I0066417

A Modern Approach to Industrial Chemistry

A Modern Approach to Industrial Chemistry

Editor: Cory Simmons

NY RESEARCH
PRESS
New York

Published by NY Research Press
118-35 Queens Blvd., Suite 400,
Forest Hills, NY 11375, USA
www.nyresearchpress.com

A Modern Approach to Industrial Chemistry
Edited by Cory Simmons

© 2019 NY Research Press

International Standard Book Number: 978-1-63238-647-2 (Hardback)

This book contains information obtained from authentic and highly regarded sources. Copyright for all individual chapters remain with the respective authors as indicated. All chapters are published with permission under the Creative Commons Attribution License or equivalent. A wide variety of references are listed. Permission and sources are indicated; for detailed attributions, please refer to the permissions page and list of contributors. Reasonable efforts have been made to publish reliable data and information, but the authors, editors and publisher cannot assume any responsibility for the validity of all materials or the consequences of their use.

Trademark Notice: Registered trademark of products or corporate names are used only for explanation and identification without intent to infringe.

Cataloging-in-Publication Data

A modern approach to industrial chemistry / edited by Cory Simmons.
    p. cm.
Includes bibliographical references and index.
ISBN 978-1-63238-647-2
1. Chemical engineering. 2. Chemistry, Technical. I. Simmons, Cory.
TP155 .M63 2019
660--dc23

Contents

Preface

I am honored to present to you this unique book which encompasses the most up-to-date data in the field. I was extremely pleased to get this opportunity of editing the work of experts from across the globe. I have also written papers in this field and researched the various aspects revolving around the progress of the discipline. I have tried to unify my knowledge along with that of stalwarts from every corner of the world, to produce a text which not only benefits the readers but also facilitates the growth of the field.

Industrial chemistry is concerned with the production of raw materials into finished industrial products by employing various chemical processes. Chemical processes employing chemical reactions, separation methods, refining techniques are commonly applied in the chemical industry for the manufacture of a wide variety of materials. All industrial chemicals are subject to quality control operations and manufacturing standards. The principles and methodologies of industrial chemistry have applications across a number of fields such as petrochemical processing, polymer manufacturing, etc. The production of various organic and inorganic chemicals, including fertilizers also fall in this domain. This book is compiled to present, in a detailed manner, the processes and systems crucial to the field of industrial chemistry. It elucidates new techniques and their applications in a multidisciplinary manner. It also presents researches that have transformed this discipline and aided its advancement. Chemical engineers, experts, researchers and students will find this book a valuable information resource.

Finally, I would like to thank all the contributing authors for their valuable time and contributions. This book would not have been possible without their efforts. I would also like to thank my friends and family for their constant support.

Editor

Synergistic effect p -phenylenediamine and n,n diphenylthiourea on the electrochemical corrosion behaviour of mild steel in dilute acid media

Roland Tolulope Loto[1] · Abimbola Patricia Popoola[2] · Akanji Lukman Olaitan[2]

Abstract Electrochemical studies of the synergistic effect of p-phenylenediamine and n,n diphenylthiourea (TPD) as corrosion inhibitor of mild steel in dilute sulphuric and hydrochloric acid through weight loss and potentiodynamic polarization at ambient temperature were performed. Experimental results showed the excellent performance of TPD with an optimal inhibition efficiency of 88.18 and 93.88 % in sulphuric and 87.42 and 87.15 % in hydrochloric acid from both tests at all concentration studied. Polarization studies show the compound to be a mixed-type inhibitor. Adsorption of deanol on the steel surface was observed to obey the Langmuir and Frumkin isotherm models. X-ray diffractometry confirmed the absence of corrosion products and complexes. Optical microscopy confirmed the selective inhibition property of TPD to be through chemical adsorption on the steel surface.

Keywords Corrosion · Inhibitor · Adsorption · Organic · Acid

Introduction

Carbon steel is extensively utilized in petrochemical plants, chemical processing plants, extractive industries, and construction and automobile industries due to its good mechanical, chemical and physical properties [1–6]. These steels are exposed to the deteriorating effect of acids in a variety of different ways resulting in corrosion. Hydrochloric acid is the most difficult of the common acids to handle from the pointview of corrosion and materials of constructions. Extreme care is required in the selection of materials to handle the acid by itself, even in relatively dilute concentrations, or in process solutions containing appreciable amounts. This acid is very corrosive to mild steel. In industries, hydrochloric acid solutions are often used in order to remove scale and salts from steel surfaces, cleaning tanks and pipelines, production of organic and inorganic compounds, regeneration of ion exchange resins, oil production, etc. Hydrochloric acid is widely used for various treatments of materials in industry. The aqueous electrolyte phase in the overhead condenser, which comes from the brine water in the crude and steam stripping, contains mostly hydrochloric acid which is released by hydrolysis of calcium chloride ($CaCl_2$) and magnesium chloride ($MgCl_2$) and also contains hydrogen sulphide (H_2S) The corrosion in this unit is mostly due to the condensed HCl. Sulphuric acid is produced more than any other chemical in the world. It has large scale uses covering nearly all industries, such as fertilizer industries, petroleum refinery, paint industry, steel pickling, extraction of non-metals and manufacture of explosives. In chemical industries, it is used for the production of dye stuffs, pharmaceuticals and fluorine. Sulphuric acid is widely used in industries such as pickling, cleaning and descaling, industrial cleaning agent and production of chemicals. Corrosion is an electrochemical process through results in the gradual deterioration of ferrous alloys through redox reactions [7]. This anomaly demands the perpetual search for more effective and versatile corrosion inhibiting compounds, due to the differential environmental conditions encountered in industry. This remains a centrepiece in corrosion prevention

✉ Roland Tolulope Loto
tolu.loto@gmail.com

[1] Department of Mechanical Engineering, Covenant University, Ota, Ogun, Nigeria

[2] Department of Chemical, Metallurgical and Materials Engineering, Tshwane University of Technology, Pretoria, South Africa

as inhibitors decelerate the electrochemical processes responsible for corrosion. The application of inhibitors is one of the most cost effective methods for corrosion control in acidic media [8]. Most common inhibitors employed in industry are compounds of organic origin whose basic constituents are nitrogen, oxygen and sulphur atoms. Inhibitors consisting of double or triple bonds facilitate the adsorption of the organic compounds onto metal surfaces, forming an impenetrable protective barrier through chemisorption reactions [9–17]. The chemical bond is formed between the electron pair and/or the pi-electron of the protonated species and the valence metal ions at the surface, thereby reducing corrosive attack in an acidic medium. Most compounds of synthetic origin are toxic, environmentally unfriendly and of high cost; thus, there is a need for low-cost, highly effective compounds [18]. Thiourea derivatives and p-phenylenedi-amine have been studied individually in previous research for corrosion inhibition properties with mixed results [19–23]; however, this research aims to study the synergistic effect of n,n diphenylthiourea and p-phenylenediamine as corrosion inhibitor for low carbon steel in 1 M sulphuric and 0.5 M hydrochloric acid.

Experimental procedure

Material specimen

Low carbon steel obtained commercially and analysed at the Advanced Materials and Tribo-Corrosion Research Laboratory, Department of Chemical and Metallurgical Engineering, Tshwane University of Technology, South Africa, gave a percentage weight composition of 0.401 % C, 0.169 % Si, 0.440 % Mn, 0.005 % P, 0.012 % S, 0.080 % Cu, 0.008 % Ni, 0.025 % Al, with the rest composed of Fe. The specimen dimension is cylindrical with 14 mm diameter.

Inhibiting compound

Combined mixture of n,n diphenylthiourea and p-phenylenediamine in equal proportions resulting in a whitish, solid powder (TPD) is the inhibitor used. n,n Diphenylth-iourea was obtained from Sigma-Aldrich, St. Louis, USA, and p-phenylenediamine was obtained from Merck Chemical, Germany. The structural formula of n,n diphenylthiourea is shown in Fig. 1a, the molecular formula is $C_{13}H_{12}N_2S$, while the molar mass is 228.312 g/mol. The molecular formula p-phenylenediamine is $C_6H_4(NH_2)_2$, while the molar mass is 108.1 g/mol. The chemical structure is shown in Fig. 1b. TPD was prepared in molar concentrations of 0.0037, 0.0074, 0.0112, 0.0149, 0.0186 and 0.0223, respectively.

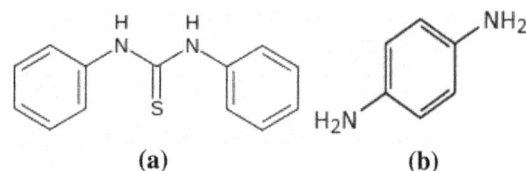

Fig. 1 a Chemical structure of n,n diphenylthiourea, **b** chemical structure of 1,4-diaminobenzene

Test solution

1 M HCl acid and 0.5 M H_2SO_4, both with 2 % recrystallized NaCl of Analar grade, were used as the corrosion test solution.

Preparation of low carbon steels

The low carbon steel (14 mm diameter) was machined into predetermined dimensions with an average length of 10 mm. The exposed ends of each steel were metallographically prepared with silicon carbide abrasive papers of 80, 120, 220, 800 and 1000 grits, washed with distilled water, rinsed with acetone, dried and stored in a dessicator for coupon analysis and potentiodynamic polarization.

Weight-loss experiments

Carbon steel samples were each immersed in 200 ml of the acid solutions (1 M HCl and 0.5 M H_2SO_4) at predetermined molar concentrations of the organic mixture (TPD for 240 h at 25 °C ambient temperature). Each sample was taken out every 48 h, washed with distilled water, rinsed with acetone, dried and re-weighed. Graphical plots of inhibition efficiency (η) versus exposure time (h) (Figs. 2, 3) for the test media were made from the obtained data (Table 1).

The corrosion rate (C) is determined from the equation below:

$$R = \left[\frac{87.6M}{DAT}\right] \quad (1)$$

M is the weight loss (mg), D is the density (g/cm^2), A is the surface area in cm^2, and T is the exposure time (h). The η was calculated from the equation below.

$$\eta = \left[\frac{C_1 - C_2}{C_1}\right] \times 100 \quad (2)$$

where C_1 and C_2 are the corrosion rates with and without of predetermined concentration of TPD. The surface coverage is calculated from the equation below:

Fig. 2 Plot of corrosion rate versus exposure time for 0–0.0037 M TPD in 0.5 M H$_2$SO$_4$

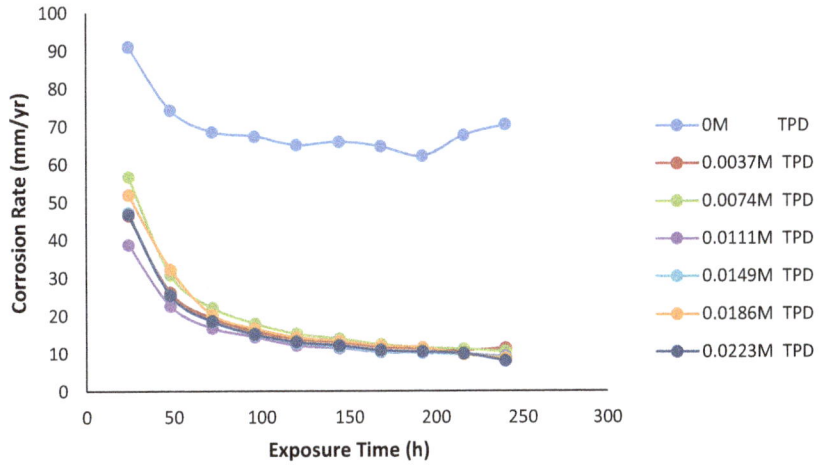

Fig. 3 Plot of inhibition efficiency versus exposure time for 0–0.0037 M TPD in 0.5 M H$_2$SO$_4$

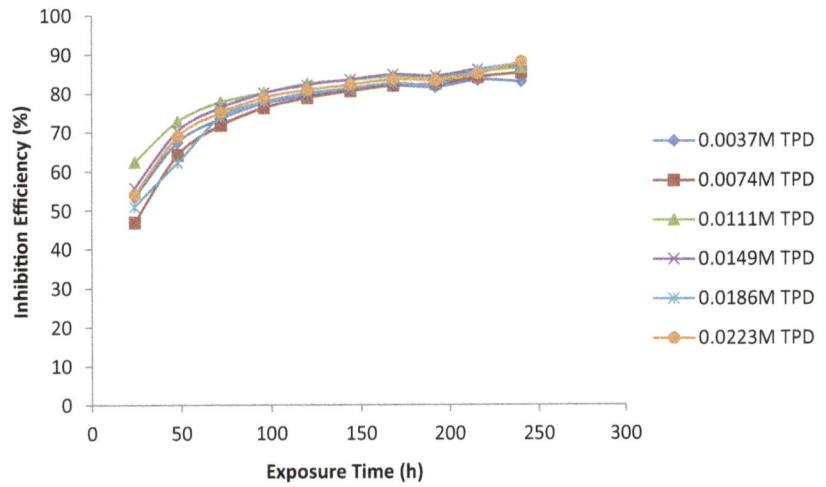

Table 1 Data obtained from weight loss analysis for low carbon steel in 0.5 M H$_2$SO$_4$ in (0.0037–0.0223 M) TPD at 240 h

Samples	Weight loss (g)	Corrosion rate (mm/year)	Inhibition efficiency (%)	TPD concentration (%)	Surface coverage (θ)	TPD concentration (mol/L)
A	2.498	70.556	0	0	0	0
B	0.422	11.459	83.12	0.0025	0.8312	0.0037
C	0.366	10.584	85.35	0.005	0.8535	0.0074
D	0.334	9.342	86.62	0.0075	0.8662	0.0111
E	0.314	9.090	87.41	0.01	0.8741	0.0149
F	0.308	8.797	87.69	0.0125	0.8769	0.0186
G	0.295	8.096	88.18	0.015	0.8818	0.0223

$$\theta = \left[1 - \frac{M_2}{M_1}\right] \quad (3)$$

where θ is the quantity of TPD adsorbed per gram (or kg) of the steel surface. M_1 and M_2 are the weight loss of carbon steel specimen in the free and inhibited test media.

Potentiodynamic polarization

Potentiodynamic polarization tests were performed with the aid of cylindrical steel samples embedded in resin mounts with exposed surface of 154 mm^2. The working electrodes were polished with differential grades of silicon carbide paper, rinsed with distilled water and dried with acetone. Polarization tests were performed at ambient temperature of 25 °C with Digi-Ivy potentiostat. A platinum rod was used as the counter electrode and silver chloride electrode (Ag/AgCl) with pH of 6.5 was used as the reference electrode. The potentials were scanned from -1.5 to $+1.5$ V at a scan rate of 0.002 V/s. The corrosion current (i_{corr}), corrosion current density (I_{corr}) and corrosion potential (E_{corr}) were determined from the Tafel plots of potential versus log I_{corr}. The corrosion rate (R), the degree of surface coverage (θ) and the percentage inhibition efficiency (%IE) were calculated from the equation below:

$$R = \frac{0.00327 \times I_{corr} \times Eq}{D} \tag{4}$$

where I_{corr} is the current density ($\mu A/cm^2$), D is the density (g/cm^3), Eq is the specimen equivalent weight (g).

The percentage inhibition efficiency (η) was calculated from the corrosion rate with the equation below:

$$\eta = 1 - \left[\frac{C_2}{C_1}\right] \times 100 \tag{5}$$

C_1 and C_2 are the corrosion rates in the absence and presence of TPD, respectively.

Optical microscopy characterization

The surface morphology of the inhibited and non-inhibited steel samples was further studied after weight-loss analysis with the aid of Nikon Eclipse LV 150 optical microscope for which micrographs were taken.

X-Ray diffraction analysis

X-ray diffraction (XRD) patterns of the film formed on the metal surface with and without TPD addition were analysed using a PANalytical X'Pert Pro powder diffractometer in θ–θ configuration with an X'Celerator detector and variable divergence and fixed receiving slits with Fe-filtered Co-Kα radiation ($\lambda = 1.789$ Å). The phases were identified using X'Pert Highscore plus software.

Results and discussion

Weight-loss measurements

Weight loss of low carbon steel during the exposure hours, with and without TPD additions in 0.5 M H_2SO_4 and 1 M HCl acid at 25 °C, was evaluated. The calculated values of weight loss (M), corrosion rate (C), surface coverage (θ) and the percentage inhibition efficiency (η) are presented in Tables 1, 2. Figures 2 and 3 show the graphical plot of corrosion rate and η versus exposure time at predetermined concentrations of TPD in H_2SO_4, while Figs. 4 and 5 show the variation of corrosion rate and η with exposure time in 1 M HCl. In Fig. 2, the corrosion rate values of samples in 0 M TPD acid solution were significantly high throughout the experimental evaluation period with slight decrease until 68-h exposure time where the corrosion rate was generally constant. This slightly contrasts the corrosion rate values in Fig. 4 for samples in 0 M TPD HCl solution where the corrosion rates though very high throughout the exposure period declined progressively. The high corrosion rates in both solutions are due to the corrosive nature of the reactive species in the acid media which rapidly destroys the steel interfacial properties and substrate metal.

Addition of specific concentrations of TPD (0.0037–0.0223 M) in both acids (1 M HCl and 0.5 M H_2SO_4) significantly decreased the corrosion rates; however, the corrosion rate values are much lower in 0.5 M H_2SO_4 than 1 M HCl. The η values are proportional to the corrosion rates. In 0.5 M H_2SO_4, the η increased from generally low percentage to very high values with time; the high values are associated with effective inhibition and protection of the steel, indicating the time-dependent effective inhibition performance of TPD in 0.5 M H_2SO_4. In 1 M HCl solution, the inhibition efficiency was generally high from the onset showing the time-independent characteristics of TPD inhibition performance in 1 M HCl. At 240-h exposure time, the corrosion data in Tables 1 and 2 show an inhibiting compound that effectively inhibits corrosion of the steel sample through adsorption onto the

Table 2 Data obtained from weight loss analysis for low carbon steel in 1 M HCl in (0.0037–0.0223 M) TPD at 240 h

Samples	Weight loss (g)	Corrosion rate (mm/year)	Inhibition efficiency (%)	TPD concentration (%)	Surface coverage (θ)	TPD concentration (mol/L)
A	3.488	126.60	0	0	0	0
B	0.559	15.97	83.97	0.0025	0.8397	0.0037
C	0.317	8.32	90.92	0.005	0.9092	0.0074
D	0.341	8.83	90.23	0.0075	0.9023	0.0111
E	0.471	12.38	86.50	0.01	0.8650	0.0149
F	0.363	11.74	89.59	0.0125	0.8959	0.0186
G	0.439	12.10	87.42	0.015	0.8742	0.0223

Fig. 4 Plot of corrosion rate versus exposure time for 0–0.0037 M TPD in 0.5 M HCl

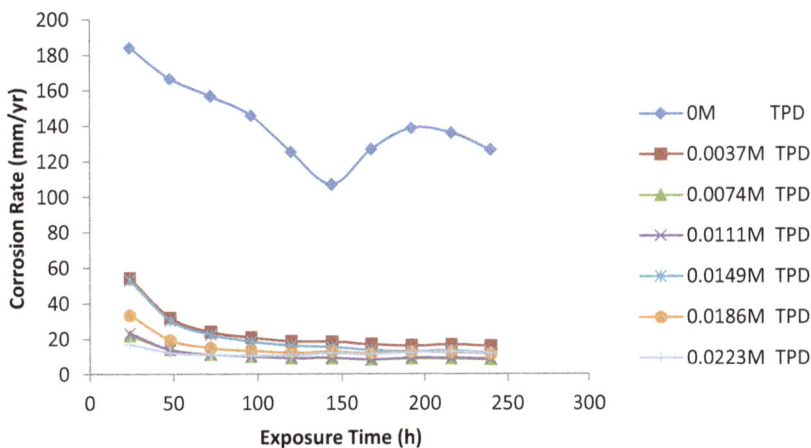

Fig. 5 Plot of inhibition efficiency versus exposure time for 0–0.0037 M TPD in 0.5 M H_2SO_4

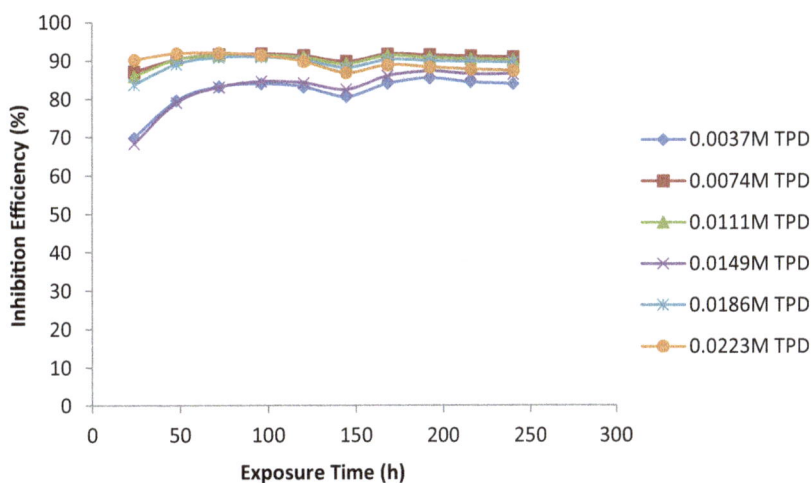

steel surface forming a protective covering through chemical reactions with the ionized steel surface [24]. TPD inhibits the redox reaction processes responsible for corrosion due to donor acceptor interactions between the pi-electrons of the heteroatoms and the vacant d orbital of steel surface atoms [25]. The electrochemical reactions are stifled over the active sites of the metal/solution interface due to the electrolytic action of TPD molecules.

Potentiodynamic polarization

The polarization data for the electrochemical influence of TPD on the corrosion behaviour of low carbon steel in 0.5 M H_2SO_4 and 1 M HCl are shown in Tables 3 and 4, while the polarization plots are shown in Figs. 6 and 7. Observation of Table 3 shows the progressive decrease in corrosion rate with increase in TPD concentration. At the highest concentration of 0.0223 M TPD, the corrosion rate is the lowest with the value of 0.344 mm/year and inhibitor efficiency is the highest at 93.88 %. This shows that the performance of TPD is subject to the inhibitor

concentration as observed from the previous discussion on weight loss. The corrosion potential varied differentially, tending towards cathodic and anodic inhibition potentials from −0.380 V (0 M TPD). This depicts the influence of TPD on inhibiting the anodic dissolution process and hydrogen evolution reactions. The influence of TPD on the steel corrosion rates in 1 M HCl is significantly smaller than in 0.5 M H_2SO_4 as shown in Table 4. Comparison of the corrosion rates shows the remarkable difference in values between Tables 3, 4. The corrosion rates are generally higher in 1 M HCl compared to 0.5 M H_2SO_4, though the rates significantly reduce in the presence of TPD in 1 M HCl, and it increased minimally with increase in TPD concentration before reducing slightly after 0.0149 M TPD. The corrosion potentials in 1 M HCl displayed similar electrochemical behaviour with the values in 0.5 M H_2SO_4. The polarization behaviours in both acid solutions are generally the same.

Study of previous research on *p*-phenylenediamine in HCl and H_2SO_4 shows that desorption occurs at very low

Table 3 Potentiodynamic polarization data for low carbon steel in 0.5 M H_2SO_4 at 0.0037–0.0223 M TPD

TPD concentration (mol/L)	Corrosion rate (mm/year)	Corrosion potential (V)	Cathodic slope (A/V)	Anodic slope (A/V)	Polarization resistance (Ω)	Corrosion current (A)	Corrosion current density (A/cm^2)	Inhibition efficiency (%)
0 M	5.631	−0.380	−7.055	7.44	34.42	7.47E−04	4.85E−04	0
0.0037 M	0.959	−0.392	−7.821	14.20	201.99	1.27E−04	8.26E−05	82.93
0.0074 M	0.594	−0.385	−7.815	13.46	326.05	7.88E−05	5.12E−05	89.42
0.0111 M	0.645	−0.346	−6.521	16.80	300.50	8.55E−05	5.55E−05	88.53
0.0149 M	0.479	−0.379	−7.902	15.12	404.61	6.35E−05	4.12E−05	91.48
0.0186 M	0.532	−0.374	−7.005	13.01	364.44	7.05E−05	4.58E−05	90.54
0.0223 M	0.344	−0.317	−8.697	9.73	563.44	4.56E−05	2.96E−05	93.88

Table 4 Potentiodynamic polarization data for low carbon steel in 0.5 M HCl at 0.0037–0.0223 M TPD

TPD concentration (mol/L)	Corrosion rate (mm/year)	Corrosion potential (V)	Cathodic slope (A/V)	Anodic slope (A/V)	Polarization resistance (Ω)	Corrosion current (A)	Corrosion current density (A/cm^2)	Inhibition efficiency (%)
0 M	5.646	−0.386	−7.243	2.966	34.26	7.50E−04	4.87E−04	0
0.0037 M	2.201	−0.381	−7.842	13.140	250.99	1.02E−04	6.65E−05	84.86
0.0074 M	2.501	−0.387	−8.512	11.120	355.98	7.22E−05	4.69E−05	89.33
0.0111 M	2.525	−0.387	−8.100	12.530	243.85	1.05E−04	6.84E−05	84.42
0.0149 M	3.102	−0.381	−7.963	9.987	313.05	8.21E−05	5.33E−05	87.86
0.0186 M	2.256	−0.370	−7.580	10.110	214.72	1.20E−04	7.77E−05	82.30
0.0223 M	2.687	−0.385	−7.719	12.450	295.76	8.69E−05	5.64E−05	87.15

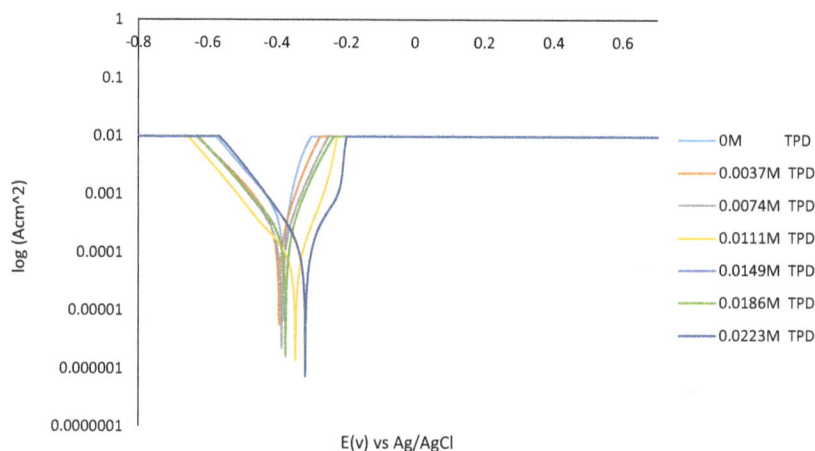

Fig. 6 Comparison plot of polarization scans for low carbon steel in 0.5 M H_2SO_4 solution at 0 M TPD −0.0223 M TPD concentration

and higher concentrations of the organic compound due to lateral repulsion between the inhibitor molecules which results in weak inhibitor covering over the steel surface and hence significant increase in corrosion rate. Observation of data from unpublished research identifies similar phenomenon for n,n diphenylthiourea in HCl. The combined action/synergistic effect of the compounds in TPD displayed remarkable improvement in corrosion inhibition and inhibition efficiency [26, 27].

Inhibition mechanism

The classification of an organic compound as anodic- or cathodic-type inhibitor depends on the displacement of the corrosion potential values within and beyond 85 mV in the anodic or cathodic direction [28, 29]. The maximum displacement value in 0.5 M H_2SO_4 is 73 mV in the anodic direction and 17 mV in the anodic direction for 1 M HCl, and thus in 0.5 M H_2SO_4 and 1 M HCl, TPD can be

Fig. 7 Comparison plot of polarization scans for low carbon steel in 0.5 M HCl solution at 0 M TPD −0.0223 M TPD concentration

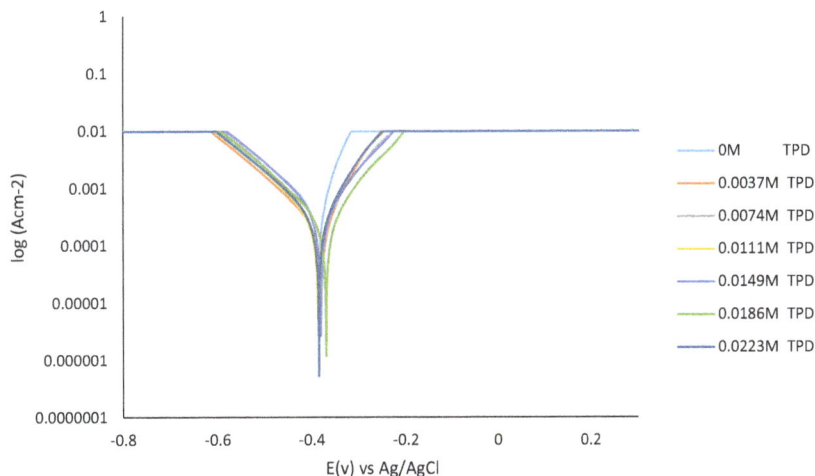

classified as mixed-type inhibitor with cathodic and anodic inhibiting characteristics. TPD has strong influence on the entire electrochemical process as the polarization resistance in Tables 3 and 4 increases with increase in inhibition efficiency till the maximum studied concentration. Organic compounds inhibit through surface coverage and adsorption which stifles the redox electrochemical process. The amines functional groups in TPD are nucleophiles that form a variety of electrophilic compounds due to the basic functionality of their nitrogen atoms and the electrophilic substitutions [30–32]. The unshared electron pair of the nitrogen atom forms a coordinate bond with a proton released from the ionized atoms at the steel surface. Amines from TPD react with acids to give salts and the valence steel electrons to produce chemical complexes responsible for the compact impenetrable barrier which strongly adsorbs to the steel and inhibits corrosion. The adsorption, however, may be considered to be due to physical and chemical reactions. The hydrophilic nature of protonated TPD functional groups and heteroatoms with their hydrophobic substituent (thiol) conforms the inhibitor molecules during adsorption process to form a polymolecular barrier layer that reliably screens the metal from the corrosive medium. The thiol molecules form a monomolecular layer by means of their strong affinity to metal and their self-assembly ability due to hydrophobic interaction.

Adsorption isotherm

The mechanism of corrosion process can be further analysed on the support of adsorption behaviour of the TPD inhibitor on the steel surface. Adsorption isotherms are

very important in determining the nature of organometallic interactions [33, 34]. Adsorption of TPD at the metal/solution interface is due to the chemical bonding through chemisorption mechanism between the inhibitor and the valence atoms at the metal surface. Langmuir and Frumkin adsorption isotherms were applied to describe the adsorption mechanism in the acid solutions, as they provided the best fits.

The isotherms are of the general form:

$$f(\theta,x)\exp(-2a\theta) = KC \qquad (6)$$

where $f(\theta, x)$ is the configuration factor which depends upon the physical model and assumptions, the basis for the derivative of the isotherm, θ is the surface coverage, C is the inhibitor concentration, x is the size ration, "a" is the molecular interaction parameter and K is the equilibrium constant of adsorption process.

The general equation for Langmuir isotherm is,

$$\left[\frac{\theta}{1-\theta}\right] = K_{ads}C \qquad (7)$$

and rearranging

$$K_{ads}C = \left[\frac{\theta}{1 + K_{ads}\theta}\right] \qquad (8)$$

where K_{ads} is the equilibrium constant of the adsorption process. The plots of TPD concentration/θ versus the TPD concentration for carbon steel in 0.5 M H_2SO_4 and 1 M HCl are continuous (Figs. 8, 9) indicating the Langmuir adsorption. The divergence of the slope in Fig. 8 from unity is as a result of the electrochemical interaction among the adsorbed TTD cations on the metal surface and changes in the values of Gibbs free energy with increase in surface coverage.

Fig. 8 Plots of TPD concentration/θ and TPD concentration for 0.5 M H_2SO_4

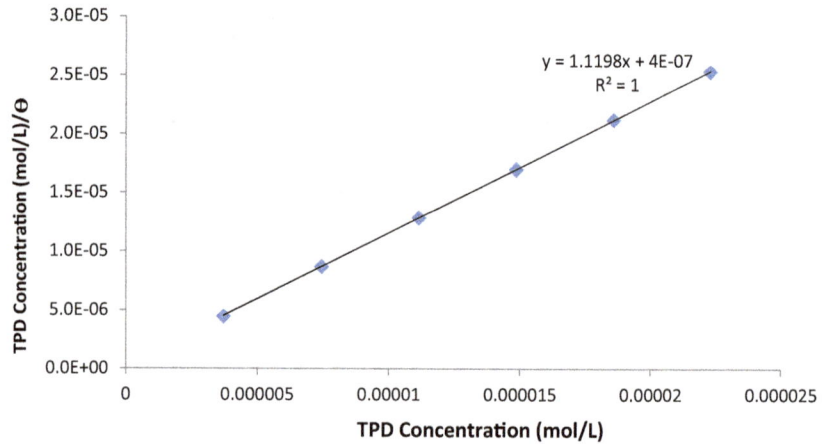

Fig. 9 Plots of TPD concentration/θ and TPD concentration for 1 M HCl

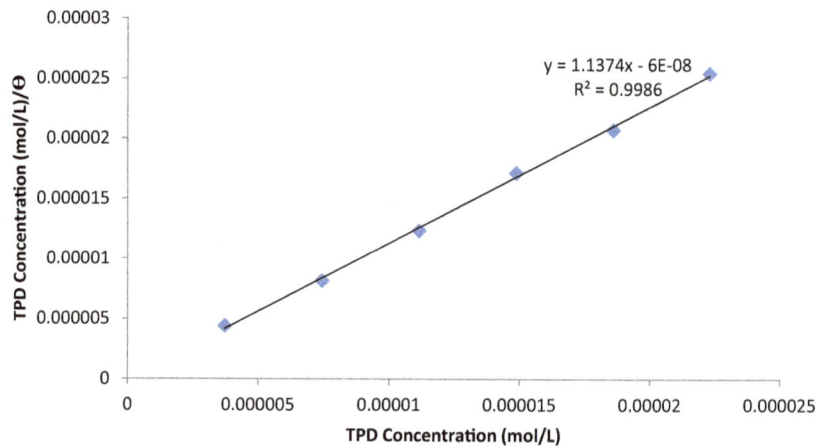

Langmuir isotherm states the following: the metal surface has a definite proportion of adsorption sites with one adsorbate and Gibbs free energy of adsorption has the same value for the sites, independent of the value of surface coverage [35].

Frumkin isotherm postulates unit coverage at high TPD concentrations and the alloy surface is not homogeneous, i.e. the effect of lateral interaction is significant. Only the active sites of the metal surface where adsorption occurs are considered. The mathematical expressions of Frumkin adsorption isotherm is

$$\log[C \times (\theta/1 - \theta)] = 2.303 \log K + 2\alpha\theta \quad (9)$$

$$KC = (\theta/1 - \theta) \exp(\theta/1 - \theta) \quad (10)$$

where K is the equilibrium constant of adsorption and α is the lateral interaction term describing the interaction between the adsorbed TPD molecules. Graphical plots of $\theta/1 - \theta$ versus TPD concentration presented in Figs. 9 and 10 are continuous. The values of lateral interaction

parameter are shown in Table 5. Increase in TPD concentration and surface coverage results in decrease in the lateral interaction term (α), suggesting that the inhibitor stifles the electrochemical process responsible for corrosion through adsorption on active sites on the steel surface. The "α" values show that the attraction between TTD molecules is weak.

Thermodynamics of the corrosion inhibition mechanism

Gibbs free energy (ΔG_{ads}) (Table 6) for the TPD corrosion inhibition was evaluated from the equilibrium constant of adsorption according to the equation.

$$\Delta G_{ads} = -2.303RT \ \log[55.5K_{ads}] \quad (11)$$

where 55.5 is the molar concentration of water in the solution, R is the universal gas constant, T is the absolute temperature and K_{ads} is the equilibrium constant of

Fig. 10 Plots of $(\theta/1 - \theta)$ and TPD concentration in 1 M HCl

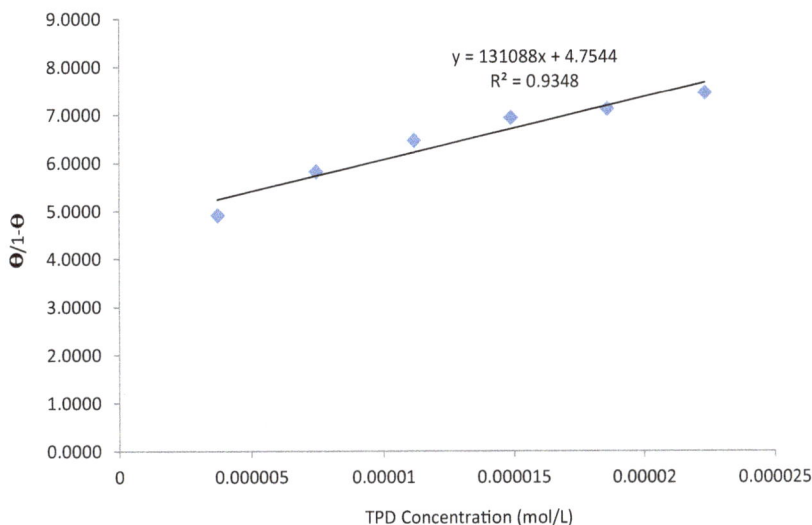

Table 5 Relationship between lateral interaction parameter, surface coverage and TPD concentration from Frumkin adsorption isotherm

Lateral interaction (α)	Equilibrium constant of adsorption (K)	Surface coverage (θ)	TPD concentration (mol/L)
−10.14	181969.17	0.8312	0.003715
−9.88	266194.77	0.8535	0.007431
−9.80	376806.42	0.8662	0.011146
−9.77	485401.77	0.8741	0.014861
−9.67	476025.26	0.8769	0.018576
−9.67	581038.31	0.8818	0.022292

Table 7 Data of Gibbs free energy, TPD concentration and inhibition efficiency for 0.5 M HCl

TPD concentration (mol/L)	Inhibition efficiency (%)	Gibbs free energy (ΔG_{ads})
0.0037	83.97	−35.09
0.0074	90.92	−34.98
0.0111	90.23	−33.77
0.0149	86.50	−32.15
0.0186	89.59	−32.33
0.0223	87.42	−31.35

Table 6 Data of Gibbs free energy, TPD concentration and inhibition efficiency for 0.5 M H_2SO_4

TPD concentration (mol/L)	Inhibition efficiency (%)	Gibbs free energy (ΔG_{ads})
0.0037	83.12	−44.88
0.0074	85.35	−43.59
0.0111	86.62	−42.84
0.0149	87.41	−42.30
0.0186	87.69	−41.81
0.0223	88.18	−41.48

adsorption. K_{ads} is related to surface coverage (θ) by the following Eq. (8).

The ΔG_{ads} data values in Tables 6 and 7 vary with change in inhibitor concentration, and this supports the Langmuir model due to the nonhomogeneous nature of the metal surface such as microscopic voids, non-metallic inclusion, and impurities. Conventionally, values of ΔG_{ads} of around −20 kJ/mol or below depict physisorption reactions and values of about −40 kJ/mol or above involve chemical reactions associated with chemisorption mechanism which tends to be more sustainable in corrosion inhibition than physisorption. The value of ΔG_{ads} for the carbon steel in 0.5 M H_2SO_4 (Table 6) shows chemisorption of TPD on the steel surface; in 1 M HCl (Table 7), the ΔG_{ads} values show physiochemical (physisorption as well as chemisorption) adsorption mechanism. The negative values of ΔGads show that TTD adsorption on the metal surface is spontaneous [35–38].

X-Ray diffraction analysis

X-ray diffraction (XRD) patterns of the carbon steel surface which gives qualitative information about the possible phases present before and after the corrosion test in 0.5 M H_2SO_4 and 1 M HCl solutions with TPD inhibitor are shown in Fig. 11. The peak values at 2θ values showed the presence of iron and carbon only on the steel surface. Observation of the diffraction peaks for the inhibited showed the absence of iron oxides and chemical

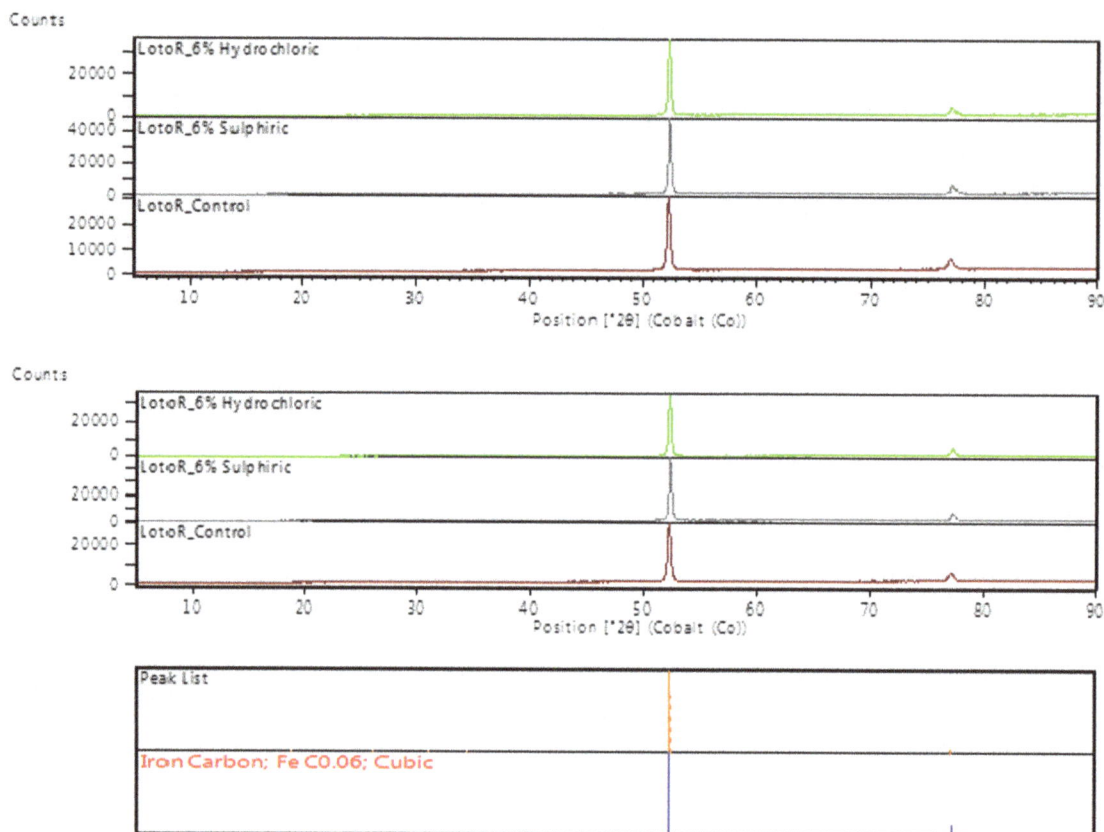

Fig. 11 XRD pattern of low carbon steel after immersion in 0.5 M H_2SO_4 and in 1 M HCl with TPD addition

compounds associated with corrosion, further proving the inhibition to be due to the electrochemical action of the functional groups of TPD and selective adsorption onto the steel

Optical microscopy analysis

The optical microscopy images of the carbon steel surfaces before and after immersion in the acid media, without the addition of TTD, are given in Fig. 12a–c. Figure 12a shows the steel sample before immersion, and the serrated surface is due to machining during sample preparation. Figure 12b, c shows the steel surfaces after 240 h of immersion in 0.5 M H_2SO_4 and 1 M HCl without TTD addition. In Fig. 12b, c, the images reveal a rough surface due to corrosion and anodic oxidation resulting from the action of SO_4^{2-} and Cl^- ions in the acid solution. The chloride ions accelerate the hydrolysis and diffusion of ionized atoms of Fe, causing rapid corrosion and deterioration of the steel sample.

The surface topography of the carbon steel (Fig. 12d, e) is as a result of the presence of TPD in the acid solution

compared with the control and uninhibited sample. The effectiveness of the protective film is clearly visible on the images. Precipitates form on the steel surface, coating the steel from corrosion and producing the micrographs below. The selective precipitation of TPD on the surface is clearly visible on the images after the electrochemical tests due to the strong adsorption of its molecules through electrolytic diffusion and electrostatic attraction onto the surface of the steel, displacing the corrosive anions through the formation of a stronger bond with the charged surface atoms of the steel. The protective film of TPD formed on the specimen surface exhibited good inhibition performance for the corrosion of carbon steel in the acid solutions and the observation is in good agreement with the weight loss and electrochemical experiments.

Conclusions

The corrosion inhibition performance of the synergistic effect of *p*-phenylenediamine and *n,n* diphenylthiourea (TPD) on mild steel in the acid solutions showed excellent results. The corrosion rate reduced sharply with TPD

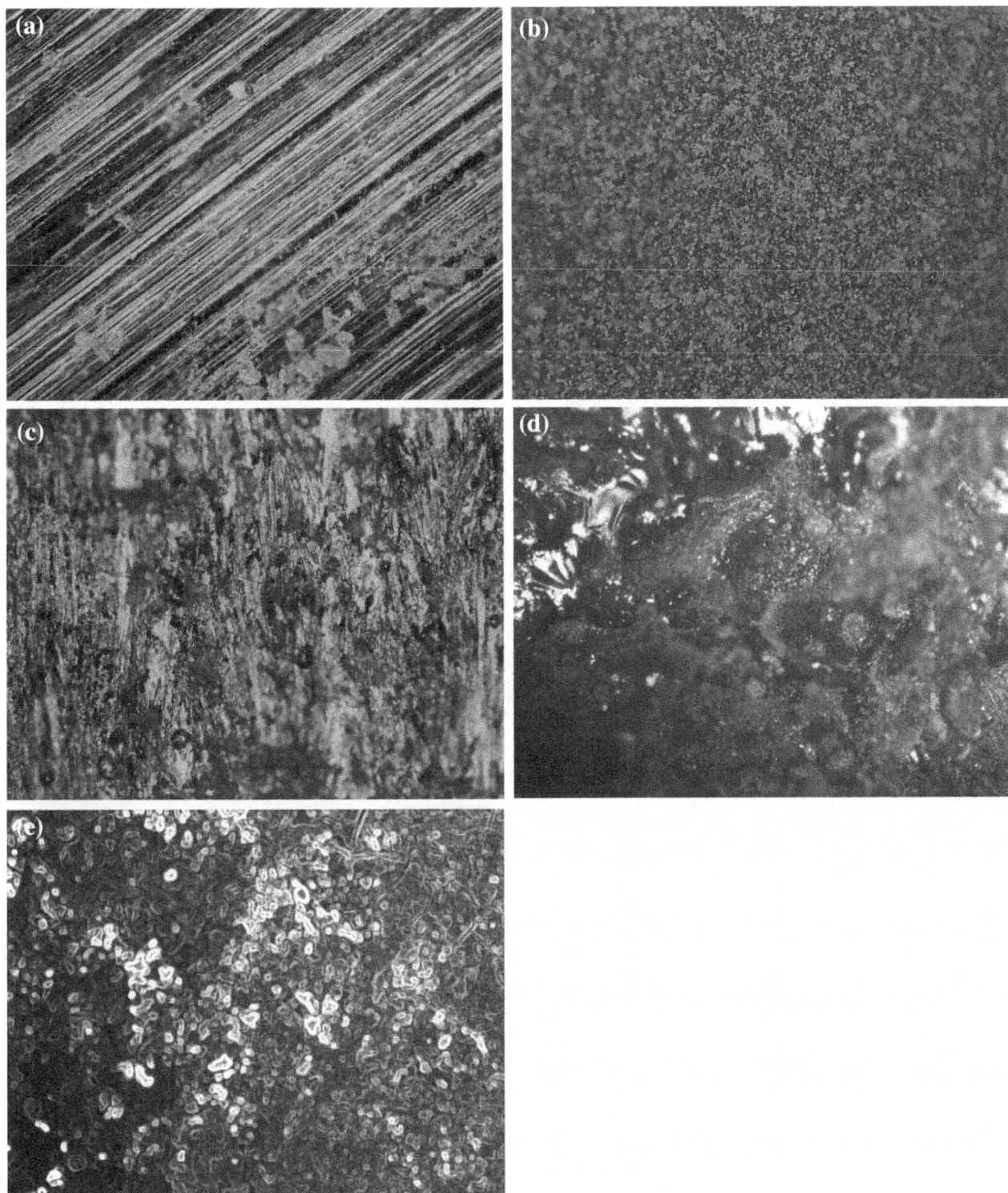

Fig. 12 Optical microscopy images at x50 of low carbon steel: **a** before immersion, **b** after immersion in 0.5 M H_2SO_4, **c** after immersion in 0.5 M HCl, **d** after immersion in 0.5 M H_2SO_4 + TTD inhibitor, **e** after immersion in 0.5 M HCl + TTD inhibitor

addition from lowest to highest concentration, significantly influencing the interfacial redox electrochemical process responsible for corrosion through adsorption on the steel surface. Mixed inhibition behaviour was observed from the corrosion potential values of the potentiodynamic analysis. Optical microscopy characterization shows the change on the surface topography and morphology of the steel surface between the inhibited and uninhibited steel samples.

Acknowledgments The authors acknowledge the Department of Mechanical Engineering, College of Engineering, Covenant University, Ota, Ogun State, Nigeria and the Department of Chemical, Metallurgical and Materials Engineering, Tshwane University of Technology, Pretoria, South Africa, for the provision of research facilities for this work.

References

1. Fouda AS, Hamdy BA (2013) Aqueous extract of propolis as corrosion inhibitor for carbon steel in aqueous solutions. Afr J Pure Appl Chem 7(10):350–359
2. Burubai W, Dagogo G (2007) Comparative study of inhibitors on the corrosion of mild steel reinforcement in concrete. Agric Eng Int CIGR E-J 9:1–10
3. Liu GQ, Zhu ZY, Ke W, Han CI, Zeng CL (2001) Corrosion. Natl Assoc Chem Eng 57(8):730
4. Ilevbare GO, Burstein GT (2003) The inhibition of pitting corrosion of stainless steel by chromate and molybdate ions. Corros Sci 45:1545–1569
5. Munoz AI, Anton JG, Nuevalos SL, Guinon JL, Herranz VP (2004) Corrosion studies of Austenitic and duplex stainless steels in aqueous lithium bromide solution at different temperatures. Corros Sci 46:2955–2974
6. Cui ZD, Wu SL, Zhu SL, Yang XJ (2006) Study on corrosion properties of pipelines in simulated produced water saturated with supercritical CO_2. Appl Surf Sci 252:2368–2374
7. Leelavathi S, Rajalakshmi R (2013) Dodonaea viscosa (L.) leaves extract as acid corrosion inhibitor for mild steel–a green approach. J Mater Environ Sci 4(5):625–638
8. Trabanelli G (1991) Inhibitors—an old remedy for a new challenge. Corrosion 47(6):410–419
9. Satapathy AK, Gunasekaran G, Sahoo SC, Amit K, Rodrigues RV (2009) Corrosion inhibition by *Justicia gendarussa* plant extract in hydrochloric acid solution. Corros Sci 51(12):2848–2856
10. Ghazoui A, Bencaht N, Al-Deyab SS, Zarrouk A, Hammouti B, Ramdani M, Guenbour M (2013) An Investigation of two novel pyridazine derivatives as corrosion inhibitor for C38 steel in 1.0 M HCl. Int J Electrochem Sci 8:2272–2292
11. Al Hamzi AH, Zarrok H, Zarrouk A, Salghi R, Hammouti B, Al-Deyab SS, Bouachrine M, Amine A, Guenoun F (2013) The role of acridin-9(10H)-one in the inhibition of carbon steel corrosion: thermodynamic, electrochemical and DFT studies. Int J Electrochem Sci 2013(8):2586–2605
12. Zarrok H, Zarrouk A, Salghi R, Oudda H, Hammouti B, Assouag M, Taleb M, Ebn Touhami M, Bouachrine M, Boukhris S (2012) Gravimetric and quantum chemical studies of 1-[4-acetyl-2-(4-chlorophenyl)quinoxalin-1(4H)-yl]acetone as corrosion inhibitor for carbon steel in hydrochloric acid solution. J Chem Pharm Res 4(12):5056–5066
13. Zarrok H, Zarrouk A, Salghi R, Ramli Y, Hammouti B, Assouag M, Essassi EM, Oudda H, Taleb M (2012) 3,7-Dimethylquinoxalin-2-(1H)-one for inhibition of acid corrosion of carbon steel J Chem. Pharm Res 4(12):5048–5055
14. Zarrouk A, Zarrok H, Salghi R, Hammouti B, Bentiss F, Touir R, Bouachrine M (2013) Evaluation of N-containing organic compound as corrosion inhibitor for carbon steel in phosphoric acid. J Mater Environ Sci 4(2):177–192
15. Zarrouk A, Zarrok AH, Salghi R, Bouroumane N, Hammouti B, Al-Deyab SS, Touzani R (2012) The Adsorption and Corrosion Inhibition of 2-[bis-(3,5-dimethyl-pyrazol-1-ylmethyl)-amino]-pentanedioic acid on carbon steel corrosion in 1.0 m HCl. Int J Electrochem Sci 7:10215–10232
16. Ben Hmamou D, Salghi R, Zarrouk A, Zarrok H, Al-Deyab SS, Benali O, Hammouti B (2012) The inhibited effect of phenolphthalein towards the corrosion of C38 steel in hydrochloric acid. Int J Electrochem Sci 7:8988–9003
17. Zarrouk A, Messali M, Aouad MR, Assouag M, Zarrok H, Salghi R, Hammouti B, Chetouani A (2012) Some new ionic liquids derivatives: synthesis, characterization and comparative study towards corrosion of C-steel in acidic media. J Chem Pharm Res 4(7):3427–3436
18. Torres VV, Amado RS, Faia de Sa C, Fernandez TL, Riehl CAS, Torres AG, D'Elia E (2011) Inhibitory action of aqueous coffee ground extracts on the corrosion of carbon steel in HCl solution. Corros Sci 53:2385–2392
19. Edrah S, Hasan SK (2010) Studies on thiourea derivatives as corrosion inhibitor for aluminum in sodium hydroxide solution. J Appl Sci Res 6(8):1045–1049
20. Fekry AM, Mohamed RR (2010) Acetyl thiourea chitosan as an eco-friendly inhibitor for mild steel in sulphuric acid medium. Electrochim Acta 55(6):1933–1939
21. Awad K (2004) Semiempirical investigation of the inhibition efficiency of thiourea derivatives as corrosion inhibitors. J Electro Chem 567(2):219–225
22. Chao CY, Lin LF, Macdonald DD (1981) A point defect model for anodic passive films. I. Film growth kinetics. J Electrochem Soc 128:1187–1194
23. Loto RT, Loto CA (2012) Effect of *P*-phenylenediamine on the corrosion of austenitic stainless steel type 304 in hydrochloric acid. Int J Electrochem Sci 7:9423–9440
24. Heckerman N, Snavely E Jr, Payne JS Jr (1966) Effects of anions on corrosion inhibition by organic compounds. J Electrochem Soc 113:677–681
25. Felicia RS, Santhanalakshmi S, Wilson SJ, John AA, Susai R (2004) Synergistic effect of succinic acid and Zn2+ in controlling corrosion of carbon steel. Bull Electrochem 20(12):561–565
26. Loto RT, Loto CA, Popoola API, Fedotova T (2015) Electrochemical effect of 1, 4-diaminobenzene on the corrosion inhibition of mild steel in dilute acid media. Der Pharma Chemica 7(5):72–93
27. Susai RS, Mary R, Noreen A, Ramaraj R (2002) Synergistic corrosion inhibition by the sodium dodecylsulphate–Zn2+ system. Corros Sci 44(10):2243–2252
28. Sahin M, Bilgiç S, Yılmaz H (2002) The inhibition effects of some cyclic nitrogen compounds on the corrosion of the steel in NaCl mediums. Appl Surf Sci 195(104):1–7
29. Ruppel DT, Dexter SC, Luther GW (2001) III role of manganese dioxide in corrosion in the presence of natural biofilms. Corrosion J 57(1):863–873
30. Gad Alla AG, Tamous HM (1990) Structural investigation of pyrazole derivatives as corrosion inhibitors for delta steel in acid chloride solutions. J Appl Electrochem 20(3):488–493
31. Quraishi MA, Sharma HK (2002) 4-Amino-3-butyl-5-mercapto-1, 2, 4-triazole: a new corrosion inhibitor for mild steel in sulphuric acid. Mater Chem Phys 78(1):18–21
32. Quraishi MA, Ansari FA, Jamal D (2004) Corrosion inhibition of Tin by some amino acids in citric acid solution. Indian J Chem Technol 11:271–274
33. Hirozawa ST (1995) Proceedings of 8th European symposium on corrosion inhibition. Ann University, Ferrara, vol 1, p 25
34. Villamil RFV, Corio P, Rubin JC, Agostinho SMI (1999) Effect of sodium dodecylsulfate on copper corrosion in sulfuric acid media in the absence and presence of benzotriazole. J Electroanal Chem 472:112–119
35. Eddy NO, Mamza PAP (2009) Inhibitive and adsorption properties of ethanol extract of seeds and leaves of *Azardirachta*

indica on the corrosion of mild steel in H_2SO_4. Port Electrochim Acta 27(4):443–456

36. Hosseini MG, Mertens SFL, Arshadi MR (2003) Synergism and antagonism in mild steel corrosion inhibition by sodium dodecylbenzenesulphonate and hexamethylenetetramine. Corros Sci 45:1473–1489

37. Solmaz R (2010) Investigation of the inhibition effect of 5-((*E*)-4-phenylbuta-1,3-dienylideneamino)-1,3,4-thiadiazole-2-thiol Schiff base on mild steel corrosion in hydrochloric acid. Corros Sci 52(10):3321–3330

38. Döner A, Solmaz R, Özcan M, Kardaş G (2011) Experimental and theoretical studies of thiazoles as corrosion inhibitors for mild steel in sulphuric acid solution. Corros Sci 53(9):2902–2913

2

Preparation of environmentally friendly activated carbon for removal of pesticide from aqueous media

Somaia G. Mohammad[1] · Sahar M. Ahmed[2]

Abstract The preparation of eco-friendly low-cost silkworm feces activated carbon (SFAC) for the removal of oxamyl pesticide from aqueous solution has been investigated in batch experiments. Structure and morphology of SFAC were characterized by Fourier transform infrared spectroscopy (FTIR), X-ray diffraction (XRD), field emission scanning electron microscopy (SEM). The specific surface area and mean pore diameter were obtained as 75.219 and 0.2035 $cm^3 g^{-1}$, respectively. The effect of different physicochemical parameters such as initial oxamyl concentrations, activated carbon dose and contact time has been studied. The results showed that the oxamyl removal on SFAC was unaffected in the pH range of 2–10. The percent removal of oxamyl onto SFAC was 99.48% from aqueous solutions. The adsorption process attained equilibrium within 120 min of contact time. Equilibrium data were analyzed by the Freundlich, Langmuir and Dubinin–Radushkevich (D-R) isotherm models. Freundlich isotherm provided the best fit to the equilibrium data. Adsorption kinetic was fitted well by the pseudo-second-order kinetic model. The results revealed that SFAC could be used a low-cost and eco-friendly alternative to other adsorbents for the oxamyl removal from aqueous solution.

Keywords Activated carbon · Silkworm feces · Removal · Oxamyl · Isotherm · Kinetics

Introduction

The common usage of pesticides (hazardous compounds) has some undesirable effects such as toxicity, carcinogenicity and mutagenicity [7, 10, 36]. The presence of pesticides in water can cause serious environmental and human health problems.

Oxamyl (methyl N'-dimethyl-N-[(methyl carbamoyl) oxy]-l-thiooxamimidate) is an oximino carbamate pesticide, systemic and active as an insecticide or a nematicide. It is used for the control of nematodes in vegetables, bananas, pineapple, peanut, cotton, soya beans, potatoes, sugar beet and other crops. Oxamyl is characterized by its high water solubility (280 g/L), has a very low soil sorption coefficient, therefore, and has high potential for movement in the soil profile. In addition, it is characterized by high acute toxicity (LD_{50} = 2.5 mg/Kg). It can easily cause contamination of both ground and surface water resources. In addition, various amounts of oxamyl have been detected in surface and ground waters not only during actual insecticide application, but also after a long period of use.

The removal of pesticides from water is one of the major environmental concerns nowadays. Photocatalytic degradation [24, 53], ultrasound combined with photo-Fenton treatment [34], advanced oxidation processes [57], ozonation [41] and adsorption [20] are different methods used to remove the pesticides from water. One of the most widely used techniques is adsorption by activated carbon (AC) coming from agricultural wastes which show greater potential for the treatment of wastewaters due to very large quantities, easy to get and very low costs [43, 39]. For the

✉ Somaia G. Mohammad
somaiagaber@yahoo.com;
sommohammad2015@gmail.com

Sahar M. Ahmed
saharahmed92@hotmail.com

[1] Central Agricultural Pesticides Laboratory, Agriculture Research Center (ARC), Dokki, Giza, Egypt

[2] Egyptian Petroleum Research Institute, Ahmed El-Zomor St., Nasr City, Cairo, Egypt

above reasons, a wide range of agricultural wastes including banana and pomegranate peels [44] and rice husk [29], sugar beet pulp [37], corn wastes [1], tea waste and rice husks [54], walnut shells [42], chestnut shells [15], orange peels [23], walnut [33] and citrus limetta peel [49]. Lit et al. [38] studied the abilities of biochars produced from six agriculture wastes (soybeans, corn stalks, rice stalks, poultry manure, cattle manure and pig manure) to remove atrazine pesticide from contaminated water.

Several authors have successfully studied a variety of low-cost, effective and locally available adsorbents for the removal of different types of pesticides, Naushad et al. [45], Tran et al. [52] and Bansal [9].

To the best of our knowledge, there is no any study devoted to the potential applicability of silkworm feces activated carbon for the removal of pesticide from an aqueous solution. Some researchers [19] prepared activated carbons from silkworm feces via chemical activation method and used them as cheap adsorbents for removal cadmium and methylene blue from aqueous solutions.

Silkworm feces are non-toxic, low-cost, effective and environmentally friendly adsorbent. They are effective in the treatment of skin diseases and possess anti-inflammatory characteristics.

The current study aims to use silkworm feces as an inexpensive and environmentally friendly precursor for the production of activated carbon (SFAC) and to test its ability to remove oxamyl pesticide from aqueous solution under different conditions. Adsorption isotherm and kinetic parameters were also calculated and discussed.

Materials and methods

Chemicals

All reagents used in this study were of analytical grade. Before each experiment, all glassware were cleaned with dilute nitric acid and repeatedly washed with deionized water.

Preparation of activated carbon (SFAC)

Silkworms' feces were kindly obtained from Sericulture Research Department, Agricultural Research Center (ARC), Egypt, were washed repeatedly with distilled water for several times to remove dirt particles and soluble impurities and were allowed to air-dry in an oven at 80 °C for 2 days. The sample was then soaked in orthophosphoric acid (H_3PO_4) with an impregnation ratio of 1:1 (w/w) for 24 h and dehydrated in an oven overnight at 105 °C. The resultant sample was activated in a closed muffle furnace to increase the surface area at 500 °C for 2 h in the absence of

air. The SFAC produced was cooled to room temperature and washed with 0.1 M HCl and successively with distilled water. Washing with distilled water was done repeatedly until the pH of the filtrate reached 6–7. The final product was dried in an oven at 105 °C for 24 h and stored in vacuum desiccators until needed [46].

Instruments

The surface morphology of prepared activated carbon was examined using a scanning electron microscopy, Quanta 250 FEG (Field Emission Gun) with accelerating voltage 30 kV. Surface area and pore size distribution were measured using BELSORP-mini II instrument. The surface functional groups and structure were studied by FTIR spectrophotometer (PerkinElmer 1720) in the range of 400–4000 cm^{-1}. The samples were examined as KBr disks. The crystallographic structure of the activated carbon sample is studied using powder X-ray diffraction analyzer (XPERT PRO, Netherland) with Cu Ka radiation (1.54 Å) at 40 kV and 40 mA, in 2θ range 4–80°.

Adsorption experiments

All chemicals used in this work were of analytical reagent grade and were used without further purification. Batch adsorption experiments were conducted for the removal of oxamyl using silkworm feces activated carbon as a function of initial pH (2–10), initial oxamyl concentration (100–2500 mg/L), adsorbent dose (0.1–1.5 g/100 mL) and contact time (5–180 min). All the experiments were carried out at room temperature.

The adsorption of oxamyl by SFAC was studied over a pH range of 2–10. The influence of the initial solution pH was studied by shaking 1 g of sorbent and 100 mL of the oxamyl solution (500 mg/L) at different pHs. The flasks were agitated for 2 h. The solution pH was adjusted at the desired value by adding a small amount of HCl (0.1 M) or NaOH (0.1 M).

After adsorption process, the adsorbent was separated from the sample by filtering and the filtrate was transferred to a separatory funnel and extracted successively three times with 20, 15 and 10 mL portions of dichloromethane. The combined extracts were dried on anhydrous sodium sulfate to remove moisture content and evaporated using a rotary evaporator on a water bath at 40 °C. All samples should be cleaned by filtration with target nylon (0.45 μm) prior to analysis in order to minimize the interference of the carbon fines with the analysis. The samples were analyzed using HPLC with DAD (diode array detector). The equilibrium and kinetics data were obtained from batch experiments.

The amount of oxamyl adsorbed (q_e) was calculated by using the following mass balance equation:

$$q_e = \frac{(C_0 - C_e)V}{W} \tag{1}$$

The removal efficiency of oxamyl was calculated as follows:

$$\text{Removal}(\%) = \frac{(C_0 - C_e)}{C_0} \times 100 \tag{2}$$

where C_0 and C_e are the initial and the equilibrium concentrations of oxamyl mg/L, respectively, V is the volume of solution (mL), and W is the amount of adsorbent used (g).

Fitness of the adsorption isotherms models

The isotherm models were evaluated for fitness of the adsorption data by calculating the sum of squared error (SEE). The SEE values were calculated by the equation:

$$\text{SEE} = \sqrt{\frac{\sum(Y - Y')2}{N}}$$

where Y is an actual score, Y' is a predicted score, and N is the number of data points. In this study, the standard error of estimate was used to confirm the best fitting. If data from the model are similar to the experimental data, error is a small number.

Analysis of oxamyl

The concentrations of oxamyl in the solutions before and after adsorption were determined using an Agilent HPLC

1260 infinity series (Agilent technologies) equipped with a quaternary pump, a variable wavelength diode array detector (DAD) and an autosampler with an electric sample valve. The column was Nucleosil C_{18} [30 × 4.6 mm (i.d) × 5 μm] film thickness. The mobile phase was 60/40 (V/V) mixture of HPLC grade acetonitrile/water. The mobile phase flow rate was 1 mL/min. The wavelength was 220 nm. The retention time of oxamyl was 2.4 min, and the injection volume was 5 μL under the conditions.

Results and discussion

Characterization of the prepared adsorbent (SFAC)

Surface morphology of SFAC

The SEM micrographs of SFAC are shown in Fig. 1a, b. The SEM image of SFAC before adsorption (Fig. 1a) illustrates the irregular size and shape of individual grains and heterogeneous surface morphology. The porous structure is appearing with different widths. After adsorption, it is obvious that the pores have been covered by the oxamyl (Fig. 1b).

Silkworm feces activated carbon (SFAC) samples were first degassed under high vacuum at 350 °C for 8 h. Nitrogen adsorption–desorption isotherms of ACs were measured at nitrogen temperature (77 K).

Figure 2 shows the N_2 adsorption–desorption isotherms of the SFAC. It is noted that the sample (SFAC) displayed isotherms of type IV, signifying that the sample contained mesopores with sizes ranging from 2 to 50 nm. According to the BET method, the specific surface area and mean pore diameter were obtained as 75.219 m^2 g^{-1} and 0.2035 cm^3 g^{-1}, respectively.

Fig. 1 SEM images of (a) silkworm feces activated carbon before adsorption of oxamyl and (b) after adsorption

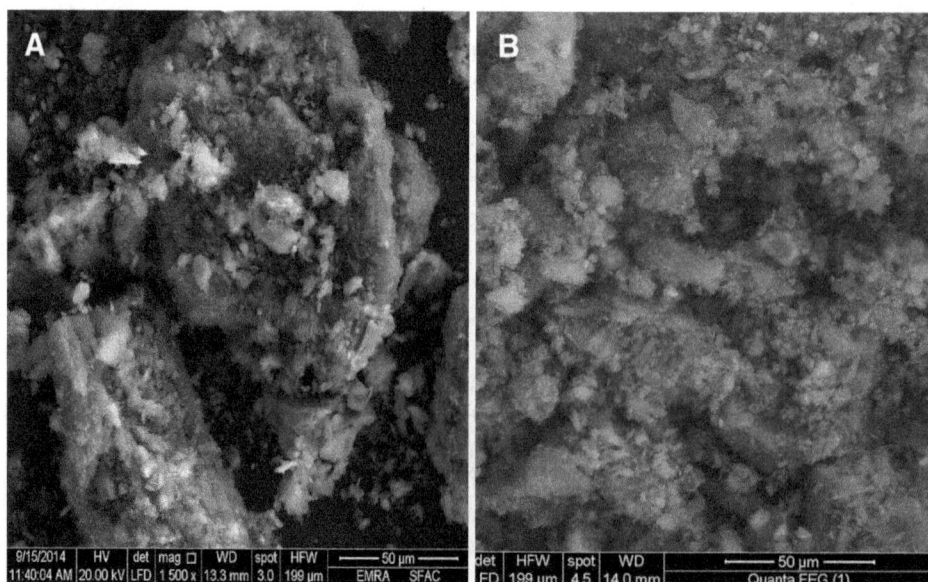

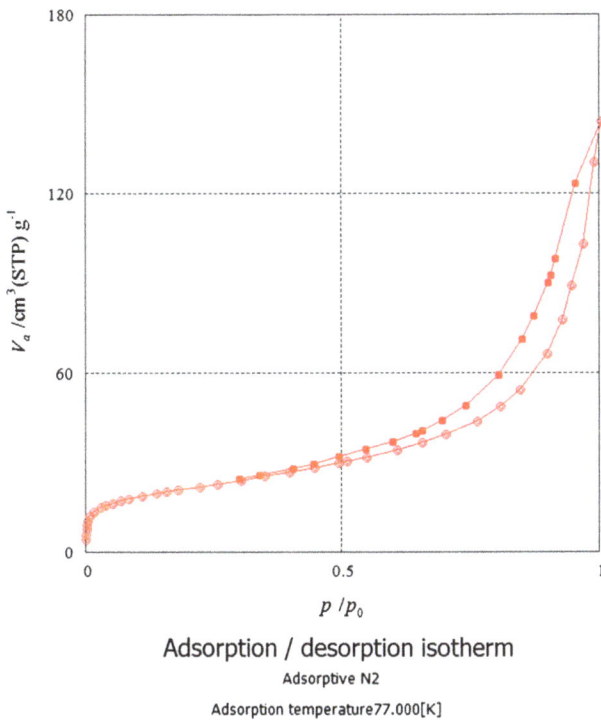

Fig. 2 N2 adsorption–desorption isotherms of the silkworm feces activated carbon

Function groups of SFAC

The FTIR spectrum of SFAC is illustrated in Fig. 3. A wide absorption band at 3431 cm^{-1} is assigned to O–H stretching vibrations of hydrogen bonded hydroxyl group of polymeric compounds, such as alcohols, phenols and carboxylic acids, as in pectin, cellulose groups on the adsorbent surface. The oxygen functional groups on the surface of SFAC greatly enhance its hydrophilic properties and also act as binding sites for the organic pollutant molecules [14].

The peaks at 2816 and 2921 cm^{-1} are attributed to the symmetric and asymmetric C–H stretching vibration of aliphatic acids. The peak observed at 1622 cm^{-1} is due to C=C stretching that can be attributed to the aromatic C–C bond, C–N and C–O– (1157 cm^{-1}). The bands located at 1077 cm^{-1} are attributed to C–H in plane. The bands around 1440 cm^{-1} are due to the symmetric bending of –CH$_3$. After loading, the most of characteristic peaks corresponding to these groups unchanged. However, there are also different changes in some peaks. The wave number of –OH group blueshifted from 3431 to 3405 cm^{-1} compared with that of SFAC. Hao et al. [30] had reported that the changes in some peaks of silkworm were seemed to be fact that –OH and–NH$_2$ participate in the adsorption process.

Adsorption of the pesticides was considered to take place mainly by dispersion forces between electrons in the oxamyl pesticide structure and electrons in the SFAC surface. The adsorption of oxamyl on SFAC may be mainly due to dispersion forces and polarization of π electrons (electron-rich portion of the adsorbate).

A covalent bond between the pesticide and the surface of the carbon is formed Al-Qodah and Shawabkah [5] and Guixia et al. [25]. Moreover, the ionization of nitrogen atoms and/or NH groups in pesticide molecules is likely to occur, leading to the adsorption of more pesticide molecules on the surface of the carbon. Figure 4 shows the interaction mechanism of oxamyl pesticide onto activated carbon prepared from silkworm feces.

X-ray diffraction

The XRD pattern of the SFAC sample (Fig. 5) showed that defined and considerably sharp peaks at 23.9° and 42° were attributed to the presence of carbon and graphite [12], while different peaks which can related to specific compounds were found in ash. Meanwhile, peaks at 25°, 27°, 31°, 35° and 42° were attributed to titanium oxide, calcium corresponding silicate magnetite and iron oxide, respectively.

The effect of the initial pH

To study the influence of pH on the adsorption capacity of SFAC for oxamyl, experiments were performed at room temperature and oxamyl initial concentration of 500 mg/L using different initial solution pH values, varying from 2 to 10. The obtained results of removal of oxamyl at different pH solution values are shown in Fig. 6. It is clear that oxamyl was unaffected by varying pH of solution. A similar trend of pH effect was observed for the adsorption of anthracene on activated carbon and P. oceanica [21] and Al-Zaben and Mekhamer [6]. Thus, medium pH was used to study the adsorption isotherms and kinetics.

The effect of adsorbent dose

Adsorbent dose is an important parameter in the determination of adsorption capacity for a given initial concentration of the adsorbate under the operating conditions. The effect of adsorbent dose on removal of oxamyl by SFAC was studied by varying the dose of adsorbent in the range of 0.1–1.5 g/100 mL solution, while all the other variables were kept constant. Batch experiments were conducted at initial concentration 500 mg/L. The results are shown in Fig. 7 for oxamyl. It is evident from the plots that the percentage removal of oxamyl from aqueous solution increases with increase in the adsorbent dose. It was observed that the removal efficiency increased from 99.18 to 99.49% for oxamyl with the adsorbent dose varying from 0.1 to 1 g, and thereafter, it reached a constant value,

Fig. 3 FTIR spectra of
(**a**) silkworm feces activated
carbon before adsorption of
oxamyl and (**b**) after adsorption

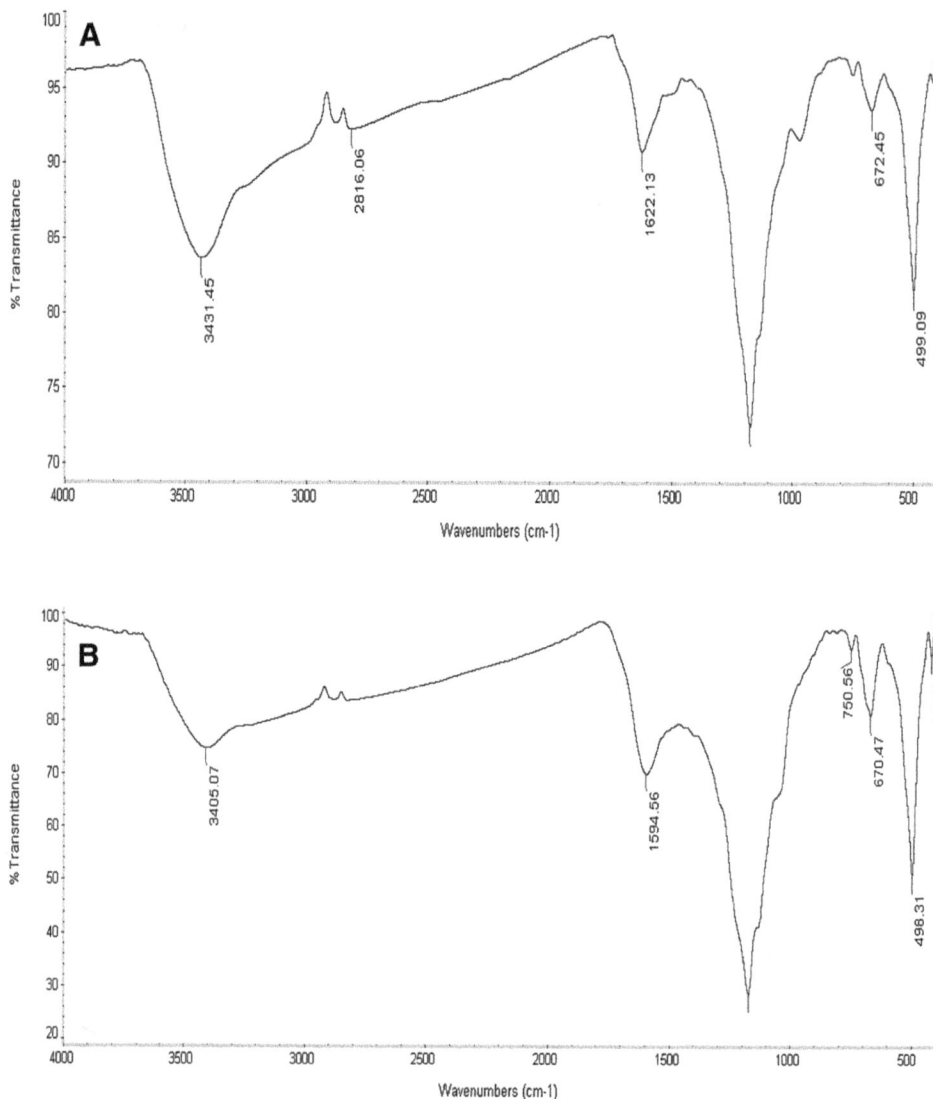

which is probably due to an increase in the number of binding sites available for biosorption [3, 48]. Further increase in adsorbent dose did not significantly change the biosorption yield. The optimum dose is 1 g/100 mL.

Effect of different concentrations

The adsorption experiments were conducted with different initial concentrations of oxamyl pesticide (100–2500 mg/L), 1 g/100 mL of SFAC for 2 h. The wide range of initial concentration of oxamyl was used to observe the adsorption performance of SFAC for oxamyl at both low and high concentrations of oxamyl. It has been found that the biosorption capacity of the adsorbent SFAC increases from 9.969 to 248.76 mg/g with increasing pesticide concentrations from 100 to 2500 mg/L, respectively. The increase in equilibrium adsorption capacity may be due to the utilization of all available active sites for adsorption at higher

oxamyl concentration, a larger mass transfer driving force and increased number of collision between pesticide molecules and SFAC, as shown in Fig. 8.

Effect of contact time

The rate at which adsorption takes place is most important during designing batch adsorption experiments. In order to determine the equilibrium time for maximum uptake of oxamyl, at a contact time study was performed with initial oxamyl concentration of 1000 mg/L, adsorbent dose (1 g for SFAC), at room temperature and contact time 5–180 min. The graphical representation of the contact time is given in Fig. 9. The adsorption of oxamyl is rapid for the first 30 min, and finally, equilibrium is established after about 120 min. The rapid pesticide adsorption at the initial stages of contact time could be attributed to the abundant availability of active sites on the surface of

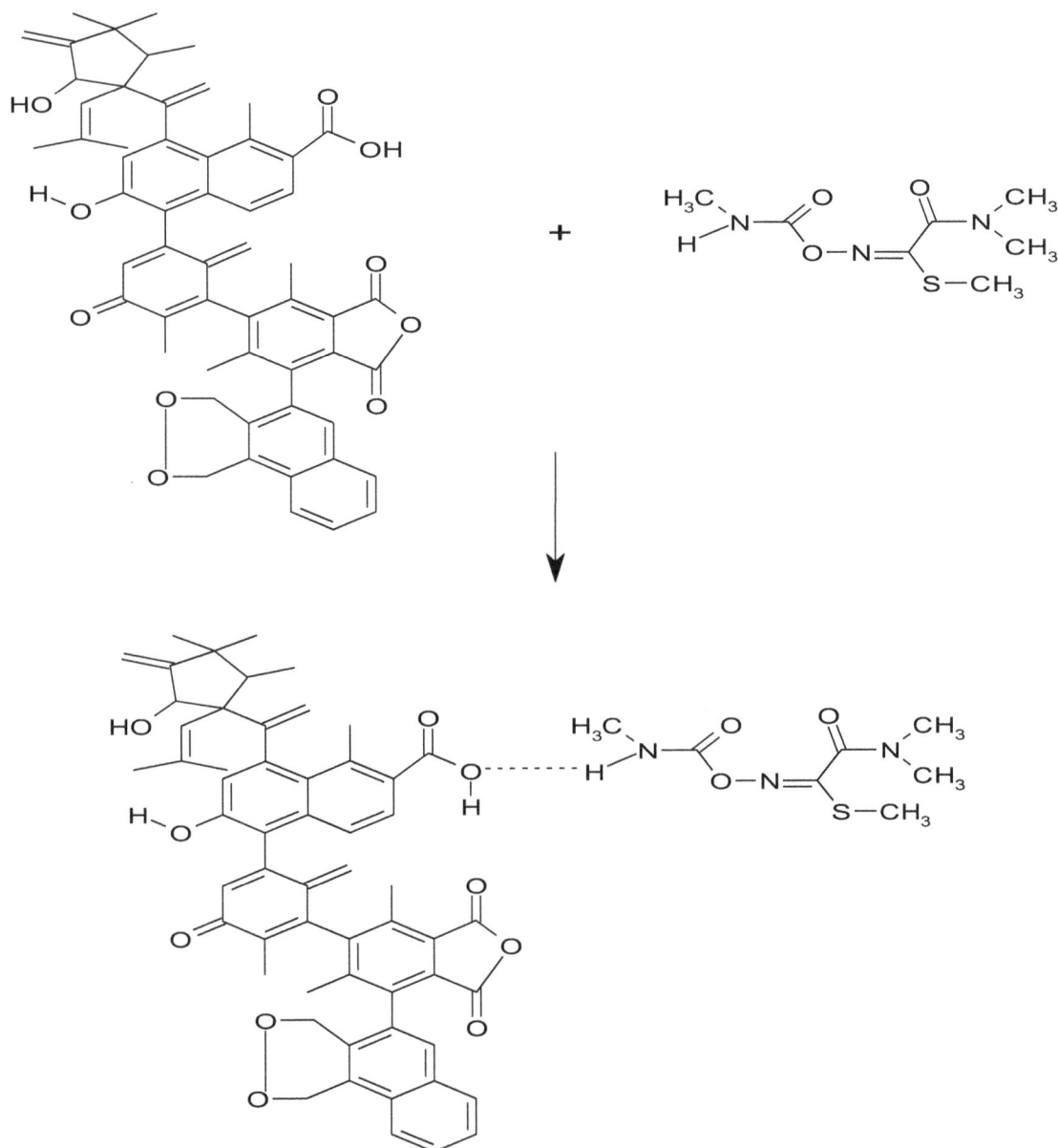

Fig. 4 Interaction mechanism of activated carbon with oxamyl

adsorbents. Afterward with the gradual occupancy of these sites, the adsorption became less efficient due to the decreased or lesser number of active sites [35]. Further increase in contact time did not enhance the adsorption, so, the optimum contact time for the adsorbent was selected as 120 min for further experiments. This is in agreement with the results obtained for hazelnut shell [16].

Biosorption kinetics

For analyzing the adsorption kinetics of oxamyl, the pseudo-first-order, pseudo-second-order and the intraparticle diffusion models were applied to the experimental data.

The pseudo-first-order rate equation is one of the most widely used equations for the adsorption of a solute from an aqueous solution and is represented as:

$$\log(q_e - q_t) = \log(q_e) - \frac{K_1}{2.303}(t) \tag{3}$$

where q_e and q_t are the amount of pesticide adsorbed (mg/g) at equilibrium and time t, respectively. K_1 is the first-order reaction rate constant (l/min). Examination of the data shows that the pseudo-first-order kinetic model is not applicable to oxamyl adsorption onto SFAC (data not shown) judged by low correlation coefficient.

The pseudo-second-order equation based on adsorption equilibrium capacity may be expressed as follows:

Fig. 5 XRD patterns of silkworm feces activated carbon

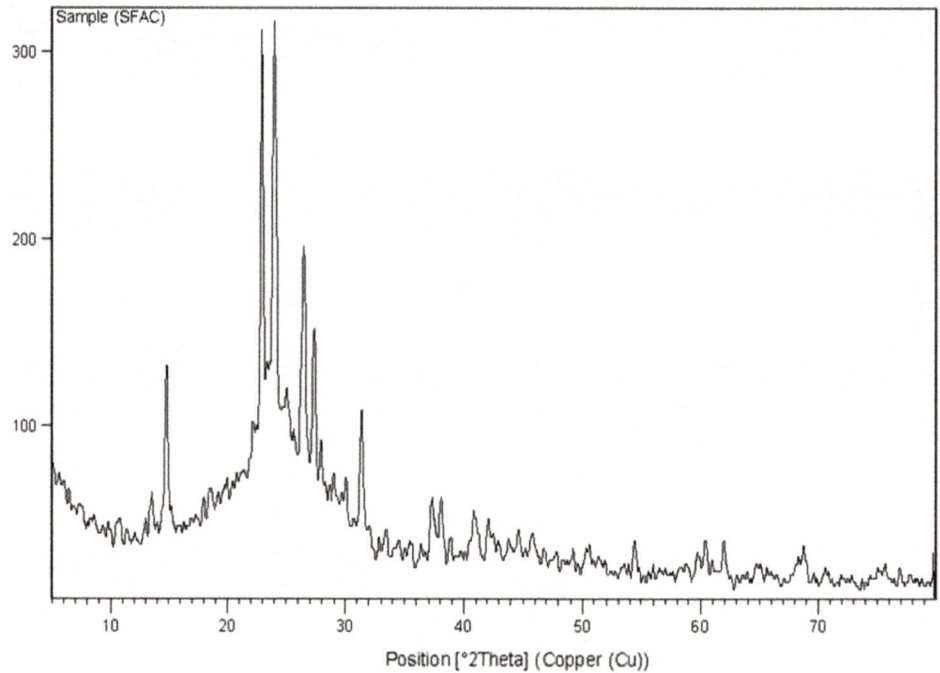

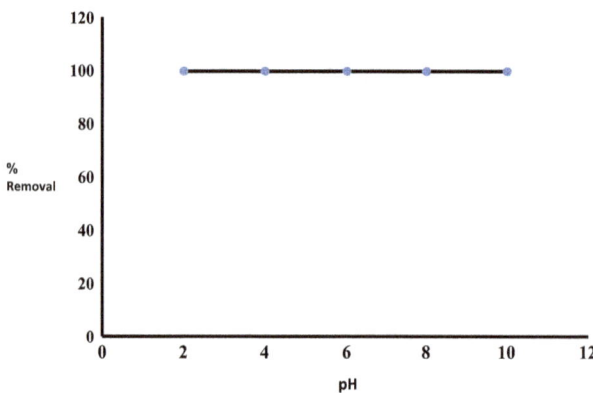

Fig. 6 Effect of pH for the removal of oxamyl using silkworm feces activated carbon (initial oxamyl concentration 500 mg/L, adsorbent dose 1 g/100 mL at room temperature)

$$\frac{t}{q_t} = \frac{1}{K_2 q_e^2} + \frac{1}{q_e}(t) \qquad (4)$$

where q_e is the equilibrium biosorption capacity and K_2 is the pseudo-second-order rate constant (g/mg min). A plot of (t/q_t) versus t gives a linear relationship for the applicability of the second-order kinetic model as shown in Fig. 10.

As seen in Table 1 due to high R^2, the pseudo-second-order model is predominant kinetic model for the oxamyl removal by silkworm feces activated carbon biosorbent. The similar kinetic result was reported for hazelnut shell, Pyracantha coccinea and drin pesticide on the surface of acid-treated olive stones [4, 18, 50]. According to the high regression coefficient, the adsorptions of oxamyl onto

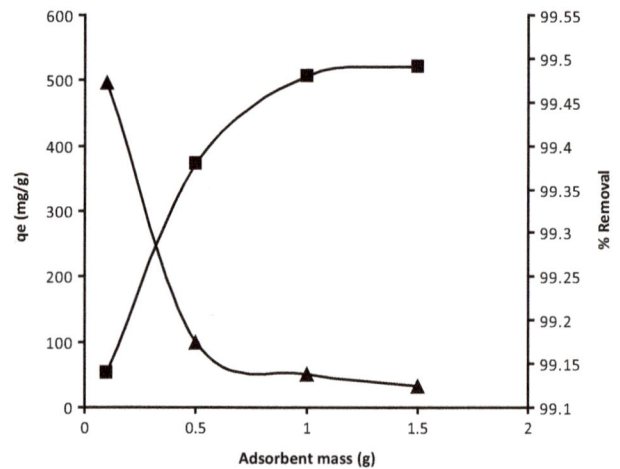

Fig. 7 Effect of different mass on removal of oxamyl using silkworm feces activated carbon (initial concentration of oxamyl 500 mg/L, adsorbent dose 0.1–1.5 g/100 mL at room temperature)

SFAC are best fitted by the pseudo-second-order kinetic model compared to pseudo-first-order kinetic model.

The adsorption mechanism

In order to identify the diffusion mechanism, the intraparticle diffusion model described by Weber and Morriss is [55].

$$q_t = K_i\, t^{1/2} + C \qquad (4)$$

where K_i is the intraparticle diffusion rate constant (mg/g min$^{1/2}$) and C (mg/g) is a constant that gives an idea about

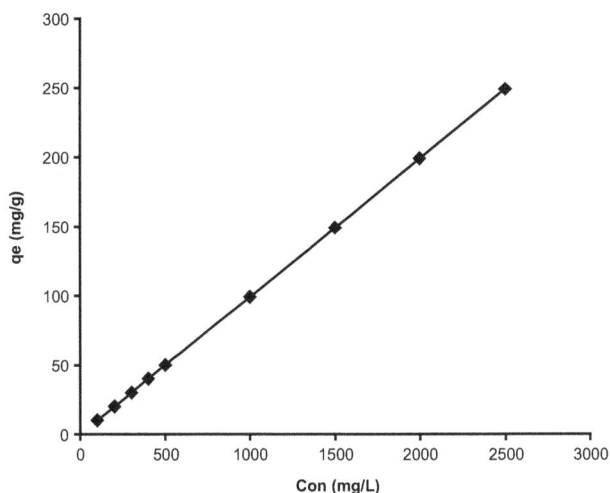

Fig. 8 Effect of initial concentrations of oxamyl (100–2500 mg/L) using silkworm feces activated carbon, adsorbent dose 1 g/100 mL at room temperature)

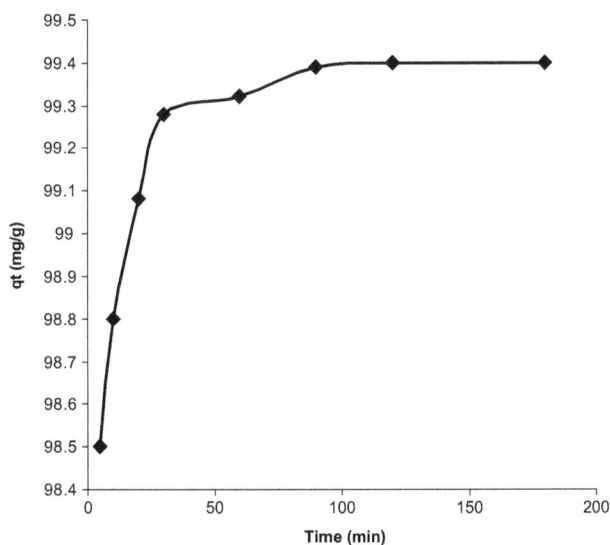

Fig. 9 Effect of contact time on the removal of oxamyl using silkworm feces activated carbon (initial oxamyl concentration 1000 mg/L, adsorbent dose 1 g/100 mL at room temperature)

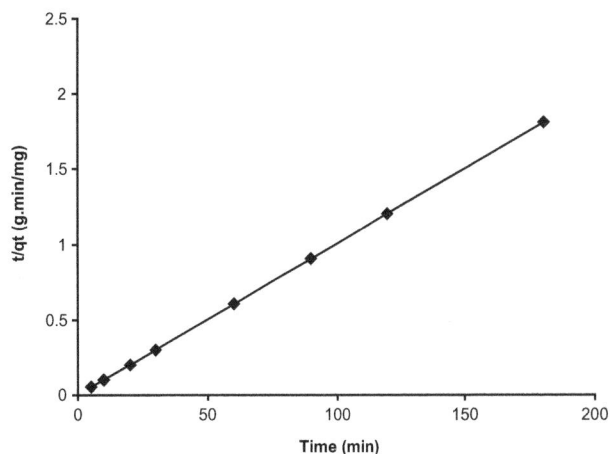

Fig. 10 Pseudo-second-order kinetic model for the removal of oxamyl using silkworm feces activated carbon (initial oxamyl concentration 1000 mg/L, adsorbent dose 1 g/100 mL at room temperature)

Table 1 Kinetic parameters for the removal of oxamyl using silkworm feces activated carbon

Kinetic model	Parameter	Value
Pseudo-second order	K_2 (g/mg min)	0.204
	R^2	1.000
	q_e (mg/g)	99.00

with the results obtained for Araucaria angustifolia [13] and garlic peel, Hameed and Ahmad [28].

Adsorption isotherm

Adsorption isotherm expresses the relationship between pesticide adsorbed onto the adsorbents and pesticide in the solution and provides important design parameters for adsorption system. Several equilibrium models have been used to describe the adsorption data. Langmuir, Freundlich and Dubinin–Radushkevich (D-R) models are the most widely used.

The Freundlich adsorption model deals with non-ideal sorption onto heterogeneous surfaces involving multilayer sorption. The linear form of the Freundlich adsorption isotherm is

$$\log q_e = \log K_f + 1/n \log C_e \qquad (5)$$

where K_f is adsorption capacity and $1/n$ is related to the degree of surface heterogeneity (smaller value indicates more heterogeneous surface, whereas value closer to or even 1.0 indicates a material with relatively homogenous binding sites). The value of n obtained as a result of plotting of log C_e versus log q_e is shown in Fig. 11. Data given in Table 2 suggest that adsorption of oxamyl onto

the thickness of the boundary layer, i.e., the larger the value of C, the greater the boundary layer effect. The value of K_i was calculated from the slope of the linear plot of q_t versus $t^{1/2}$. According to this model, if the regressions of q_t versus $t^{1/2}$ is linear and pass through the origin, then intraparticle diffusion is the rate controlling step. Examination of the data showed that the regression was linear, but the plot did not pass through the origin (data not shown), suggesting that removal of oxamyl on silkworm feces activated carbon involved intraparticle diffusion but not the only rate controlling step. Other kinetic models may control the adsorption rate.

It could be stated that this process is complex and may involve more than one mechanism. This is in accordance

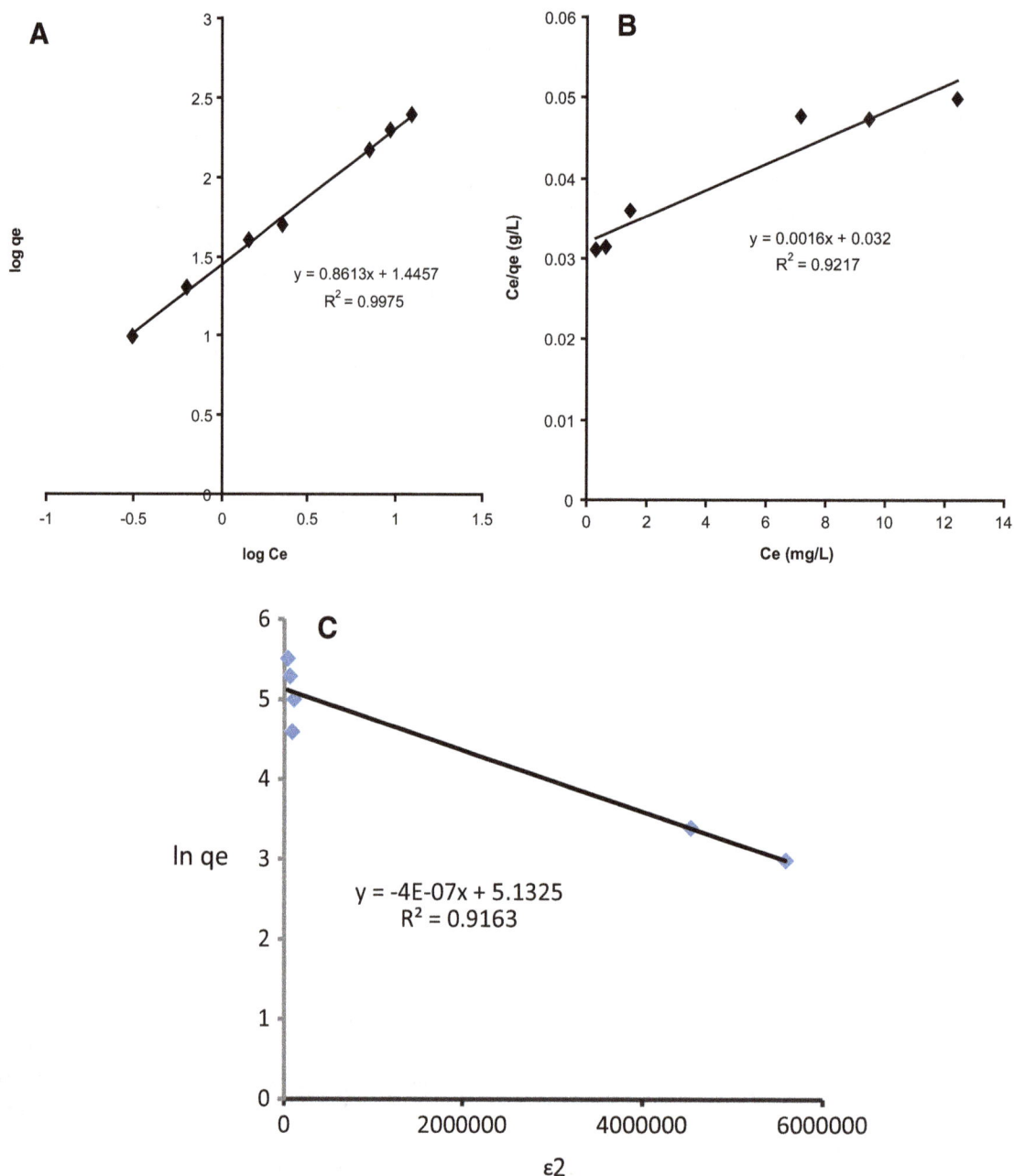

Fig. 11 Freundlich isotherm (**a**), Langmuir (**b**) and Dubinin–Radushkevich (D-R) isotherm (**c**) for removal of oxamyl using silkworm feces activated carbon (initial oxamyl concentration 100–2500 mg/L, adsorbent dose 1 g/100 mL at room temperature)

Table 2 Comparison of various isotherm constants for the removal of oxamyl using silkworm feces activated carbon

Freundlich		Langmuir		D-R	
K_f	27.906	q_m (mg/g)	625.00	q_e (mg/g)	169.44
n	1.161	b (L/mg)	0.05	E (KJ/mol)	1.11
R^2	0.9975	R^2	0.9217	R^2	0.9163
SEE	0.02	SEE	0.005	SEE	0.88

SFAC is well represented by Freundlich isotherm model and support the assumption that adsorption takes place on heterogeneous surfaces (H.M.F 1906).

The values of 'n' indicate the degree of nonlinearity between solution concentration and adsorption. If the value of n is equal to unity, the adsorption is linear; if the value is below unity, this implies that adsorption process is chemical; if the value is above unity, adsorption is a favorable physical process [27, 40].

The Langmuir model suggests that uptake occurs on a homogeneous surface by monolayer sorption without interaction between the adsorbed molecules [56]. The linear form of Langmuir adsorption isotherm is

$$\frac{C_e}{q_e} = \frac{1}{Q_m b} + \frac{C_e}{Q_m} \quad (6)$$

where q_e is the amount adsorbed onto adsorbent at equilibrium, b is the Langmuir constant, and q_m is the monolayer adsorption capacity. The plot of C_e/q_e versus C_e is employed to generate the intercept value of $1/bq_m$ and slope of $1/q_m$ as shown in Fig. 11.

One of the essential characteristics of this model can be expressed in terms of the dimensionless separation factor for equilibrium parameter, R_L, defined by [22]

$$R_L = \frac{1}{1 + bC_0} \quad (7)$$

The value of R_L indicates the type of isotherm to be irreversible ($R_L = 0$), favorable ($0 < R_L < 1$), linear ($R_L = 1$) or unfavorable ($R_L > 1$). The value of R_L in the present investigation was found to be 0.007, indicating that the adsorption of oxamyl on SFAC is favorable. Adsorption capacity from the present study was compared with other adsorbents from previous studies [8] which is shown in Table 2. q_{max} of oxamyl in silkworm feces was almost 4 times greater than that in apricot stone which is shown in Table 2 according to the source of the activated carbon. Also the starting material significantly influences physical and chemical properties of activated carbon as a result.

Based on the correlation coefficient (R^2) shown in Table 2, the adsorption isotherms of oxamyl on SFAC can be slightly better described by the Freundlich equation. Thus, the results of the present study indicate that biosorption of oxamyl onto SFAC is heterogeneous in nature. However, the sum of squared error SEE test (Table 2) for Langmuir isotherm model exhibited lower values than for Freundlich isotherm model. Similar results have been reported for the adsorption of phosphate ions by pine cone [11].

Dubinin–Radushkevich (D-R) proposed another equation used in the analysis of isotherms. D-R model was applied to estimate the porosity apparent free energy and the characteristic of adsorption [17]. The D-R isotherm does not assume a homogeneous surface or constant sorption potential, and it has commonly been applied in the following form Eq. (8) and its linear form can be shown in Eq. (9):

$$q_e = q_m \exp\left(-K\varepsilon^2\right) \quad (8)$$

$$\ln q_e = \ln q_m - \beta\varepsilon^2 \quad (9)$$

Table 3 Adsorption capacities of oxamyl by various adsorbents

Adsorbent	q_{max} (mg/g)	Reference
Apricot stone activated carbon	147.05	[29]
Silkworm feces activated carbon	625.0	This work

where K is a constant related to the adsorption energy, q_e (mg/g) is the amount of pesticide adsorbed per g of adsorbent, q_m represents the maximum adsorption capacity of adsorbent, β (mol^2/J^2) is a constant related to adsorption energy, while ε is the Polanyi potential that can be calculated from Eq. (10):

$$\varepsilon = RT \ln\left[1 + \frac{1}{C_e}\right] \quad (10)$$

The values of β and q_m can be obtained by plotting $\ln q_e$ vs. ε^2. The mean free energy of adsorption (E, J/mol), defined as the free energy change when one mole is transferred from infinity in solution to the surface of the sorbent, is calculated using the following relation Eq. (11):

$$E = 1 \left/ \sqrt{2\beta} \right. \quad (11)$$

The calculated values of D-R parameters are given in Fig. 11 and Table 2. The saturation adsorption capacities q_m obtained using D-R isotherm model for adsorption of oxamyl SFAC are 169.44 mg/g. The mean energy of adsorption is the free energy change when one mole of pesticide is transferred to the surface of the solid from infinity in the solution. The value of this parameter can give information about adsorption mechanism. When one mole of pesticide is transferred, its value in the range of 1–8 kJ/mol suggests physical adsorption [47], the value of E is between 8 and 16 kJ/mol, which indicates the adsorption process, follows by ion-exchange [31], while its value in the range of 20–40 kJ/mol is indicative of chemisorption [51]. So, the values of E calculated is 1.11 kJ/mol, indicating that weak physical forces such as van der Waals and hydrogen bonding affect the adsorption process of oxamyl onto SFAC. These E values are in agreement with [2] for the adsorption of dyes by loofa activated carbon and drin pesticides onto acid-treated olive stones [32].

Adsorption capacity from the present study was compared with other adsorbents from previous studies [8] are shown in Table 3. The maximum adsorption capacity qmax of oxamyl by silkworm feces was almost 4 times greater than that in Apricot stone in Table 3 according to the source of the activated carbon. Also the starting material significantly influences physical and chemical properties of activated carbon as a result.

Conclusion

The present study focused on removal of oxamyl pesticide from aqueous solution using eco-friendly silkworm feces-based activated carbon. The adsorption has been examined with the variations in the parameters of activated carbon dose, initial oxamyl concentration and contact time. The experimental data were analyzed using Langmuir, Freundlich and Dubinin–Radushkevich (D-R) isotherm models. The Freundlich model provides the best correlation of the experimental equilibrium data. The adsorption system obeys the pseudo-second-order kinetic model. The results indicated that the SFAC could be a promising biosorbent for the removal of oxamyl from aqueous solution.

Acknowledgements Many thanks to all the members of Central Agricultural Pesticides Laboratory (CAPL), Agricultural Research Center, Dokki, Giza, Egypt, for their valuable assistance and facilities they provided.

References

1. Abdel-Ghani N, El-Chaghaby G, Zahran E (2015) Pentachlorophenol (PCP) adsorption from aqueous solution by activated carbons prepared from corn wastes. Int J Environ Sci Tech 12(1):211–222
2. Abdelwahab O (2008) Evaluation of the use of loofa activated carbons as potential adsorbents for aqueous solutions containing dye. Desalination 222:357–367
3. Ahmad R (2009) Studies on adsorption of crystal violet dye from aqueous solution onto coniferous pinus bark powder (CPBP). J Hazard Mater 171:767–773
4. Akar T, Celik S, Akar ST (2010) Biosorption performance of surface modified biomass obtained from *Pyracantha coccinea* for the decolonization of dye contaminated solutions. Chem Eng J 160(2):466–472
5. Al-Qodah Z, Shawabkah R (2009) Production and characterization of granular activated carbon from activated sludge. Braz J Chem Eng 26:127–136
6. Al-Zaben MI, Mekhamer WK (2013) Removal of 4-chloro-2-methyl phenoxy acetic acid pesticide using coffee wastes from aqueous solution. Chem, Arabian J. doi:10.1016/j.arabjc.2013.05.003 **(In press, corrected proof)**
7. Ayranc JE, Hoda N (2005) Adsorption of phthalic acid and its esters onto high-area activated carbon-cloth studied by in situ UV-spectroscopy. Chemosphere 60:1600–1607
8. Ahmed SM, Mohammad SG (2014) Egyptian apricot stone (Prunus armeniaca) as a low cost and eco-friendly biosorbent for oxamyl removal from aqueous solutions. Am J Exp Agri 4(3):302–321
9. Bansal OP (2004) Kinetics of interaction of three carbamate pesticides with Indian soils: aligarh district. Pest Manag Sci 60(11):1149–1155
10. Becker DL, Wilson SC (1980) Carbon adsorption Handbook In: Chereminisoff PN, Ellebush F (eds) Ann Harbor Science Publishers, Michigan. 167–212
11. Benyoucef S, Amrani M (2011) Adsorption of phosphate ions onto low cost *Aleppo pine* adsorbent. Desalination 275:231–236
12. Bouchelta C, Medjram MS, Bertrand O, Bellat JP (2008) Preparation and characterization of activated carbon from date stones by physical activation with steam. J Anal Appl Pyrolysis 82:70–77

13. Calvete T, Lima EC, Cardoso NF, Dias SLP, Pavan FA (2009) Application of carbon adsorbents prepared from the Brazilian-pine fruit shell for removal of Procion Red MX 3B from aqueous solution-kinetic, equilibrium and thermodynamic studies. Chem Eng J 155:627–636
14. Chen H, Wang X, Li J, Wang X (2015) Cotton derived carbonaceous aerogels for the efficient removal of organic pollutants and heavy metal ions. J Mater Chem A 3:6073–6081
15. Cobas M, Meijide J, Sanromán MA, Pazos M (2016) Chestnut shells to mitigate pesticide contamination. J Taiwan Inst Chem Eng 61:166–173
16. Dogan M, Abak H, Alkan M (2009) Adsorption of methylene blue onto hazelnut shell: kinetics, mechanism and activation parameters. J Hazard Mater 164(1):172–181
17. Dubinin MM, Zaverina ED, Radushkevich LV (1947) Sorption and structure of active carbons. I. Adsorption of organic vapors. ZhFizKhim. 21:1351–1362
18. El Bakouri H, Usero J, Morillo J, Ouassini A (2009) Adsorptive features of acid-treated olive stones for drin pesticides: equilibrium, kinetic and thermodynamic modeling studies. Bioresour Technol 100:4147–4155
19. ElShafei GMS, ElSherbiny IM, Darwish AS, Philip C (2014) Silkworms' feces-based activated carbons as cheap adsorbents for removal of cadmium and methylene blue from aqueous solutions. Chem Eng Res Des 92(3):461–470
20. ElShafei GMS, Nasr IN, Ayman SM, Mohammad SG (2009) Kinetics and thermodynamics of adsorption of cadusafos on soils. J Hazard Mater 172:1608–1616
21. El Khames M, Ramzi K, Elimame E, Younes M (2014) Adsorption of anthracene using activated carbon and *Posidonia oceanica*. Arabian J Chem 7:109–113
22. Farooq U, Kozinski JA, Khan MA, Athar M (2010) Biosorption of heavy metal ions using wheat based biosorbents-a review of the recent literature. Bioresour Technol 101:5043–5053
23. Fernandez ME, Nunell GV, Bonelli PR, Cukierman AL (2014) Activated carbon developed from orange peels: batch and dynamic competitive adsorption of basic dyes. Ind Crops Products 62:437–445
24. Gong J, Yang C, Zhang W (2011) Liquid phase deposition of tungsten doped TiO$_2$ films for visible light photoelectrocatalytic degradation of dodecyl benzenesulfonate. Chem Eng J 167:190–197
25. Guixia Z, Lang J, Yudong H, Li Jiaxing, Huanli D, Xiangke W, Wenping H (2011) Sulfonated graphene for persistent aromatic pollutant management. Adv Mater 23:3959–3963
26. Freundlich HMF (1906) Uber die adsorption in lasungen. Z Phys Chem 57(1906):385–470
27. Hadi M, Samarghandi MR, McKay G (2010) Equilibrium two parameter isotherms of acid dyes sorption by activated carbons: study of residual errors. Chem Eng J 160:408–416
28. Hameed BH, Ahmad AA (2009) Batch adsorption of methylene blue from aqueous solution by garlic peel, an agricultural waste biomass. J Hazard Mater 164:870–875
29. Han R, Ding D, Xu Y, Zou W, Wang Y, Li Y, Zou L (2008) Use of rice husk for the adsorption of Congo red from aqueous solution in column mode. Bioresour Technol 99:2938–2946
30. Hao C, Jie Z, Guoliang D (2011) Silkworm exuviae—A new non-conventional and low-cost adsorbent for removal of methylene blue from aqueous solutions. J Hazard Mater 186:1320–1327
31. Helfferich F (1962) Ion Exchange. McGraw-Hill Book Co, New York
32. El Hicham, Jose U, Jose M, Abdelhamid O (2009) Adsorptive features of acid-treated olive stones for Drin pesticides: equilibrium, kinetic and thermodynamic modeling studies. Bioresour Tech. 100:4147–4155

33. Heibati B, Rodriguez-Couto S, Amrane A, Rafatullah M, Hawari A, Al-Ghouti MA (2014) Uptake of Reactive Black 5 by pumice and walnut activated carbon: chemistry and adsorption mechanisms. J Ind Eng Chem 20(5):2939–2947

34. Katsumata H, Kobayashi T, Kaneco S, Suzuki T, Ohta K (2011) Degradation of linuron by ultrasound combined with photo-Fenton treatment. Chem Eng J 166:468–473

35. Kannan N, Karrupasamy K (1998) Low cost adsorbents for the removal of phenyl acetic acid from aqueous solution. Indian J Environ Prot 18:683–690

36. Kouras A, Zouboulis A, Samara C, Kouimtzis T (1998) Environ Pollut 103:193–202

37. Li D, Yan J, Liu Z, Liu Z (2016) Adsorption kinetic studies for removal of methylene blue using activated carbon prepared from sugar beet pulp. Int J Environ Sci Technol 13:1815–1822

38. Naa Liu, Alberto BC, Chih HW, Xiaoling Y, Feng D (2015) Characterization of biochars derived from agriculture wastes and their adsorptive removal of atrazine from aqueous solution: a comparativestudy. Bioresource Tech. 198:55–62

39. Mahmoud MS, Ahmed SM, Mohamamd SG, Abou Elmagd AM (2014) Evaluation of Egyptian Banana Peel (*Musa* sp.) as a green sorbent for groundwater treatment. Int J Eng Tech 4(11):648–659

40. Mahmoodi NM, Arami M (2008) Modeling and sensitivity analysis of dyes adsorption onto natural adsorbent from colored textile wastewater. J Appl Polym Sci 109:4043–4048

41. Maldonado MI, Malato S, Perez-Estrada LA, Gernjak W, Oller I, Domenech X (2006) Partial degradation of five pesticides and an industrial pollutant by ozonation in a pilot-plant scale reactor. J Hazard Mater 38:363–369

42. Memon GZ, Moghal M, Memon JR, Memon NN, Bhanger M (2014) Adsorption of selected pesticides from aqueous solutions using cost effective walnut shells. J Eng 4:43–56

43. Mohammad SG (2013) Biosorption of pesticide onto a low cost carbon produced from Apricot Stone (*Prunus armeniaca*): equilibrium, kinetic and thermodynamic studies. J Appl Sci Res 9(10):6459–6469

44. Mohammad SG, Ahmed SM, Badawi AM (2015) A comparative adsorption study with different agricultural waste adsorbents for removal of oxamyl pesticide. Desalination Wat Treat 55(8):2109–2120

45. Naushad M, Alothman ZA, Khan MR (2013) Removal of malathion from aqueous solution using De-Acidite FF-IP resin and determination by UPLC-MS/MS: equilibrium, kinetics and thermodynamics studies. Talanta 15(115):15–23

46. Njoku VO, Hameed BH (2011) Preparation and characterization of activated carbon from corncob by chemical activation with H_3PO_4 for 2,4-dichlorophenoxyacetic acid adsorption. Chem Eng J 173:391–399

47. Onyango MS, Kojima Y, Aoyi O, Bernardo EC, Matsuda H (2004) Adsorption equilibrium modeling and solution chemistry dependence of fluoride removal from water by trivalent cation exchanged zeolite F-9. J Colloid Interface Sci 279:341–350

48. Saeed A, Sharif M, Iqbal M (2010) Application potential of grapefruit peel as dye sorbent: kinetics, equilibrium and mechanism of crystal violet adsorption. J Hazard Mater 179:564–572

49. Sadia S, Nasar Abu (2016) Removal of methylene blue dye from artificially contaminated water using citrus limetta peel waste as a very low cost adsorbent. J Taiwan Inst Chem Eng 66:154–163

50. Safa Y, Bhatti HN (2011) Kinetic and thermodynamic modeling for the removal of Direct Red-31 and Direct Orange-26 dyes from aqueous solutions by rice husk. Desalination 272(1–3):313–322

51. Tahir SS, Rauf N (2006) Removal of cationic dye from aqueous solutions by adsorption onto bentonite clay. Chemosphere 63:1842–1848

52. Tran VS, Ngo HH, Guo W, Zhang J, Liang S, Ton-That C, Zhang X (2015) Typical low cost biosorbents for adsorptive removal of specific organic pollutants from water. Bioresour Technol 182:353–363

53. Uğurlu M, Karaoğlu MH (2011) TiO_2 supported on sepiolite: preparation, structural and thermal characterization and catalytic behaviour in photocatalytic treatment of phenol and lignin from olive mill wastewater. Chem Eng J 166:859–867

54. Vithanage M, Mayakaduwa SS, Herath I, Sik YO, Dinesh M (2016) Kinetics, thermodynamics and mechanistic studies of carbofuran removal using biochars from tea waste and rice husks. Chemosphere 150:781–789

55. Weber WJ, Morriss JC (1963) Kinetics of adsorption on carbon from solution. J Sanit Eng Div Am Soc Civil Eng 9:31–60

56. Zainal IG (2010) Biosorption of Cr(VI) from aqueous solution using new adsorbent: equilibrium and thermodynamic study. E J Chem 7:S488–S494

57. Zhou T, Lim TT, Chin SS, Fane AG (2011) Treatment of organics in reverse osmosis concentrate from a municipal wastewater reclamation plant: feasibility test of advanced oxidation processes with/without pretreatment. Chem Eng J 166:932–939

3

Saccocalyx satureioides as corrosion inhibitor for carbon steel in acid solution

M. Benahmed[1] · N. Djeddi[1] · S. Akkal[2] · H. Laouar[3]

Abstract The inhibitory effect of the crude ethyl acetate extract of *Saccocalyx satureioides* was estimated on the corrosion of carbon steel (X52) in 1 M HCl solution sing weight loss measurement, potentiodynamic polarization, electrochemical impedance spectroscopy and scanning electron microscopy techniques. Potentiodynamic polarization curves indicated that the plants extract behaves as a mixed-type inhibitor. The adsorption of the inhibitor on the carbon steel surface was found to follow Freundlich adsorption isotherm. The free energies, enthalpies and entropies for the adsorption and dissolution process were discussed in details. The inhibiting action increases with increasing concentration of the extract. The results obtained show that the extract of the aerial parts of *Saccocalyx satureioides* could serve as an effective corrosion inhibitor of carbon steel (X52) in hydrochloric acid medium.

Keywords Carbon steel X52 · *Saccocalyx satureioides* · Inhibition

✉ M. Benahmed
  riad43200@yahoo.fr

1   Laboratoire des Molécules Bioactives et Applications, Université Larbi Tébessi, Route de Constantine, 12000 Tébessa, Algeria

2   Laboratoire de Phytochimie et Analyses physicochimiques et Biologiques, Département de Chimie, Faculté de Sciences exactes, Université Mentouri Constantine, Route d'Ain el Bey, 25000 Constantine, Algeria

3   Laboratoire de Valorisation des Ressources Naturelles Biologiques, Département de Biologie, Université Ferhat Abbas de Sétif, Sétif, Algeria

Introduction

Metals and alloys are frequently used in industrial applications due to their convenience and low cost. Among these, carbon steel is one of the most important alloys which are continually used. Unfortunately, it suffers serious acid corrosion.

To prevent carbon steel against acid corrosion, inhibitors are often used. Unfortunately, some of them are not environmentally friendly. Therefore, the uses of eco-friendly and biodegradable products are preferred. Recently, many published works shown that some naturally occurring substances have corrosion inhibitor properties for different metals in various environments [1–6].

Saccocalyx satureioides (Lamiaceae) is an Algerian endemic species, which grows in the dunes of the pre-desert area. This plant is 20–100 cm high. Its flowers can be white, rose or crimson [7]. Various studies reported the chemical composition of *S. satureioides* oils obtained from the same plant but collected in North and Northeast of Algeria. Some of these investigation reported that this species reveal a strong antimicrobial activity against *Staphylococcus aureus*, *Escherichia coli* and *Klebsiella* [8–10]. Moreover, a research work reported by Mohamadi et al. [11] revealed the identification of seventeen compounds from *S. satureioides*: piceol, vanillin, ferulic aldehyde, 3,3′-bis (3,4-dihydro-4-hydroxy-6,8-dimethoxy-2H-1-benzopyran), 3,3-bis (3,4-dihydro-4-hydroxy-6-methoxy-2H-1-benzopyran), dimethylcaffeic acid, balanophonin, 7-methyl-sudachitin, caffeicacid, *p*-coumaric acid, isoscutellarein-7-*O*-[β-D-allopyranosyl-(1 → 2)]β-D-glucopyranoside, isoscutellarein-7-*O*-[β-D-allopyranosyl-(1 → 2)]-6″-*O*-acetyl-β-D-lucopyranoside, isoscutellarein-7-*O*-[6‴-*O*-acetyl-β-D-allopyranosyl-(1 → 2)]-β-D-glucopyranoside, quercetin, isoscutellarein-7-*O*-[6‴-*O*-acetyl- β-Dallo pyranosyl-(1 → 2)]-6''-O-acetyl–D β-glucopyranoside, apigenin-7-O-[6''-trans-p-coumaroyl]-β-D-glucopyranoside and sideritiflavone.

In this work, we describe the testing of ethyl acetate extract parts S. satureioides as corrosion inhibitor for carbon steel (X52) in HCl solution, using weight loss and electrochemical techniques, such as potentiodynamic polarization and electrochemical impedance spectroscopy (EIS).

Experimental

Plant material

The aerial part of Saccocalyx satureioides was picked up during June 2011, in El Masrane near Djelfa, Algeria. After identification, a voucher specimen was deposited in the Herbarium of the Museum of natural history of Nice city (Voucher number B-6309).

Extraction and isolation

The air-dried aerial parts of S. satureioides (800 g) were extracted three times with boiling methyl alcohol (70 %). The hydro-alcoholic solutions were concentrated in vacuum to aridness and the residue was dissolved in hot water and kept overnight at room temperature. After filtration, the aqueous solution was successively treated with ethyl acetate and n-butanol, and then their extracts were concentrated to dryness [12, 13].

Materials

Carbon steel (X52), with the chemical composition displayed in Table 1, was used in this study. Coupons of carbon steel (X52) with dimensions $1.0 \times 0.8 \times 0.4$ cm were used for weight loss measurements. For electrochemical measurements, a steel cube embedded in epoxy resin, leaving an exposed surface area of 0.8 cm^2 as a working electrode. The coupons were abraded with different grade of emery papers, (600, 800, 1200 and 2000), degreased with acetone and rinsed with distillated water, before its immersion in 1 M HCl with and without the addition of different concentrations of Saccocalyx satureioides ethyl acetate extract (SSE).

Corrosion tests

Weight loss measurements

Weight loss measurements were conducted under total immersion in stagnant aerated condition using 250 mL capacity beakers containing 150 ml test solution at 293–323 K maintained in a thermostated water bath. Carbon steel (X52) coupons were weighed and suspended in the previous beakers. After 6 h, the coupons were taken out, washed in distilled water, dried and then weighed. From the weight loss values, corrosion rates (Cr), surface coverage and the inhibition efficiency IE (%) were calculated using the expressions [14]:

$$Cr = \frac{\Delta W}{At} \quad (1)$$

where ΔW is the weight loss, A is the sectional area of carbon steel (X52), t is the exposure time.

$$\theta = \frac{Cr_0 - Cr_i}{Cr_0} \quad (2)$$

$$IE(\%) = \theta \times 100 \quad (3)$$

where θ the surface coverage, Cr_0 and Cr_i are the corrosion rates of the carbon steel (X52) coupons in absence and presence of inhibitor, respectively.

Potentiodynamic polarization measurement

All electrochemical measurements were carried out using a computer controlled Voltalab PGZ 301 instrument with Voltamaster software at room temperature, without and with the addition of different concentrations of SSE to 1 M HCl solution. Open circuit potential (E_{ocp}), was measured for 60 min to allow stabilization of the steady state potential. The potential of the potentiodynamic polarization curves ranged from a cathodic potential of -100 mV to an anodic potential of $+100$ mV versus OCP at a scan rate of 1 mV s^{-1}. The inhibition efficiency IE (%) was calculated using the relation:

$$IE\ (\%) = \frac{i_{corr} - i_{corr(i)}}{i_{corr}} \times 100 \quad (4)$$

where i_{corr} and $i_{corr(i)}$ are the corrosion current density without and with the addition of the inhibitor, respectively.

Table 1 Chemical composition of carbon steel (X52)	Composition	C	Mn	Si	S	P	Fe
	Wt%	0.1039	0.9710	0.1261	0.0021	0.0020	Remainder

Electrochemical impedance spectroscopy

Electrochemical impedance spectroscopy measurements were carried out using the same instrument described above with the open circuit potential. Every sample was immersed for 30 min over a frequency range of 100 kHz–0.01 Hz with applied potential signal amplitude of 5 mV. The percentage inhibition efficiency IE (%) was calculated from,

$$IE(\%) = \frac{R_i - R_0}{R_i} \times 100 \qquad (5)$$

where R_0 and R_i are charge transfer resistance of carbon steel (X52) in uninhibited and inhibited solutions, respectively.

Scanning electron microscope (SEM)

The specimens used to examine the surface morphology were prepared as described above then immersed in 1 M HCl in the absence and presence of 900 mg L^{-1} SSE at room temperature. SEM images were taken for specimens that have been submerged for 6 h. The SEM used was a JEOL model JSM 6390LV.

Results and discussion

Weight loss measurements

Corrosion rate and inhibition efficiency of carbon steel (X52) in 1 M HCl solution at 293–323 K in the absence and presence SSE were shown in Table 2. From the values obtained, it is clear that the presence of inhibitor leads to a reduction of the corrosion rate. It is indicated that inhibition efficiency of carbon steel (X52) rises with increasing SSE concentration up to 87 % at 293 K (Fig. 1a). Several scientists in their studies (Table 3) reported similar results [2, 4, 6, 15–19], which confirm that the SSE act as effective inhibitors. The evolution in inhibition efficiency may be attributed to the adsorption of the particles at the carbon steel (X52) surfaces [20]. As the temperature increases, the corrosion rate (Cr) increases and inhibition efficiency decreases (Fig. 1b). This suggests a probable desorption of some adsorbed inhibitor from the steel surface at higher temperature. This behavior indicates that the phytochemical components are physically adsorbed on the carbon steel surface [14].

Adsorption isotherm

The degrees of surface coverage (θ) for different inhibitor concentrations were calculated by weight loss data. The

Table 2 Corrosion rate data of carbon steel (X52) in 1 M HCl for various concentrations of SSE

C (mg L^{-1})	Temperature											
	293 K			303 K			313 K			323 K		
	Cr (mg cm^{-2} h^{-1})	θ	IE (%)	Cr (mg cm^{-2} h^{-1})	θ	IE (%)	Cr (mg cm^{-2} h^{-1})	θ	IE (%)	Cr (mg cm^{-2} h^{-1})	θ	IE (%)
0	0.53	–	–	1.23	–	–	1.87	–	–	2.21	–	–
200	0.28	0.4717	47.17	0.72	0.4146	41.46	1.25	0.3315	33.15	1.65	0.2534	25.34
400	0.23	0.5660	56.60	0.62	0.4959	49.59	1.10	0.4118	41.18	1.43	0.3529	35.29
600	0.17	0.6792	67.92	0.49	0.6016	60.16	0.91	0.5134	51.34	1.19	0.4615	46.15
800	0.11	0.7924	79.24	0.34	0.7236	72.36	0.65	0.6524	65.24	0.99	0.5520	55.20
900	0.06	0.8679	86.79	0.23	0.8130	81.30	0.50	0.7326	73.26	0.86	0.6108	61.08

surface coverage values were fitted to different adsorption isotherm models, including Langmuir, Frumkin, Freundlich and Temkin isotherms, and the best results

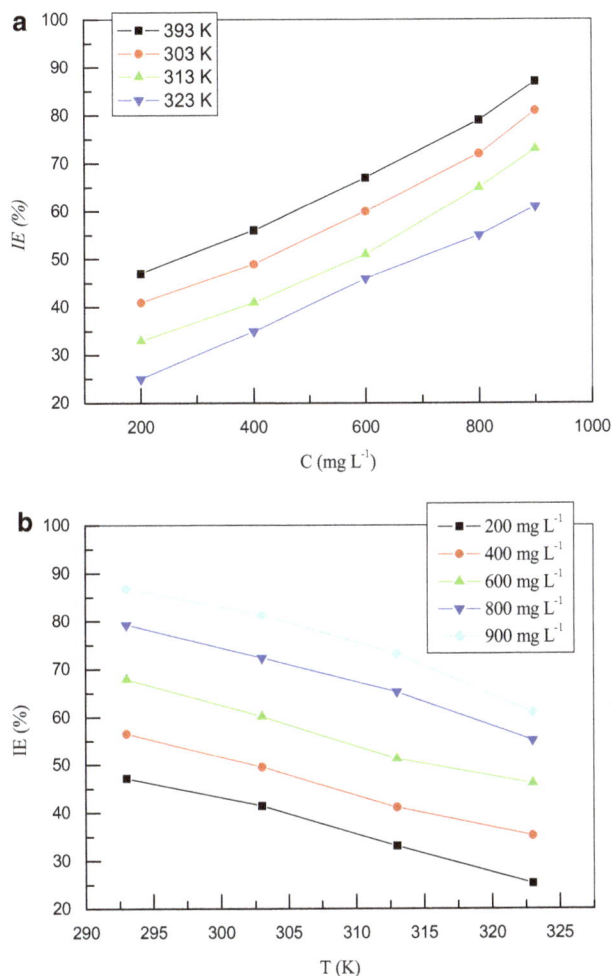

Fig. 1 Variation of inhibition efficiency of SSE in 1 M HCl with: **a** extract concentration, **b** temperature

revealed by the correlation coefficient (r^2) were obtained with the Freundlich adsorption isotherm for all considered temperatures. Figure 2 shows the plot of $\log\theta$ versus $\log C$ for SSE at 293–323 K.

Freundlich adsorption isotherm could be represented using the following equation:

$$\log\theta = \log K_{ads} + n\log C \qquad (6)$$

where $0 < n < 1$; θ is the surface coverage, C is the inhibitor concentration and K_{ads} is the adsorption constant. The K_{ads} values presented in the Table 4 were calculated from the intercept lines on the $\log\theta$ axis.

Table 4 shows that K_{ads} values decrease with increasing temperature. This result can be interpreted in a way that the temperature rise up induce desorption of some adsorbed components of the extracts from the surface. It is indeed consistent with the proposed physisorption mechanism [14].

Thermodynamic parameters

The standard adsorption free energy (ΔG^0_{ads}) can be calculated on the basis equation [21, 22]:

$$\Delta G^0_{ads} = -RT\ln(C_{H_2O}\cdot K_{ads}) \qquad (7)$$

where R is the gas constant, T the absolute temperature (K) and C_{H2O} is the concentration of water expressed in mg L^{-1} with an approximate value of 10^6. It should be noted that the unit of C_{H2O} lies in that of K_{ads} [23].

The standard adsorption enthalpy (ΔH^0_{ads}) could be calculated using Van't Hoff equation [25]:

$$\frac{\delta\ln K_{ads}}{\delta T} = \frac{\Delta H^0_{ads}}{RT^2}. \qquad (8)$$

Equation (6) can be rewritten as:

$$\ln K_{ads} = \frac{-\Delta H^0_{ads}}{RT} + I \qquad (9)$$

Table 3 Critical concentration and percentage inhibition efficiency for different plants extracts

Natural products	Acidic media	Metal exposed	Optimum concentration	Highest of IE (%)
Diethyl ether extract of *Ptychotis verticillata* [15]	1 M HCl	Mild steel	0.25 g L^{-1}	75.00
Ethyl acetate extract of *Ptychotis verticillata* [15]			0.5 g L^{-1}	86.00
Pelargonium extract [16]	1 M HCl	Mild steel	10 mL L^{-1}	76.79
Geissospermum laeve extract [2]	1 M HCl	C38 steel	100 mg L^{-1}	92.00
Mentha pulegium [17]	1 M HCl	Steel	2.76 g L^{-1}	78.00
Henna extract [18]	1 M HCl	C-Steel	3000 mg L^{-1}	88.42
Alkaloid extract of *G. ouregou* [4]	0.1 M HCl	Low carbon steel	250 mg L^{-1}	88.00
Alkaloid extract of *S. tinctoria* [4]				90.00
Alkaloids extract of *Oxandra asbeckii* [19]	1 M HCl	C38 steel	100 mg L^{-1}	87.00
Bark extract of *Neolamarckia cadamba* [6]	1 M HCl	Mild steel	5 mg L^{-1}	82.00
Leaves extract of *Neolamarckia cadamba* [6]				84.00
Pure alkaloid of *Neolamarckia cadamba* [6]				83.00

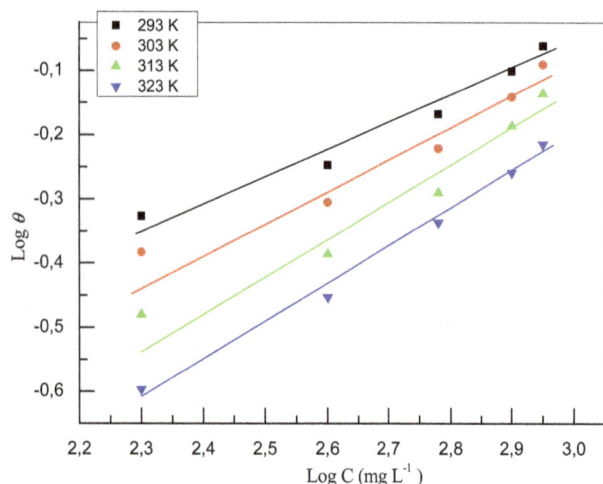

Fig. 2 Freundlich adsorption plot for carbon steel (X52) in 1 M HCl containing different concentration of SSE

Table 4 Parameters of the linear regression

Temperature (K)	r^2	k_{ads} (L mg^{-1})
293	0.982	0.0543
303	0.972	0.0387
313	0.973	0.0198
323	0.996	0.0112

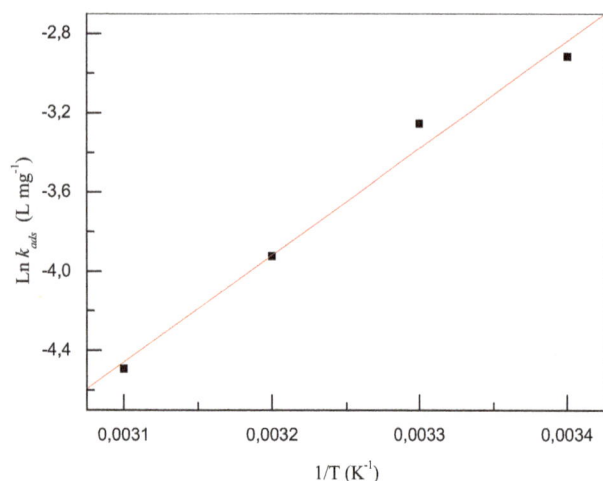

Fig. 3 Straight line of lnK_{ads} versus $1/T$ for SSE

where I is a constant of integration. Figure 3 presents straight lines of ln K_{ads} versus $1/T$ with good linear relationship (the linear correlation coefficients is 0.9970). ΔH^0_{ads} values calculated from the slope ($-\frac{\Delta H^0_{ads}}{R}$) and listed in Table 5.

From the obtained values of ΔG^0_{ads} and ΔH^0_{ads} parameters, the standard adsorption entropy (ΔS^0_{ads}) can be

Table 5 Standard thermodynamic parameters of the adsorption of SSE in 1 M HCl solution

T (K)	ΔG^0_{ads} (kJ mol^{-1})	ΔH^0_{ads} (kJ mol^{-1})	ΔS^0_{ads} (J mol^{-1} K^{-1})
293	−26.56	−44.94	−62.73
303	−26.61		−60.49
313	−25.74		−61.34
323	−25.04		−61.61

calculated using the thermodynamic Gibbs–Helmholtz equation:

$$\Delta S^0_{ads} = \frac{\Delta H^0_{ads} - \Delta G^0_{ads}}{T} \tag{10}$$

All thermodynamic parameters are recorded in Table 5. The negative values of ΔG^0_{ads} specify the spontaneity of the adsorption process [20]. In the present work, the calculated value of the adsorption free energy lies between −25.04 and −26.61 kJ mol^{-1}. It is less than the −40 kJ mol^{-1} required for chemical adsorption. Allows with the decrease of IE (%) with increasing temperature confirm the mechanism of physical adsorption [18]. The negative sign of ΔH^0_{ads} shows that the adsorption of inhibitor is an exothermic process [25, 26]. For physisorption, ΔG^0_{ads} is in the order of 40 kJ mol^{-1} while for chemisorption, ΔH^0_{ads} approaches 100 kJ mol^{-1} [27].

Activation parameters of the corrosion process

The temperature has a great effect on the rate electrochemical corrosion of metal. It is apparent from the Fig. 1b, that inhibition efficiency declines with increasing temperature. The reliance of corrosion rate on temperature can be stated by the Arrhenius equation [19]:

$$\ln Cr = -\frac{Ea}{RT} + \ln D \tag{11}$$

Ea is the apparent activation energy of the carbon steel (X52) dissolution and D is the Arrhenius pre-exponential factor. The logarithm of the Cr versus $1/T$ can be characterized by straight lines and the activation energy values were calculated from Arrhenius plots: (Fig. 4a).

The values of activation energy enumerated in Table 6 are higher in the presence of the SSE than in its absence. It may be attributed to the geometric blocking effect of adsorbed inhibitive species on the metal surface [28]. Further inspection of Table 6 also revealed that Ea increases with increase in SSE concentration. This means that the corrosion reaction will further be pushed to surface sites that are characterized by progressively higher values of Ea in the presence of extracts [29].

Saccocalyx satureioides as corrosion inhibitor for carbon steel in acid solution

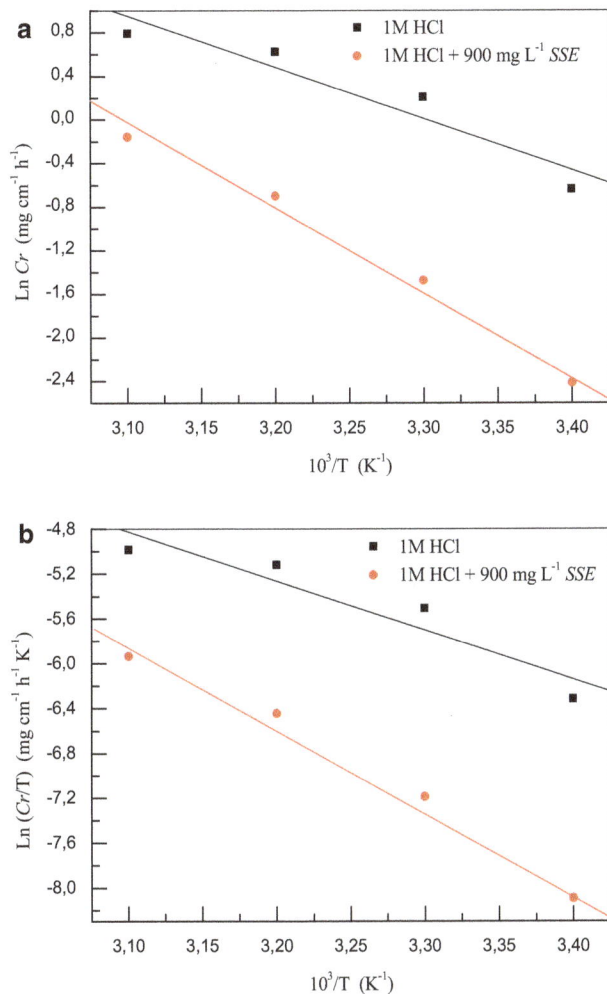

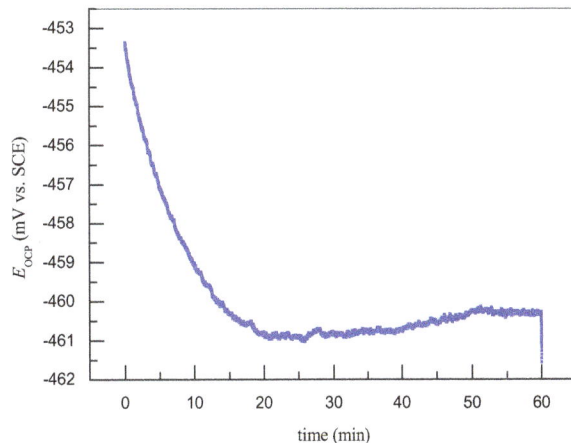

Fig. 5 E_{OCP}–time curve for carbon steel (X52) in 1 M HCl

$$\ln Cr = \frac{RT}{h N_a} \exp \frac{\Delta S_a}{R} \exp(-\frac{\Delta H_a}{RT}) \qquad (12)$$

where h is the Planck's constant, N_a is Avogadro's number. A plot of $\frac{\ln Cr}{T}$ vs. $\frac{1}{T}$ gave a straight line (Fig. 4b) with a slope of $-\frac{\Delta H}{RT}$ and an intercept of $\ln \frac{R}{N_a h} + \Delta S_a$, from which the values of ΔS_a^0 and ΔH_a were calculated and registered in Table 6. The positive signs of enthalpies indicate the endothermic nature of the dissolution process [30].

The values of ΔS_a^0 given in Table 6 are positive and get improved in the presence of extracts. This behavior can be explicated as a result of the replacement process of water molecules during adsorption of SSE on the steel surface [27]. This observation is in agreement with many studies that have been reported by several authors [31–35].

Fig. 4 Arrhenius plots related to the corrosion rate for carbon steel (X52) in 1 M HCl without and with 900 mg L^{-1} of SSE. **a** lnCr vs. $1/T$, **b** ln(Cr/T) vs. $1/T$

Table 6 Activation parameters Ea, ΔH_a^0 and ΔS_a^0 for carbon steel (X52) in 1 M HCl at different concentrations of SSE at different temperatures

Conc. (mg L^{-1})	Ea (kJ mol^{-1})	ΔH_a^0 (kJ mol^{-1})	10^{-1} ΔS_a^0 (J mol^{-1} K^{-1})
0	39.10	36.39	27.02
200	50.93	48.22	30.53
400	50.35	47.65	30.16
600	53.68	50.97	31.06
800	60.18	57.47	32.90
900	62.75	60.06	33.53

The enthalpy, ΔH_{ads}^0 and entropy, ΔS_{ads}^0 of activation for the corrosion process, the alternative formulation of Arrhenius equation was used [28, 29]:

Open circuit potential measurement

Figure 5 presents the variation of the open circuit potential (E_{OCP}) of the carbon steel (X52) vs time in 1 M HCl solution. Apparently, E_{OCP} remains almost unchanged, which indicate that the working electrode (WE) attains the steady state.

Polarization measurement

The anodic and cathodic polarization curves of carbon steel (X52) in 1 M HCl solution devoid and containing increasing concentrations of SSE are shown in Fig. 6. The respective kinetic parameters derived from the above plots are given in Table 7. It was demonstrated from the data that the addition of SSE shrank the corrosion current density (i_{corr}). The decrease may be due to the adsorption of the inhibitor on metal/acid interface [35].

In existence of SSE both cathodic and anodic Tafel slopes (β_c, β_a) decrease. This result indicates that the SSE

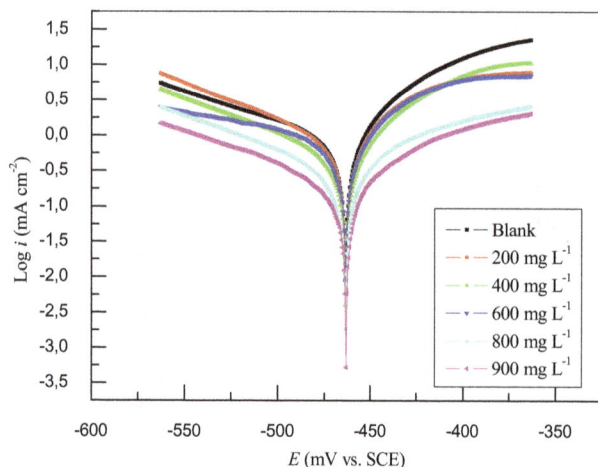

Fig. 6 Potentiodynamic polarization curves for carbon steel (X52) in 1 M HCl containing different concentrations of SSE

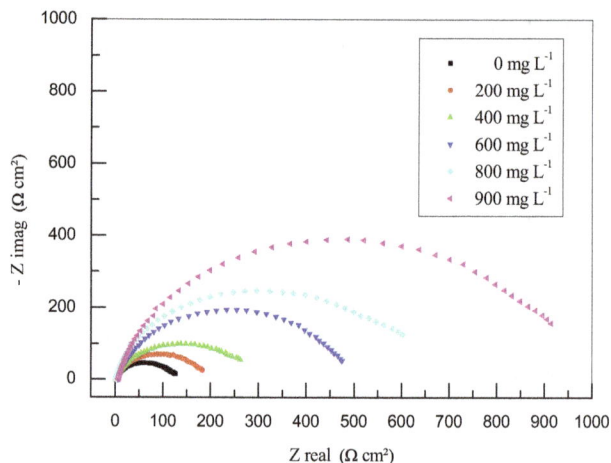

Fig. 7 Nyquist plots for carbon steel (X52) in 1 M HCl containing different concentrations of SSE

is a mixed-type inhibitor acting on both hydrogen evolution reaction and metal dissolution [36]. E_{corr} nearly does not change in the presence of SSE. According to Li and Moretti [24, 37], the inhibition category belongs to geometric blocking effect. That is, the inhibition action manifest through the reduction of the reaction area on the surface of the corroding metal [23].

Electrochemical impedance spectroscopy

The corrosion of carbon steel (X52) in 1 M HCl solution in the presence of plant extract was investigated by EIS at room temperature after an exposure period of 30 min. Impedance graphs (Fig. 7) are obtained for frequency range 100 kHz–0.01 Hz at the open circuit potential for steel in 1 M HCl in absence and presence of SSE, and the values of charge transfer resistance (R_{ct}) and double-layer capacitance (C_{dl}) are presented in Table 8. Figure 7 showed that the impedance spectra display only one capacitive loop, which indicated that the corrosion of steel is mainly controlled by a charge transfer process [29, 38]. It is also clear that these are not impeccable semicircles and this difference has been attributed to frequency dispersion and the heterogeneity of the metal surface [19]. On the other hand, it is clear from the plots that the size of these

loops raises on increasing extract concentration. This suggests that the formed inhibitive film was strengthened by the addition of plant extract [39].

The impedance spectra for Nyquist plots were analyzed by fitting to equivalent circuit model (Fig. 8), which was used to describe carbon steel/solution interface. A simple electrical equivalent circuit (EEC) has been proposed to model the experimental data. In this EEC, R_1 is the solution resistance, R_{ct} is the charge transfer resistance and Q is identified with the capacity. Generally, when a non-ideal frequency reply is present, it is usually accepted to replace the double-layer capacitance by constant phase element (CPE) [36]. Excellent fit with this model was obtained for all experimental data. As an example, the Nyquist plots for 1 M HCl alone and at 900 mg L^{-1} are presented in Figs. 9 and 10, respectively.

The values of the C_{dl} can be calculated from CPE parameter and R_{ct} according to the following equation [21]:

$$C_{dl} = R_{ct}^{\frac{1-n}{n}} Q^{\frac{1}{n}} \tag{13}$$

n is the deviation parameter of the CPE: $0 \leq 0 \leq 1$, for $n = 1$, Eq. (13) agrees to the impedance of an ideal capacitor, where Q is identified with the capacity.

The electrochemical parameters including R_{ct}, Q and n, obtained from fitting are listed in Table 8. In the Table 8, are also given the calculated "double-layer capacitance"

Table 7 Potentiodynamic polarizations parameters of carbon steel (X52) in 1 M HCl for various concentrations of SSE

Conc. (mg L^{-1})	$-E_{corr}$ (mV/SCE)	i_{corr} (mA cm^{-2})	β_a (mV Dec^{-1})	$-\beta_c$ (mV Dec^{-1})	IE (%)
0	463.8	1.5671	102.4	151.0	–
200	463.5	0.7924	83.2	119.7	49
400	463.8	0.6471	73.4	113.0	59
600	465.3	0.5238	50.4	114.8	66
800	462.6	0.3263	48.7	108.0	79
900	462.9	0.2012	45.3	92.7	87

Table 8 Electrochemical impedance spectroscopy parameters of carbon steel (X52) in 1 M HCl for various concentrations of SSE

Conc. (mg L^{-1})	R_{ct} (Ω cm^2)	10^5 Q (S^n Ω^{-1} cm^{-2})	n	C_{dl} (μF cm^{-2})	IE (%)
0	127.8 ± 0.47	50	0.83 ± 0.51	289.4	–
200	190.3 ± 0.70	41	0.84 ± 0.51	255.6	33
400	288.8 ± 1.00	19	0.86 ± 0.50	124.3	56
600	480.5 ± 0.77	16	0.78 ± 0.50	80.9	73
800	638.0 ± 0.93	10	0.82 ± 0.50	54.8	80
900	952.5 ± 0.73	9	0.76 ± 0.50	40.5	86

Fig. 8 Equivalent circuit used to fit the capacitive loop

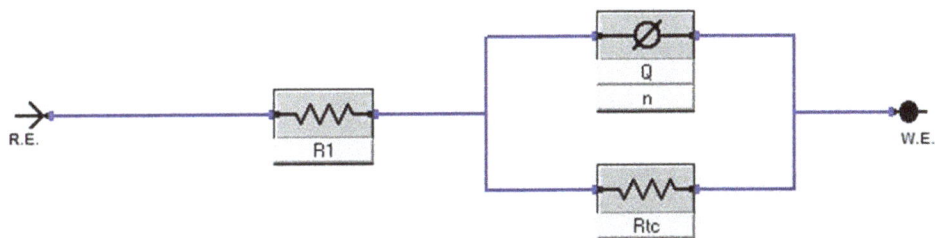

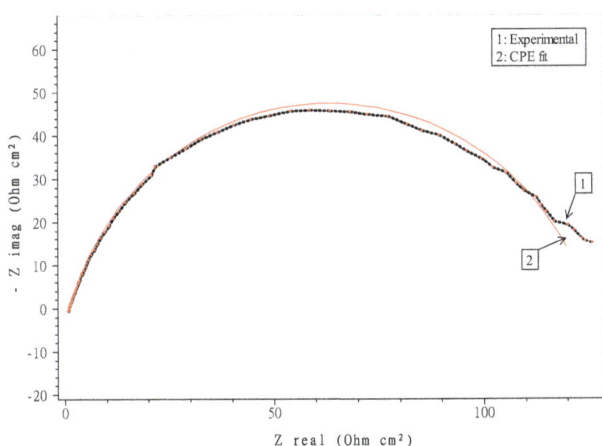

Fig. 9 Nyquist plot of carbon steel (X52) in 1 M HCl

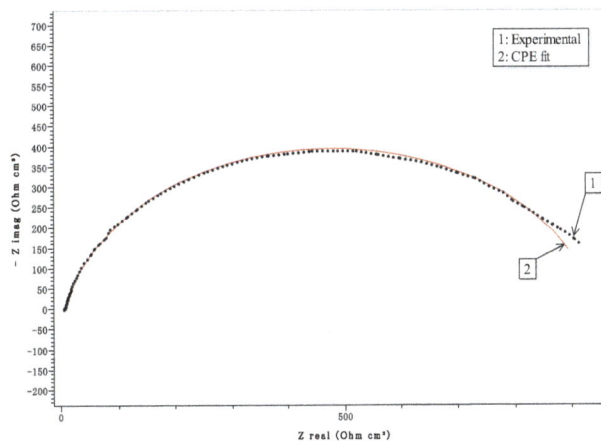

Fig. 10 Nyquist plot of carbon steel (X52) in 1 M HCl + 900 mg L^{-1} of SSE

values, derived from the CPE parameters. Furthermore, the estimates of the margins of error calculated for the parameters are presented in Table 8.

From the impedance parameters (Table 8), it is clear that the R_{ct} values increase with inhibitor concentration, and thus the inhibition efficiency rises to 86 % at 900 mg L^{-1}. The values of double-layer capacitance are also brought down to the maximum extent in the presence of SSE and the decrease in the values of C_{dl} follows the order similar to that obtained for i_{corr} in this study. The decrease in C_{dl} is due to the adsorption of the SSE on carbon steel (X52) surface leading to the formation of a film from the acidic solution [33].

Scanning electron microscope (SEM)

Scanning electron microscope (SEM) images were taken to study the surface morphology of carbon steel (X52) in the absence and presence of SSE as an inhibitor. SEM images are given in Fig. 11. The carbon steel (X52) sample dipped in the inhibitor solutions has apparently smooth surface when compared with that of rough corroded specimen immersed in HCl alone. This improvement in surface morphology indicates the formation of a protective layer on the steel surface which decreases the material degradation [40].

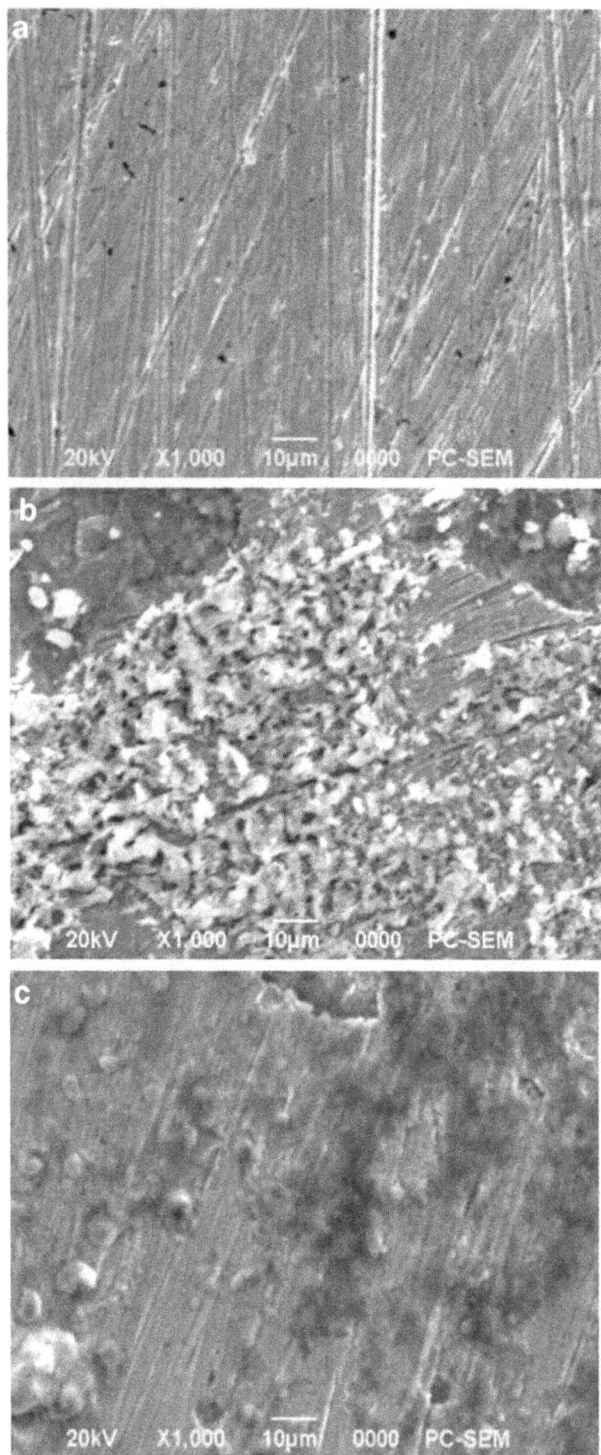

Fig. 11 SEM images of, **a** abraded carbon steel (X52), **b** carbon steel (X52) immersed in 1 M HCl solution, **c** carbon steel (X52) immersed in 1 M HCl containing 900 mg L^{-1} of SSE

Inhibition mechanism

Most of the *Saccocalyx satureioides* extract constituents are hydroxy aromatic compounds, such as piceol, vanillin, ferulic aldehyde and flavonoids [11]. The corresponding chemical structures of these compounds are shown in Fig. 12. Inspection of the figure reveals that these compounds contain O in functional groups, aromatic rings and O-heterocyclic rings, which encounter the general characteristics of typical corrosion inhibitors.

In acidic solution, the oxygen atom of these chemical compounds can be protonated easily, leading to positively charged inhibitor species. The charge of the metal surface is determined by the value of $Ecorr - Eq = 0$. The $Eq = 0$ of iron is -530 mV vs. SCE in HCl [38]. In this system, the value of $Ecorr$ is 463 mV vs. SCE. The steel surface charges positive in 1 M HCl solution because of the value of $Ecorr - Eq = 0 > 0$. Thus, in acid solution, the metal surface is negatively charged due to the specifically adsorbed chloride anions on the metal surface. Then the adsorption can occur between positively charged inhibitor molecules and negatively charged metal surface leading to physisorption of the inhibitor molecules. Further, when the protonated molecules are adsorbed on the steel surface, a coordinate band may be formed by partial transference of electrons from oxygen atoms to vacant d orbits of iron.

Conclusion

From the global experimental results the next conclusions can be summarized:

- Results obtained over weight loss measurements and electrochemical tests revealed that the ethyl acetate extract of *Saccocalyx satureioides* acts as efficient corrosion inhibitors of the carbon steel in acid solution.
- Potentiodynamic polarization measurements exhibit that SSE acts as a mixed-type inhibitor.
- The adsorption of SSE on the carbon steel (X52) surface follows Freundlich adsorption isotherm.
- The negative values of free energy of adsorption (ΔG_{ads}^0) specify that the adsorption process is spontaneous physically adsorbed on the carbon steel (X52) surface.

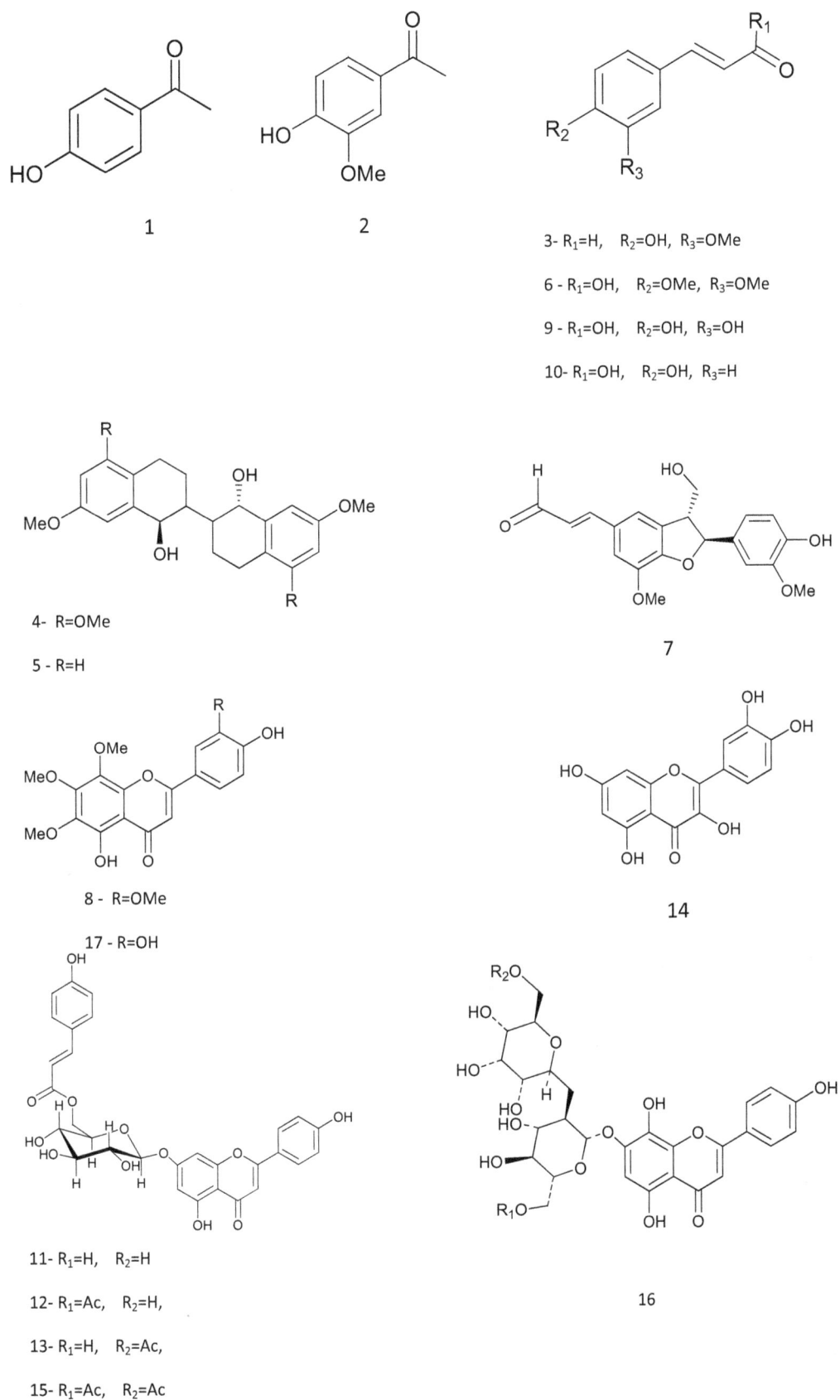

Fig. 12 Structures of the isolated compounds from *Saccocalyx satureioides*

Acknowledgments The authors want to thank the MESRES Algeria (Ministère de l'Enseignement Supérieur et de la Recherche Scientifique) for the financial support.

References

1. Ji Gopal, Anjum Shadma, Sundaram Shanthi, Prakash Rajiv (2015) *Musa paradisica* peel extract as green corrosion inhibitor for mild steel in HCl solution. Corros Sci 90:107–117

2. Faustin M, Maciuk A, Salvin P, Roos C, Lebrini M (2015) Corrosion inhibition of C38 steel by alkaloids extract of *Geissospermum* laeve in 1 M hydrochloric acid: electrochemical and phytochemical studies. Corros Sci 92:287–300

3. Umoren Saviour A, Obot Ime B, Madhankumar A, Gasem Zuhair M (2015) Performance evaluation of pectin as ecofriendly corrosion inhibitor for X60 pipeline steel in acid medium: experimental and theoretical approaches. Carbohydr Polym 124:280–291

4. Lecante A, Robert F, Blandinières PA, Roos C (2011) Anticorrosive properties of *S. tinctoria* and *G. ouregou* alkaloid extracts on low carbon steel. Curr Appl Phys 11:714–724

5. Pereira S, Pêgas M, Fernandez T, Magalhaes M, Schntag T, Lago D, Ferreira L, D'Elia E (2012) Inhibitory action of aqueous garlic peel extract on the corrosion of carbon steel in HCl solution. Corros Sci 65:360–366

6. Raja PB, Qureshi AK, Rahim AA, Osman H, Awang K (2013) *Neolamarckia cadamba* alkaloids as eco-friendly corrosion inhibitors for mild steel in 1 M HCl media. Corros Sci 69:292–301

7. Quezel P, Santa S (1963) Nouvelles flores de l'Algérie et des régions désertiques méridionales. T. II, CNRS, Paris

8. Laouer H, Akkal S, Debarnot C, Canard B, Meierhenrich UJ, Baldovini N (2006) Chemical composition and antimicrobial activity of the essential oil of *Saccocalyx satureioides* Coss. Et Dur. Nat Prod Commun 8:645–650

9. Bendahou M, Benyoucef M, Muselli A, Desjobert JM, Paolini J, Bernardini A (2008) Antimicrobial activity and chemical composition of *Saccocalyx satureioides* Coss. Et Dur. Essential oil and extract obtained by microwave extraction. Comparison with hydrodistillation. J Essent Oil Res 20:174–178

10. Zerroug MM, Laouer H, Strange RN, Nicklin J (2011) The effect of essential oil of *Saccocalyx* satureioides Coss. Et Dur. On the growth of and the production of solanapyrone A by *Ascochyta Rabiei* (Pass.) Labr. J Adv Environ Biol 5:501–506

11. Mohamadi S, Zhao M, Amrani A, Marchioni E, Zama D, Benayache F, Benayache S (2015) On-line screening and identification of antioxidant phenolic compounds of *Saccocalyx satureioides* Coss. et Dur. Ind Crops Prod 76:910–919

12. Akkal S, Louaar S, Benahmed M, Laouer H, Duddeck H (2010) A new isoflavone glycoside from the aerial parts of *Retama sphaerocarpa*. Chem Nat Compd 46:719–721

13. Benahmed M, Akkal S, Louaar S, Laouer H, Duddeck H (2006) A new furanocoumarin glycoside from *Carum montanum* (Apiaceae). Biochem Syst Ecol 34:645–647

14. Djeddi N, Benahmed M, Akkal S, Laouer H, Makhloufi E, Gherraf N (2014) Study on methylene dichloride and butanolic extracts of *Reuteralutea* (Apiaceae) as effective corrosion inhibitions of carbon steel in HCl solution. Res Chem Intermed. doi:10.1007/s11164-014-1555-3

15. El ouariachi E, Hammouti B, Paolini J, Bouyanzer A, Desjobert J-M, Majidi L, Salghi R, Costa J (2015) Inhibition of corrosion of mild steel in 1 M HCl by the essential oil or solvent extracts of *Ptychotis verticillata*. Res Chem Intermed 41:935–946

16. El Ouadi Y, Bouyanzer A, Desjobert J-M, Hammouti B, Ben Hadda T, Costa J, Jodeh S, Majidi L, Chetouani A, Warad I, Paolini J, Mabkhot Y (2014) Evaluation of *Pelargonium* extract and oil as eco-friendly corrosion inhibitor for steel in acidic chloride solutions and pharmacological properties. Res Chem Intermed. doi:10.1007/s11164-014-1802-7

17. Bouyanzer A, Hammouti B, Majidi L (2006) Pennyroyal oil from *Mentha pulegium* as corrosion inhibitor for steel. Mater Lett 60:2840–2843

18. Hamdy A, El-Gendy NSh (2013) Thermodynamic, adsorption and electrochemical studies for corrosion inhibition of carbon steel by henna extract in acid medium. Egypt J Petrol 22:17–25

19. Lebrini M, Robert F, Lecante A, Roos C (2011) Corrosion inhibition of C38 steel in 1 M hydrochloric acid medium by alkaloids extract from *Oxandra asbeckii* plant. Corros Sci 53:687–695

20. Tang Y, Zhang F, Hu S, Cao Z, Wu Z, Jing W (2013) Novel benzimidazole derivatives as corrosion inhibitors of mild steel in the acidic media. Part I: gravimetric, electrochemical, SEM and XPS studies. Corros Sci 74:271–281

21. El Bribri A, Tabyaoui M, Tabyaoui B, El Attari H, Bentiss F (2013) The use of *Euphorbia falcata* as eco-friendly corrosion inhibitor of carbon steel in hydrochloric acid solution. Mater Chem Phys 141:240–247

22. Yaro AS, Khadom AA, Wael RK (2013) Apricot juice as green corrosion inhibitor of mild steel in phosphoric acid. Alex Eng J 52:129–135

23. Li X, Deng S, Fu H (2012) Inhibition of the corrosion of steel in HCl, H_2SO_4 solutions by *bamboo* leaf extract. Corros Sci 62:163–175

24. Li L, Zhang X, Lei J, He J, Zhang S, Pan F (2012) Adsorption and corrosion inhibition of *Osmanthus fragran* leaves extract on carbon steel. Corros Sci 63:82–90

25. Deyab MA, Abd El-Rehim SS (2013) Effect of succinic acid on carbon steel corrosion in produced water of crude oil. J Taiwan Inst Chem Eng. doi:10.1016/j.jtice.2013.09.004

26. Ostovari A, Hoseinieh SM, Peikari M, Shadizadeh SR, Hashemi SJ (2009) Corrosion inhibition of mild steel in 1 M HCl solution by henna extract: a comparative study of the inhibition by henna and its constituents (Lawsone, gallic acid, α-D-glucose and tannic acid). Corros Sci 51:1935–1949

27. Pournazari Sh, Moayed MH, Rahimizadeh M (2013) In situ inhibitor synthesis from admixture of benzaldehyde and benzene-1,2-diamine along with $FeCl_3$ catalyst as a new corrosion inhibitor for mild steel in 0.5 M sulphuric acid. Corros Sci 71:20–31

28. Tebbji K, Faska N, Tounsi A, Oudda H, Benkaddour M, Hammouti B (2007) The effect of some lactones as inhibitors for the corrosion of mild steel in 1 M hydrochloric acid. Mater Chem Phys 106:260–267

29. Behpour M, Ghoreishi SM, Khayatkashani M, Soltani N (2012) Green approach to corrosion inhibition of mild steel in two acidic solutions by the extracts of *Punicagranatum peel* main constituents. Mater Chem Phys 131:621–633

30. Hodaifa G, Ochando-Pulido JM, Alami SBD, Rodriguez-Vives S, Martinez-Ferez A (2013) Kinetic and thermodynamic parameters of iron adsorption onto olive stones. Ind Crop Prod 49:526–534

31. Zarrok H, Zarrouk A, Hammouti B, Salghi R, Jama C, Bentiss F (2012) Corrosion control of carbon steel in phosphoric acid by purpald—weight loss, electrochemical and XPS studies. Corros Sci 64:243–252

32. Bouklah M, Benchat N, Hammouti B, Aouniti A, Kertit S (2006) Thermodynamic characterization of steel corrosion and inhibitor adsorption of pyridazine compounds in 0.5 M H_2SO_4. Mater Lett 60:1901–1905

33. Fouda AS, Wahed HA (2011) Corrosion inhibition of copper in HNO_3 solution using thiophene and its derivatives. Arab J Chem. doi:10.1016/arabjc2011.02.014

34. Singh AK, Quraishi MA (2010) Effect of cefazolin on the corrosion of mild steel in HCl solution. Corros Sci 52:152–160

35. Ahamad I, Prasad R, Quraishi MA (2010) Adsorption and inhibitive properties of some new *Mannich* bases of *Isatinderivaties* on corrosion of mild steel in acidic media. Corros Sci 52:1472–1481

36. Bobina M, Kellenberger A, Millet JP, Muntean C, Vaszilcsin N (2013) Corrosion resistance of carbon steel in weak acid solutions in the presence of L-histidine as corrosion inhibitor. Corros Sci 69:389–395

37. Moretti G, Guidi F, Fabris F (2013) Corrosion inhibition of the mild steel in 0.5 M HCl by 2-butyl-hexahydropyrrolo [1,2-b] [1, 2] oxazole. Corros Sci 76:206–218

38. Deng S, Li X (2012) Inhibition by Ginkgo leaves extract of the corrosion of steel in HCl and H_2SO_4 solutions. Corros Sci 55:407–415

39. Hazwan Hussin M, Jain Kassim M (2011) The corrosion inhibition and adsorption behavior of *Encaria gambir* extract on mild steel in 1 M HCl. Mater. Chem Phys 125:461–468

40. Gualdrón AF, Becerra EN, Peña DY, Gutiérrez JC (2013) Inhibitory effect of *Eucalyptus* and *Lippia Alba* essential oils on the corrosion of mild steel in hydrochloric acid. J Mater Environ Sci 4:143–158

4

Levulinic acid from corncob by subcritical water process

Chynthia Devi Hartono[1] · Kevin Jonathan Marlie[1] · Jindrayani Nyoo Putro[2] ·
Felycia Edi Soetardjo[1] · Yi Hsu Ju[2] · Dwi Agustin Nuryani Sirodj[3] ·
Suryadi Ismadji[1]

Abstract The productions of levulinic acid from corncob were carried out by subcritical water process in a temperature range of 180–220 °C, reaction time of 30, 45, and 60 min. The acid modified zeolite was used as the catalyst in the subcritical water process. The ratio between the mass of zeolite and volume of hydrochloric acid in the modification process were 1:5, 1:10 and 1:15. The optimum values of the process variables in the subcritical water process for the production of levulinic acid from corncob were: Temperature of 200 °C; 1:15 zeolite to acid ratio; and reaction time of 60 min. The maximum levulinic acid concentration obtained in this study was 52,480 ppm or 262.4 mg/g dried corncob.

Keywords Levulinic acid · Subcritical water · Modified zeolite

✉ Suryadi Ismadji
suryadiismadji@yahoo.com

Felycia Edi Soetardjo
felyciae@yahoo.com

[1] Department of Chemical Engineering, Widya Mandala Surabaya Catholic University, Kalijudan 37, Surabaya 60114, Indonesia

[2] Department of Chemical Engineering, National Taiwan University of Science and Technology, No. 43, Sec. 4, Keelung Rd, Taipei 106, Taiwan, People's Republic of China

[3] Department of Industrial Engineering, Widya Mandala Surabaya Catholic University, Kalijudan 37, Surabaya 60114, Indonesia

Introduction

Levulinic acid (4-oxopentanoic acid or γ-ketovaleric acid) is an organic compound with a short-chain fatty acids containing carbonyl group of ketones and carboxylic acids. Levulinic acid is an important chemical platform for the production of various organic compounds. It can be used for the production of polymers, resins, fuel additives, flavors, and others high-added organic substances. This chemical can be produced through several routes [1–7] and one of the most promising processes is the dehydrative treatment of biomass or carbohydrate with various kinds of acids.

Biomass can be used as the precursor to produce levulinic acid and other organic chemicals. The use of biomass as the raw material for the production of levulinic acid in commercial scale was developed by Biofine renewables [3, 7]. The Biofine process consists of two different stages of processes, the first stage of the process is the production of 5-hydroxymethylfurfural (HMF) while the second stage is the production of levulinic acid [3].

Several studies have reported that various types of homogeneous as well as heterogeneous catalysts have been used for the preparation of levulinic acid from lignocellulosic biomass [2–4, 7–9]. Usually, the homogeneous catalysts are more effective than some of heterogeneous catalysts; however, the drawbacks of the use of homogeneous catalysts for levulinic acid production are associated with the corrosion of the equipment, environmental problem, and re-use of the catalyst. One of the advantages of using heterogeneous catalyst for the production of levulinic acid is the heterogeneous catalyst can be easily recovered and reused [3].

Zeolites have been used as catalysts or catalyst supports in many reaction systems. The properties of zeolites, such

as porosity, types and the amount of surface acidity, and the type of the structure greatly influence the selectivity and catalytic performance of these materials. A number of synthetic zeolites have been used as the catalyst for the levulinic acid production, however, zeolites with low acidity and porosity gave a poor catalytic performance on the conversion of sugars into levulinic acid [3]. Zeolite-type materials, such as faujasite and modernite, have been used for the synthesis of levulinic acid from C_6 sugars and cellulose [6, 8, 10, 11].

Some of agricultural wastes and other lignocellulosic materials have the potential application as the precursors for levulinic acid production [12]. The production of levulinic acid from agricultural waste materials involves two critical steps of processes; the first process is hydrolysis, in the hydrolysis process the hemicellulose and cellulose are converted into C_5 and C_6 sugars. The second process is dehydration process, in this process the C_5 and C_6 sugars are dehydrated into levulinic acid and furan derivatives [12].

In this study, the production of levulinic acid from corncob was conducted on subcritical water condition using acid modified zeolite as heterogeneous catalyst. Subcritical water (SCW) process is an environmentally friendly method, which can be applied in various applications, such as extraction, hydrolysis, and wet oxidation of organic compounds. Subcritical water is defined as the hot compressed water (HCW) or hydrothermal liquefaction at a temperature between 100 and 374 °C under conditions of high pressure to maintain water in the liquid form [13]. At this subcritical condition, water acts as solvent and catalyst for the hydrolysis of cellulose and hemicellulose in the corncob. The use of acid modified zeolite increases the acidity of the system lead to the increase of the hydrolysis and dehydration rate of reactions and subsequently increases the yield of levulinic acid.

To the best of our knowledge, there is no single study used the subcritical water process combined with acid modified zeolite as the catalyst in the production of levulinic acid from lignocellulosic waste material (corncob). The optimum condition for the production of levulinic acid from corncob was determined by Response Surface Methodology (RSM).

Experimental

Materials

Corncobs used in this study were obtained from a local market in Surabaya, East Java, Indonesia. Prior to use, the corncobs were repeatedly washed with tap water to remove dirt. Subsequently the corncobs were dried in an oven

(Memmert, type VM.2500) at 110 °C for 4 h. The dried corncobs were pulverized into powder (20/60 mesh) using a JUNKE & KUNKEL hammer mill. The ultimate analysis of the corncob was determined using a CHNS/O analyzer model 2400 from Perkin-Elmer, while the proximate analysis was conducted according to the procedure of ASTM. The results of ultimate and proximate analyses of the corncob are summarized in Table 1.

Natural zeolite used in this research was obtained from Ponorogo, East Java, Indonesia. The purification of natural zeolite was conducted using hydrogen peroxide solution (H_2O_2) at room temperature (30 °C) to remove organic impurities. The purified zeolite then was pulverized to particle size of 40/60 mesh. The chemical composition of the purified natural zeolite was SiO_2 (60.14 %), Al_2O_3 (12.52 %), CaO (2.51 %), Fe_2O_3 (2.49 %), Na_2O (2.44 %), K_2O (1.28 %), MgO (0.49 %), H_2O (14.40 %), and loss on ignition (3.73 %).

All chemicals used in this study, such as sodium hydroxide (NaOH), hydrochloric acid (HCl), hydrogen peroxide (H_2O_2), the standard reference of levulinic acid, etc., were purchased from Sigma Aldrich Singapore and directly used without any further purification.

Natural zeolite modification

The natural zeolite was modified using hydrochloric acid solution (2 N). The ratio between the zeolite powder and hydrochloric acid were 1:5, 1:10, and 1:15 (weight/volume). Thirty grams of zeolite powder were mixed with a certain volume of HCl solution and transferred into a round bottom flask. Subsequently the mixture was heated at 70 °C under reflux and continuous stirring at 500 rpm for 24 h. After the modification completed, the acid modified zeolite was separated from the mixture by vacuum filtration system. The solid was repeatedly washed with distilled

Table 1 Proximate and ultimate analysis of corncob and its pretreated form

Component	Corncob, wt%	NaOH pretreated corncob, wt%
Ultimate analysis (dry basis)		
Carbon	54.1	53.8
Hydrogen	6.8	6.9
Nitrogen	0.3	0.2
Sulfur	0.1	0.1
Oxygen	38.7	39.0
Proximate analysis (dry basis)		
Moisture content	10.4	10.1
Volatile matter	67.1	71.8
Fixed carbon	19.4	15.2
Ash	3.1	2.9

water to remove the excess HCl solution. The acid modified zeolite was dried in oven at 110 °C for 24 h to remove free moisture content. Then, modified zeolite was calcined in a furnace at a temperature of 400 °C for 4 h.

Delignification process

Delignification process was carried out by soaking of corncob powder into 20 % of NaOH solution. The ratio between solid and solution was 1:10 (weight/volume). The delignification process was conducted at a temperature of 30 °C under constant stirring (500 rpm). After the process completed (24 h), the treated corncob was separated from the liquid using vacuum filtration system. The biomass was repeatedly washed with distilled water until the pH of the washing solution around 6.5–7. Subsequently the treated corncob was dried at 110 °C for 24 h.

Conversion of corncob to levulinic acid

The preparation of levulinic acid from corncob was conducted in a subcritical reactor system. The subcritical reactor system consists of 150 ml high pressure stainless steel vessel, a pressure gage, an external electrical heating system, type K thermocouple, and M8 screws for tightening the reactor with its cap. The maximum allowable temperature and pressure of the vessel are 250 °C and 100 bar, respectively. The reaction experiments were conducted at a pressure of 30 bar and three different temperatures (180, 200, and 220 °C). The typical reaction experiment is briefly described as follows: 20 g of corncob powder were mixed with 100 ml of distilled water; subsequently 0.5 g of acid modified zeolite was added into the mixture. The mixture was heated until the desired temperature was reached, and during the heating process, the nitrogen gas was introduced to the system to maintain the water in the liquid condition. During the reaction process, the mixture was stirred at 300 rpm. After the hydrolysis time was reached (30, 45, and 60 min), the reactor was rapidly cooled to room temperature. The solid was separated from the liquid by centrifugation at 3000 rpm. The concentrations of levulinic acid and other organic substances, such as sugars, organic acids and HMF, were determined by high performance liquid chromatography (HPLC) analysis.

Characterization of corncob and zeolite

The chemical composition of the corncob and delignified corncob was determined using Thermal gravimetric Analysis (TGA). The analysis was performed on a TGA/DSC-1 star system (Mettler-Toledo) with ramping and

cooling rate of 10 °C/min from room temperature to 800 °C under continuous nitrogen gas flow at a flowrate of 50 ml/min. The mass of the sample in each measurement was 10 mg.

The surface topography of the corncob and zeolite catalysts was characterized using a field emission Scanning Electron Microscope (SEM), JEOL JSM 6390 equipped with backscattered electron (BSE) detector at an accelerating voltage of 15 and 20 kV at a working distance of 12 mm. Prior to SEM analysis, an ultra-thin layer of conductive platinum was sputter-coated on the samples using an auto fine coater (JFC-1200, JEOL, Ltd., Japan) for 120 s in an argon atmosphere.

The X-ray powder diffraction (XRD) analysis of the samples was performed on a Philips PANalytical X'Pert powder X-ray diffractometer with a monochromated high intensity Cu Kα_1 radiation ($\lambda = 1.54056$ Å). The XRD was operated at 40 kV, 30 mA, and a step size of 0.05°/s from the 2θ angle between 5 and 90°.

The surface acidity of the zeolite acid activated zeolite was determined by amine adsorption analysis. A brief description of the method is as follows: a known amount of air dried zeolite or acid activated zeolite (50 mg) were added into a series of test tubes. Subsequently, different volumes (20–50 ml) of n-butylamine solution in benzene (0.01 M) were added to the test tubes. The test tubes then tightly stoppered and stores at 30 °C. After the equilibrium condition was achieved, the remaining n-butylamine in the solution was determined by titration using 0.016 M trichloroacetic acid solution in benzene, and 2,4 dinitrophenol was used as the indicator.

HPLC analysis

The organic compounds in the aqueous phase of the product from subcritical water process was analyzed using a Jasco chromatographic separation module consisting of a model PU-2089 quaternary low pressure gradient pump, a model RI-2031 refractive index detector and a model LC-NetII/ADC hardware interface system. Prior to the injection in the HPLC system, all of the liquid samples were filtered through a 0.22 μm PVDF syringe filter. The analysis of monomeric sugars was conducted with an Aminex HPX-87P sugar column (Bio-Rad, 300 × 7.8 mm) using degassed HPLC-grade water isocratically flowing at a rate of 0.60 ml/min. The column was operated at 85 °C. For the analysis of organic compounds, a Bio-Rad Aminex HPX-87H column (300 × 7.8 mm) was used as the separating column. The isocratic elution of sulfuric acid aqueous solution (5 mM) was used as the mobile phase with the flow rate of 0.6 ml/min. The column oven was set at 55 °C. Details of the procedure can be seen elsewhere [12].

Results and discussion

To determine the chemical composition of corncob and sodium hydroxide treated corncob, the thermal gravimetric Analysis (TGA) was conducted under the nitrogen environment. The TGA curves of both samples are given in Fig. 1. At temperature between 50 and 200 °C, the weight loss of corncob and the pretreated corncob mainly due to the evaporation of both free moisture content and bound water. From Fig. 1 it can be seen that a gradual thermal decomposition process with a significant weight loss for both samples (more than 60 %) are observed at a range of temperature from 250 to 400 °C. This significant weight loss of the biomasses mainly due to the thermal decomposition of hemicellulose (200–300 °C) and cellulose (300–360 °C) into smaller molecular weight compounds, such as water, carbon dioxide, carbon monoxide, methane, and other organic compounds. Some of lignin also degraded at this range of temperatures, which mainly due to the breakdown of chemical bonds with low activation energy [12, 14]. The breakdown of more stable bonds in the lignin occurred in temperature range from 400 to 500 °C. At higher temperature (above 500 °C), the weight loss of both biomasses was insignificant as seen in Fig. 1. The chemical compositions of corncob and its pretreated form which were determined by TGA method are listed in Table 2. Because the corncob contains high cellulose, this material is suitable as the raw material for levulinic acid production.

The SEM images of natural zeolite and acid modified zeolite are shown in Fig. 2. The modification using acid did not change the surface morphology of zeolite as indicated in Fig. 2. The XRD analysis was used to determine the crystalline structure of zeolite. In general, the modification using hydrochloric acid did not change or alter the crystalline structure of zeolite as shown in Fig. 3. The total

surface acidity of natural zeolite was 0.517 mg n-butylamine/g and after modification using hydrochloric acid solution, the total surface acidity increased to 0.815 mg n-butylamine/g. The increase of surface acidity of acid modified zeolite due to the removal of some exchangeable cations (Ca^{2+}, Fe^{3+} and Al^{3+}) from the framework of zeolite and replaced by H^+.

The production of levulinic acid from lignocellulosic materials involves several complex reaction mechanisms which also producing several intermediate products. In the hydrolysis process, the cellulose is converted into glucose, while the hemicellulose is converted into hexose (glucose, mannose, and galactose) and pentose (xylose and arabinose). In the dehydration process, hexose will be converted into 5-hydroxy-methylfurfural (HMF) and pentose will be converted into furfural. The decomposition of HMF produces levulinic acid and formic acid. A byproduct produced during the process is humin, black insoluble polymeric materials.

The subcritical water process has unique behavior and has been known as a green process for several applications [13, 15, 16]. Under high temperature and pressure, the water dissociates into H_3O^+ and OH^- ions, and the presence of these excess ions indicates that the water can act as an acid or base catalyst. The subcritical water hydrolysis of pretreated corncob were conducted either with or without solid acid catalyst additions. The subcritical water hydrolysis products are summarized in Table 3. Without addition of solid acid catalyst, the breakdown of cellulose and hemicellulose into monomeric sugars significantly low as indicated in Table 3.

At subcritical condition the ion products (H_3O^+ and OH^-) in water will make the water slightly acidic and at this condition the water become a good solvent for converting cellulose and hemicellulose to sugar monomers. The yield of monomeric sugars (calculated as the amount monomeric sugar/L solution) in the subcritical water process hydrolysis without the presence of catalyst increased with the increase of temperature from 180 to 220 °C (from 1.54 to 2.62 g/L) as seen in Table 3.

At constant pressure, the increase of temperature will decrease the dielectric constant of water and increase the ionization of water into H_3O^+ and OH^- leading to more acidic of the system. The presence of H_3O^+ (hydroxonium) in the system represents the nature of the proton in aqueous solution and this proton subsequent attacks β-1,4-glycosidic bonds as the linking bonds of several monomeric D-glucose units in the long chain polymer of cellulose, and resulting C_6 sugars as the product. The attack of hydroxonium ions into the linking bond of the hemicellulose chain, resulting C_5 sugars as the product. With the increasing of temperature, the amount of hydroxonium ions also increase, therefore the breakdown of linking bonds of

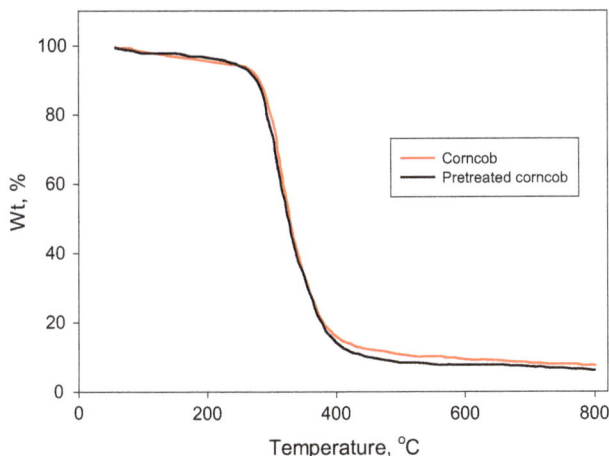

Fig. 1 Thermogravimetric curve of corncob and NaOH pretreated corncob

Table 2 Chemical composition of corncob and its pretreated form

Component	Corncob, wt%	NaOH pretreated corncob, wt%	TGA temperature, °C
Water	4.3	2.8	40–200
Hemicellulose	13.1	11.3	200–300
Cellulose	54.4	62.2	300–360
Lignin	20.1	18.2	360–500
Ash + carbon	8.1	5.5	>500

Fig. 2 SEM images of **a** natural zeolite, **b** modified zeolite (1:5), **c** modified zeolite (1:10), and **d** modified zeolite (1:15)

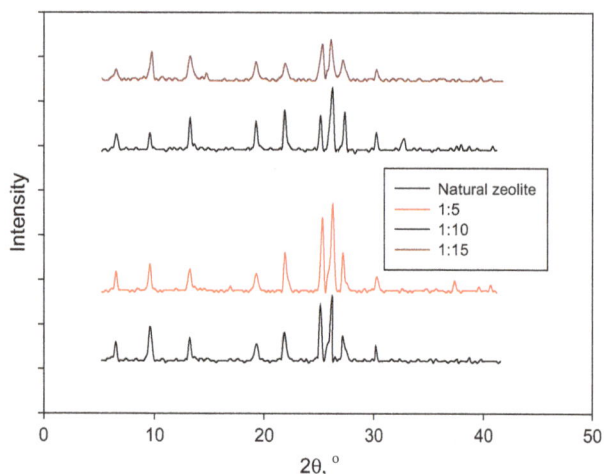

Fig. 3 X-Ray diffraction pattern of natural and acid modified zeolite

the cellulose and hemicellulose became increase leading to the increase of yield of sugars.

The addition of solid acid catalyst (modified zeolite) into the system significantly enhanced the breakdown of cellulose and hemicellulose into monomeric sugars (clearly seen in the temperature range of 180°–220°). The addition of the acid modified zeolite increased the number of protons (hydroxonium ions from subcritical water and H^+ from the surface of acid modified zeolite), with the excess number of protons in the solution, the breakdown of linking bonds of the cellulose and hemicellulose became significantly increasing and as the results the yield of monomeric sugars also increases as seen in Table 3.

In the levulinic acid production process, the C_6 sugars were dehydrated to HMF, this intermediate product subsequently converted into LA and formic acid. The C_5

Table 3 Monomeric sugars in subcritical water hydrolysis product

Temperature, °C	Acid activated zeolite, g	Yield mg/g dried corncob			
		Glucose	Xylose	Galactose	Arabinose
180	0	2.40	4.05	0.85	0.40
	0.5	55.65	40.20	27.40	14.55
200	0	6.05	4.60	1.05	0.55
	0.5	82.55	57.75	42.55	36.10
220	0	7.60	4.55	0.55	0.40
	0.5	120.65	76.90	47.60	42.20

sugars were converted to furfural, and the later was further degraded into formic acid and other insoluble products [17]. In the first step of dehydration of glucose, the isomerization reaction of glucose-fructose occurred and subsequently it further dehydrated to HMF and the later converted rapidly to LA and formic acid. The temperature plays important role in the dehydration process of glucose into LA, since all the reactions were endothermic process, the increase of temperature also increases the rate of reaction and the yield of products also increase. At temperature above 180 °C, the isomerization reaction of glucose-fructose occurred much faster, and more HMF was produced during the process, however, based on the kinetic parameters for the hydrolysis of sugarcane bagasse proposed by Girisuta et al. [17], the formation of LA or dehydration of HMF is much faster than other reactions. As soon as the HMF formed it was instantaneously converted to LA.

To obtain optimum process parameters for the levulinic acid production from corncob using catalytic subcritical water process, the response surface methodology (RSM) was employed to analyze the experimental data. The following polynomial equation was fitted to the response resulted from RSM by the LSM (least square method):

$$Y = \alpha_o + \sum_{i-1}^{k} \alpha_i X_i + \sum_{i-1}^{k} \alpha_{ii} X_i^2 + \sum_{i=1}^{k-1} \sum_{j=i+1}^{k} \alpha_{ij} X_i X_j \quad (1)$$

where Y is the concentration of levulinic acid (C_{LA}) in the product, α_o is a constant coefficient, α_I are the linear coefficients, α_{ij} are the interaction coefficients, and α_{ii} are the quadratic coefficients. X_i and X_j are the codec values of the variables. The independent variables used in this study were ratio of zeolite and acid (R), temperature (T, °C), and reaction time (t, min). The regression model was calculated using Minitab 16.1.1 Statistical software to estimate the response of dependent variables. The analysis of variance (ANOVA) was employed to confirm the adequacy of the model parameters. The suitability of the model to represent the data was determined by the value of R^2.

The full quadratic model that describes the relationship between the effects of ratio of zeolite and acid (R), temperature (T, °C), and reaction time (t, min) on the concentration of levulinic acid is given as follow

$$C_{LA} = 37102.7 + 5393.3\,R + 3893.3\,T \\ + 6040.8\,t - 254.5\,R^2 - 14713.5\,T^2 + 1485.8\,t^2 \\ + 689.4\,RT + 1993.5\,Rt - 1039.5\,Tt. \quad (2)$$

p value of the quadratic model (<0.0001) was significant at the probability level of 5 % ($R^2 = 0.9614$). The first order effect of variables R, T, and t on the output parameter (C_{LA}) were significant at the confidence level of 95 %. However, the second order effect of R and t as well as the interactions between R and t, R and T, T and t were insignificant as indicated in Table 4. Re-arrangement of Eq. (2) with the inclusion only the significant parameters give the following result:

$$C_{LA} = 37102.7 + 5393.3\,R + 3893.3\,T \\ + 6040.8\,t - 14713.5\,T^2. \quad (3)$$

The effects of ratio of zeolite and acid (R), temperature (T) and time (t) of subcritical water hydrolysis on the concentration of levulinic acid are plotted as surface plots in Figs. 4, 5 and 6. Both of these parameters have positive effects on the yield of levulinic acid (concentration). As mentioned before that temperature play important role both in hydrolysis and hydration processes, by increasing temperature the formation of levulinic acid or dehydration of HMF is much faster than other reactions. However, if the temperature is too high and the activation energy of the formation of humin is achieved, the degradation of HMF into humin is faster than the dehydration of HMF into levulinic acid and this phenomenon decreases the yield of levulinic acid. By increasing the subcritical hydrolysis time, the contact between the cellulose and hemicellulose with the ionic product of water (H_3O^+ and OH^-) become more intense and longer, and more of the cellulose and hemicellulose molecules were hydrolyzed and converted into monomeric sugars and subsequently dehydrated into HMF and levulinic acid. The ratio of zeolite and

Table 4 Analysis of variance for concentration of levulinic acid as a function of ratio of zeolite and acid (R), temperature (T, °C), and reaction time (t, min)

Source	DF	Seq SS	Adj SS	Adj MS	F	p value
Regression	9	1,498,173,667	1,498,173,667	166,463,741	39.05	0.000
Linear	3	645,895,733	645,895,733	215,298,578	50.51	0.000
R	1	232,702,558	232,702,558	232,702,558	54.59	0.001
T	1	121,263,058	121,263,058	121,263,058	28.45	0.000
t	1	291,930,117	291,930,117	291,930,117	68.48	0.000
Square	3	830,159,318	830,159,318	276,719,773	64.91	0.000
R^2	1	5,369,842	239,105	239,105	0.06	0.822
T^2	1	816,638,331	799,339,635	799,339,635	187.51	0.000
t^2	1	8,151,145	8,151,145	8,151,145	1.91	0.225
Interaction	3	22,118,617	22,118,617	7,372,872	1.73	0.276
RT	1	1,900,814	1,900,814	1,900,814	0.45	0.534
Rt	1	15,895,770	15,895,770	15,895,770	3.73	0.111
Tt	1	4,322,033	4,322,033	4,322,033	1.01	0.360
Residual error	5	21,314,232	21,314,232	4,262,846		
Lack-of-fit	3	20,009,560	20,009,560	6,669,853	10.22	0.090
Pure error	2	1,304,673	1,304,673	652,336		
Total	14	1,519,487,900				

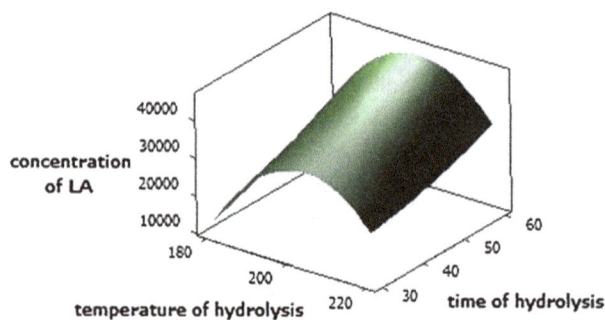

Fig. 4 Surface plot of concentration levulinic acid as a function of temperature and time of subcritical water hydrolysis

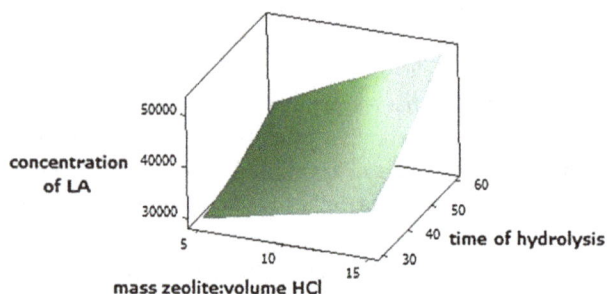

Fig. 6 Surface plot of concentration levulinic acid a function of ratio of zeolite and acid, and temperature of subcritical water hydrolysis

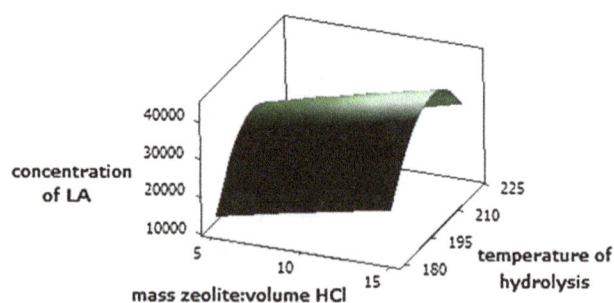

Fig. 5 Surface plot of concentration levulinic acid as a function of ratio of zeolite and acid, and time of subcritical water hydrolysis

hydrochloric acid also had a positive effect on the concentration of levulinic acid, by increasing of the ratio of acid, the ion exchange between some metal cations with H^+ also increased. Subsequently, with the increased of H^+ in the surface of zeolite catalyst also increased the number of protons in the solution leading to the increase of the breakdown of linking bonds of the cellulose and hemicellulose to produce monomeric sugars. These monomeric sugars under acidic condition and high temperature were dehydrated into levulinic acid. The experimental results of the effects of temperature, reaction time, and the ratio of zeolite and hydrochloric acid (activation of zeolite) on the yield of levulinic acid are given in Table 5.

To obtain the maximum yield or concentration of levulinic acid is an important point in this study to establish an efficient process. This objective can be achieved through the setting of all significant parameters at optimum conditions. The optimum condition of the production of levulinic acid from corncob through subcritical water process is depicted in Fig. 7. RSM indicates the optimum conditions for the variable of ratio of zeolite and acid was coded 1, variable of hydrolysis temperature was coded 0.1111 and hydrolysis time was coded 1. These units

Table 5 The effect of temperature and reaction time on the yield of levulinic acid

Temperature, °C	Time of hydrolysis, min	Ratio zeolite: volume HCl, g:ml	Yield of levulinic acid, mg/g dried corncob
180	30	1:10	73.8
180	45	1:5	69.5
180	45	1:15	114.7
180	60	1:10	129.3
200	30	1:5	138.4
200	30	1:15	174.3
200	45	1:10	181.9
200	60	1:5	194.2
200	60	1:15	269.9
220	30	1:10	119.8
220	45	1:5	104.9
220	45	1:15	163.9
220	60	1:10	154.4

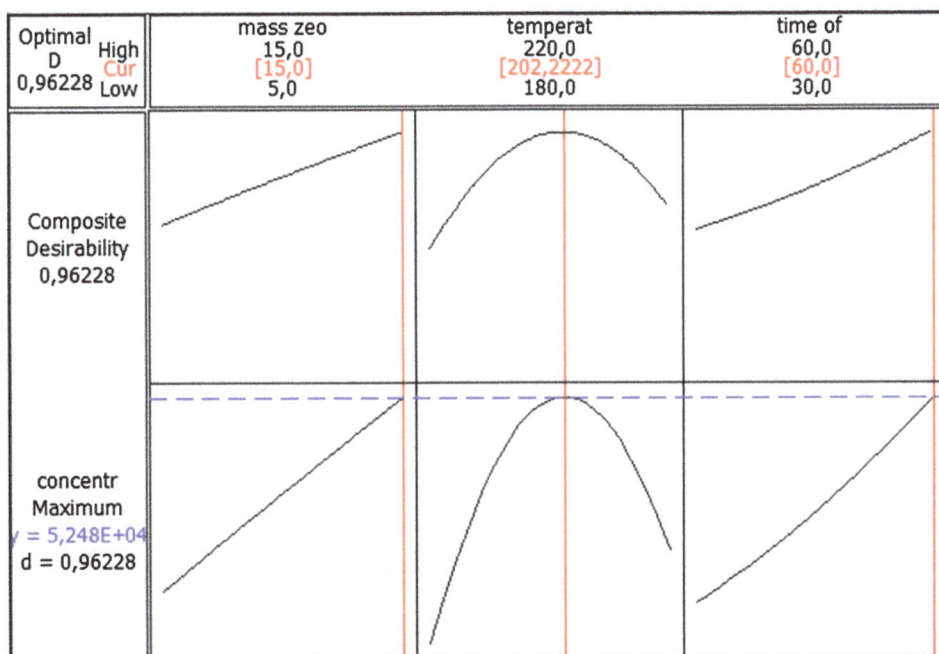

Fig. 7 Independent factor optimization during subcritical water hydrolysis and hydration processes of corncob

correlate to zeolite and an acid ratio of 1:15, reaction temperature of 200 °C, and reaction time of 60 min and correspond to the optimum concentration of levulinic acid of 52,480 ppm (262.4 mg/g dried corncob). To test the validity of the optimum condition obtained from the RSM, an experiment has also been conducted using process variables values from the RSM, and as the result the concentration of levulinic acid of 53,989.7 ppm (269.9 mg/g) was obtained. Since the difference between the experiment and the optimize value from RSM only 2.8 %, therefore, these theoretical optimum values obtained from RSM are considered to be appropriate.

The stability and reusability of the heterogeneous catalyst are crucial issues for industrial application. To examine the stability and reusability of acid modified zeolite, the catalyst was recovered from the reaction mixture, re-calcined at 400 °C for 4 h, and reused five times. The reaction temperature of 200 °C, reaction time of 60 min, and zeolite to acid ratio of 1:15 were used as the reaction parameters to study of the reusability of catalyst. The reusability results of the spent catalyst are depicted in Fig. 8. This figure clearly shows that the yield of levulinic acid gradually decrease after the first run. This phenomenon indicates that the catalyst has gradually deactivated during the reaction

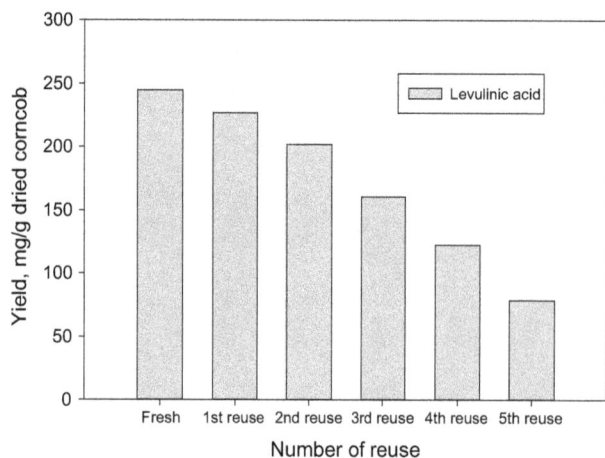

Fig. 8 The stability and reusability of spent catalyst

cycle. The activation of catalyst during the reaction cycle due to the leaching of surface acid sites (the acidity of fresh catalyst was 0.815 mg n-butylamine/g and after 5th cycle was 0.423 mg n-butylamine/g) and the formation of humin in the active sites of the catalyst.

Conclusion

Corncob had been successfully used as the new raw material for levulinic acid production. The production of levulinic acid was conducted in subcritical condition with the presence of acid modified zeolite as catalyst. The yield of levulinic acid in the final product was strongly influenced by the ratio of zeolite and acid, reaction temperature, and reaction time. The optimum yield of levulinic acid was 262.4 mg/g dried corncob, and was obtained at temperature of 200 °C, reaction time of 60 min, and zeolite to acid ratio of 1:15.

Acknowledgments The financial support from The World Academy of Science Research Grant 2015/2016 with contract no 14-095/RG/CHE/AS-1; UNESCO FR:325028591 and LPPM Widya Mandala Surabaya Catholic University through Outstanding Lecturer Research Grant 2014/2015 with contract number 845a/WM01.5/N/2014 is gratefully acknowledged.

Authors' contributions CDH, KJM and JNP conducted the experiments, while DANS performed the statistical analysis, FES and YHJ drafting the manuscript, SI performed the experiment design and corrected the manuscript.

Compliance with ethical standards

Conflict of interest The authors declare that they have no competing interests.

References

1. Cavinato G, Toniolo L (1990) Levulinic acid synthesis via regiospecific carbonylation of methyl vinyl ketone or of its reaction products with hydrochloric acid or an aleanol or of a mixturjz of acetone with a formaldehyde precursor catalyzed by a highly active Pd-HCl system. J Mol Catal 58:251–267
2. Mukherjee A, Dumont MJ, Raghavan V (2015) Review: sustainable production of hydroxymethylfurfural and levulinic acid: challenges and opportunities. Biomass Bioenergy 72:143–183
3. Ramli NAS, Amin NAS (2015) Fe/HY zeolite as an effective catalyst for levulinic acid production from glucose: characterization and catalytic performance. Appl Catal B Environ 163:487–498
4. Chamnankid B, Ratanatawanate C, Faungnawakij K (2014) Conversion of xylose to levulinic acid over modified acid functions of alkaline-treated zeolite Y in hot-compressed water. Chem Eng J 258:341–347
5. Lourvanij K, Rorrer GL (1994) Dehydration of glucose to organic acids in microporous pillared clay catalysts. Appl Catal A 109:147–165
6. Jow J, Rorrer GL, Hawley MC, Lamport DTA (1987) Dehydration of D-fructose to levulinic acid over LZY zeolite catalyst. Biomass 14:185–194
7. Girisuta B, Danon B, Manurung R, Janssen LPBM, Heeres HJ (2008) Experimental and kinetic modelling studies on the acid-catalysed hydrolysis of the water hyacinth plant to levulinic acid. Bioresour Technol 99:8367–8375
8. Ya'aini N, Amin NAS, Asmadi M (2012) Optimization of levulinic acid from lignocellulosic biomass using a new hybrid catalyst. Bioresour Technol 116:58–65
9. Deng W, Zhang Q, Wang Y (2014) Catalytic transformations of cellulose and cellulose-derived carbohydrates into organic acids. Catal Today 234:31–41
10. Lourvanij K, Rorrer GL (1993) Reactions of aqueous glucose solutions over solid-acid Y-zeolite catalyst at 110–160 °C. Ind Eng Chem Res 32:11–19
11. Zeng W, Cheng DG, Zhang H, Chen F, Zhan X (2010) Dehydration of glucose to levulinic acid over MFI-type zeolite in subcritical water at moderate conditions. React Kinetics Mech Catal 100:377–384
12. Putro JN, Kurniawan A, Soetaredjo FE, Lin SY, Ju YH, Ismadji S (2015) Production of gamma-valerolactone from sugarcane bagasse over TiO2-supported platinum and acid-activated bentonite as co-catalyst. RSC Adv 5:41285–41299
13. Ahmed IN, Nguyen PLT, Huynh LH, Ismadji S, Ju YH (2013) Bioethanol production from pretreated *Melalueca leucadendron* shedding bark—simultaneous saccharification and fermentation at high solid loading. Bioresour Technol 136:213–221
14. Yang H, Yan R, Chen H, Lee DH, Zheng C (2007) Characteristics of hemicellulose, cellulose and lignin pyrolysis. Fuel 86:1781–1788
15. Tsigie YA, Huynh LH, Ismadji S, Engida AM, Ju YH (2012) Insitu biodiesel production from wet *Chlorella vulgaris* under subcritical condition. Chem Eng J 213:104–108
16. Go AW, Sutanto S, Nguyen PLT, Ismadji S, Gunawan S, Ju YH (2014) Biodiesel production under subcritical condition using subcritical water treated whole *Jatropha curcas* seed kernels and possible use of hydrolysates to grow *Yarrowia lipolytica*. Fuel 120:46–52
17. Girisuta B, Dussan K, Haverty D, Leahy JJ, Hayes MHB (2013) A kinetic study of acid catalysed hydrolysis of sugar cane bagasse to levulinic acid. Chem Eng J 217:61–70

5

Mitigation of corrosion of carbon steel in acidic solutions using an aqueous extract of *Tilia cordata* as green corrosion inhibitor

A. S. Fouda[1] · A. S. Abousalem[1] · G. Y. EL-Ewady[1]

Abstract The effectiveness of using *Tilia cordata* extract as a green corrosion inhibitor for carbon steel in 1 M hydrochloric acid solutions was demonstrated by employing some chemical and electrochemical techniques. The surface morphology of C-steel specimens was examined. The results showed that *Tilia cordata* has corrosion inhibition characteristics with efficiency of 96% as the concentration of *Tilia cordata* extract increased to 300 mg L^{-1}. Charge transfer resistance (R_{ct}) value increases while both the capacitance of the double layer (C_{dl}) and corrosion current (i_{corr}) values decrease with increasing the extract concentration. The effect of temperature was studied in the range 30–60 °C. Some thermodynamic parameters were calculated and discussed. The adsorption of extract on the C-steel surface was found to obey Langmuir adsorption isotherm. Polarization results showed that the investigated extract acts as mixed type inhibitor. All the different used techniques gave similar results.

Keywords Corrosion inhibition · HCl · C-steel · *Tilia cordata* extract · Adsorption

Introduction

Corrosion is the process whereby metals and alloys lose their useful properties as a result of a reaction with the surrounding environment. The definition of corrosion is not limited to metals and alloys, but the term, corrosion, can be extended to include ceramics and other nonmetallic materials. Corrosion leads to major problems in most industrial fields [1]. In the recent years, corrosion becomes one of the most challenging topics for scientists and the engineering society. From the standpoint of the importance of metals in industry, carbon steel is considered to be the metal of choice due to its distinctive characteristics, besides the wide application aspects. In several industrial processes such as acid cleaning, acid well acidizing, and acid pickling, rust and contaminated scales were generally removed using acid solutions. Moreover, hydrochloric acid can be produced as a byproduct of crude oil desalting process and some oil refinery treatments [2]. Hydrochloric and sulphuric acids are the most widely used mineral acids in the pickling processes of metals [3]. It is generally known that the spontaneous dissolution of iron in acid solution produces Fe^{2+} ions, which corresponds to the anodic reaction, accompanied by discharging of electrons by hydrogen evolution at cathodic sites on the metal surface. Because of the general destructive attack of acid solutions, the use of inhibitors to control the aggressiveness of acid environment was found to have widespread applications in many industries [4]. The feasibility of protection methods depend mainly on the conditions and surrounding environment that materials, particularly metals, experienced during service conditions. The use of corrosion inhibitor to minimize corrosion rate in closed service system is well-established and generally accepted [5–8]. In literature, several organic compounds have been reported as potential corrosion inhibitor for different metals, but on the other hand, ecological and healthy problems have been arisen because of using such synthetic compounds. In recent years, considerable amount of effort devoted to find low cost and efficient corrosion inhibitors from natural resources such as

✉ A. S. Abousalem
ashraf.abousalem@gmail.com

[1] Chemistry Department, Faculty of Science, Mansoura University, Mansoura 35516, Egypt

plant extracts that can be used as promising alternative sources for corrosion inhibitors [9–12]. Several studies reports the use of plant extracts as potential agents to reduce corrosion in various industrial solutions [13–15]. Inhibitors extracted from plants are renewable resources, readily available, acceptable and friendly for human and environment [11]. To our knowledge, there seems no results have been reported in the literature for using *Tilia cordata* as corrosion inhibitor. The present work provide an investigation on using an aqueous extract, simply prepared from the leaves of *Tilia cordata*, as a green corrosion inhibitor to control the corrosion behavior of carbon steel in hydrochloric acid solutions.

Materials and methods

Specimen preparation

The carbon steel used in the present study was brought from Nile Delta Fields, Petrobel Company, Abu-Mady in Egypt. The chemical composition of the material (wt%) is as follows; C (0.17–0.20), Mn (0.35), P (0.025), Si (0.003) and Fe (balance). For weight loss measurements, a rectangle sheet of carbon steel was mechanically cut to prepare seven specimens, each of identical dimensions, $20 \times 20 \times 1$ mm. A small hole with a diameter 2 mm was punched at one top corner of each specimen. Glass hooks were used to hold the specimens in the test solutions.

Solution preparation

Analar grade reagents and bi-distilled water were used for the acid solution preparation. The test solution was 1 M HCl. The hydrochloric acid was purchased from Al-Gomhoria Co. for chemicals, in Egypt. A volume of 100 ml of HCl solutions was freshly prepared and used as test solution before each experiment.

Preparation of plant extract

The investigated plant was purchased from "Al Nakyti" a local plant supplier in Egypt. The studied aqueous extract was prepared from the leaves of *Tilia cordata*. The leaves were air-dried, grinding down to small pieces. A sample of 150 g of grinded leaves was added to 250 ml of bi-distilled water in a conical flask of a volume 1000 ml, the mixture was boiled for 30 min, and then cooled in dark place, after the extraction process, the crude extract was filtered using Whatman Filter Papers to remove undesirable solid residues and contamination. A volume of 10 ml of the crude extract is taken out and desiccated at fixed temperature. The resultant desiccated solid residue is weighted to aid in determining the

concentration of the aqueous extract. A stock solution of 1000 mg L^{-1} was prepared by taking 70 ml of the crude extract and completed to 1000 ml with bi-distilled water. The range of the plant extract concentration was as follows; 50–300 mg L^{-1}. Table 1 lists the proposed chemical structure of some phytochemical constituents isolated from a crude extract of *Tilia cordata* [16–19].

Weight loss measurements

The surface of carbon steel specimens were abraded, to a mirror finish, using different grades of emery paper starting with coarser type 80 to finer one 1200, degreased with acetone, rinsed with bi-distilled water and gently dried using filter papers. The procedure of the experiment was as follows; first, the mass of specimens was measured precisely to 0.0001 digits by using a high sensitive electronic balance, and then the specimens were immersed into 100 ml of 1 M Hydrochloric acid solution in absence and presence of various concentrations of *Tilia cordata* extract at 30 °C. For weight loss experiment, the immersion time intervals were (30 min until 180 min). After each interval time, the tested specimens were taken out of the solution, rinsed with bi-distilled water, thoroughly dried and the mass after immersion is precisely reweighted and recorded. The corrosion rate (C_R) in mg cm^{-2} min^{-1} was calculated from the value of weight loss divided by the total surface area (cm^2) and immersion time (min).

Electrochemical technique

Three different electrochemical techniques were conducted using three electrode system assembled in a glass cell as follows; carbon steel specimen as working electrode (1 cm^2), saturated calomel electrode (SCE) acts a reference electrode, and platinum wire serves as an auxiliary electrode. In the present work, the working electrode was made of squared specimen of carbon steel welded with copper rod from one side and totally encapsulated into a glass rod, of larger diameter (5 mm) so that only one face of the carbon steel specimen, of dimension (1 cm × 1 cm), was left to be exposed to the test solution and act as the active working surface. The reference electrode was connected to a Luggin capillary and the tip of the Luggin capillary is set to be very close to the surface of the working electrode in order to partially eliminate error originated from IR drop. All the measurements performed were subjected to stagnant conditions. Before starting electrochemical experiments, the working electrode was prepared in the same manner of weight loss method and the electrode potential was stabilized for 20 min. All electrochemical results were obtained using Gamry Instrument (PCI4/750) with a Gamry system based on the ESA400 and computerized frameworks include DC105 software for potentiodynamic

Table 1 List of some phytochemical constituents isolated from a crude extract *Tilia cordata*

Chemical constituents of *Tilia cordata* extract	
Quercetin	2-(3,4-dihydroxyphenyl)-3,5,7-trihydroxy-4H-chromen-4-one
Kaempherol	3,5,7-Trihydroxy-2-(4-hydroxyphenyl)-4H-chromen-4-one
P-coumaric acid	(E)-3-(4-hydroxyphenyl)-2-propenoic acid

polarization measurements, EIS300 software for electrochemical impedance (ac) spectroscopy, and EFM140 software for electrochemical frequency modulation measurements. In addition, Echem Analyst 6.03 software was used for data fitting, graphing and plotting.

Potentiodynamic polarization

Tafel curves obtained from potential polarization scan by automatically sweeping the working electrode potential with a scan rate of 5 mVs^{-1} from (−700 to 700 mV vs. SCE) at open circuit potential. The corrosion current density was calculated by the extrapolation of cathodic and anodic Tafel lines (β_c and β_a) to an intersection that gives log i_{corr} and the relevant corrosion potential (E_{corr}) for the free acid and each concentration of the investigated plant extract. The inhibition efficiency (%IE) and surface coverage (θ) were calculated using the obtained values of i_{corr}.

Electrochemical impedance spectroscopy

EIS experiments were carried out at open circuit potential, alternative current signals, in frequency range from 100 kHz to 0.5 Hz with amplitude of 10 mV peak-to-peak,

were applied to measure the impedance of the corrosion process. An equivalent electrical circuit was tested to explain the impedance results.

Electrochemical frequency modulation

Two different frequencies 2 and 5 Hz with base frequency equals to 0.1 Hz [20–22] were applied to obtain the intermodulation spectra of the electrochemical frequency modulation. It is necessary for the value of lower frequency to be not greater than a half of the higher frequency. The higher frequency must also be sufficiently slow that the charging of the double layer does not contribute to the current response. The current responses, obtained from EFM spectra, assigned for intermodulation and harmonical current peaks. The large peaks [23] function to calculate the corrosion current density (i_{corr}), the slopes of Tafel curves (β_a and β_c) and corresponding causality factors [CF-2 and CF-3].

Surface analysis

Three squared pieces of carbon steel specimens, of dimensions (20 × 20 × 1 mm), were used in this analysis, the first piece of metal was used to represent the carbon steel surface without neither exposure to acid nor treatment with extract inhibitor, the second specimen was completely immersed for 12 h into the free acid (1 M HCl) and without exposure to plant extract inhibitor. The third sample of metal was immersed for 12 h in a test solution of 1 M HCl containing 300 mg L^{-1} of *Tilia cordata* extract. The surface of these specimens was scanned and examined using scanning electron microscope (SEM, JOEL, JSM-T20, Japan) and the elemental composition of carbon steel surface was detected using energy dispersive X-ray spectroscopy (EDX) Type: Philips X-ray diffractometer (pw-1390) equipped with Cu-tube (Cu Ka1, 1 = 1.54051 Å).

Results and discussion

Weight loss measurements

Weight loss in mg per cm^2 was determined in laboratory after equal time interval of immersion into a test solution of 1 M HCl without and with treatment of various concentrations of the investigated additive. As shown in Fig. 1, curves for additive-containing systems fall below of the free acid. This suggests that the weight loss of carbon steel is a function of both the type and the concentration of the additive. The surface coverage, likewise, the inhibition efficiency increases when the concentration of the additive plant extract increases as indicated by the drop occurs in weight loss per cm^2, while the inhibition efficiency decreases as the temperature of the test solution increases, this suggests electrostatic mode of adsorption (physical

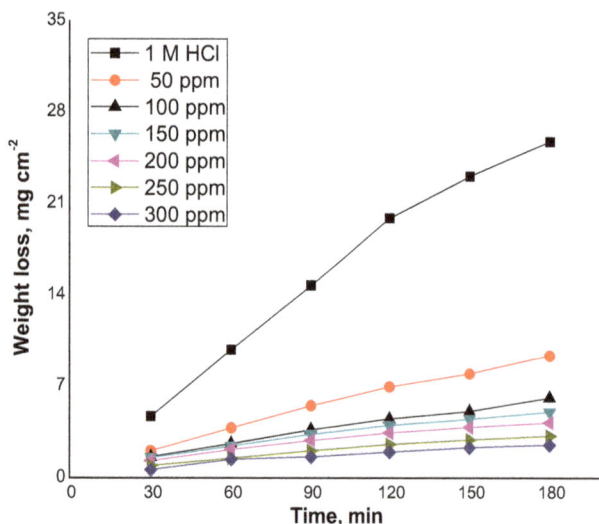

Fig. 1 Plots of mass loss vs. immersion time for carbon steel corrosion in 1 M HCl without and with the treatment of *Tilia cordata* extract at 30 °C

adsorption) generating on the metal surface by the action of plant extract constituents. Table 2 lists the inhibition efficiency (%IE) and the surface coverage (θ) obtained from weight loss calculation. The inhibition efficiency (%IE) and the degree of surface coverage (θ) were calculated from Eq. (1):

$$\%IE = \theta \times 100 = \left[(C'_R - C_R) / C'_R\right] \times 100, \qquad (1)$$

where θ is the surface coverage, C'_R, C_R represent the corrosion rates in absence and presence of extract inhibitor, respectively.

Adsorption isotherm

Different mathematical representations were tested to obtain the best fit representing the adsorption isotherms. The studied corrosion inhibition system corresponds to Langmuir adsorption isotherm given by the following equation [24]:

$$C/\theta = 1/k_{ads} + C, \qquad (2)$$

where θ represents the surface coverage, C is the concentration of the plant extract inhibitor and K_{ads} is the equilibrium constant of adsorption related to the free energy of adsorption ΔG°_{ads} by the following [25]:

$$K_{ads} = 1/5.55 \exp\left(-\Delta G^{\circ}_{ads}/RT\right). \qquad (3)$$

Straight lines obtained when plotting C/θ vs. C for the adsorption of the plant additive on the surface of carbon steel in hydrochloric acid at 30 °C is shown in Fig. 2 and the adsorption of thermodynamic parameters were calculated and listed in Table 3. It can be seen that, ΔG°_{ads} become less negative, in other meaning, increase when rising the temperature. This suggests that the adsorption process occurred by the electrostatic attraction. The negative values of ΔG°_{ads} suggest that the adsorbed layer on the carbon steel surface is stable and the adsorption process is spontaneous [26]. The values of ΔG°_{ads} were less negative than -20 kJ mol^{-1} indicating that the adsorption mechanism of the investigated extract on carbon steel in 1 M HCl solution is consistent with physisorption [27, 28]. The heat of adsorption (ΔH°_{ads}) could be calculated according to the Vant Hoff equation [29]:

$$Log\ K_{ads} = \left(-\Delta H^{\circ}_{ads}/2.303\ RT\right) + constant. \qquad (4)$$

Log K_{ads} was plotted against 1/T as shown in Fig. 3 to calculate the heat of adsorption ΔH°_{ads}. The straight lines were obtained with slope equal to $(-\Delta H^{\circ}_{ads}/2.303\ R)$. In consistence with the following equation [30]:

$$\Delta G^{\circ}_{ads} = \Delta H^{\circ}_{ads} - T\ \Delta S^{\circ}_{ads}. \qquad (5)$$

Table 2 The effect of different concentration of *Tilia cordata* extract on the corrosion rate (C_R) (mg cm^2 min^{-1}) and inhibition efficiency (%IE) of carbon steel in 1 M HCl solution at different temperatures

$[C_{inh}]$ (mg L^{-1})	30 °C		40 °C		50 °C	
	C_R (mg cm^{-2} min^{-1}) × 10^{-3}	%IE	C_R (mg cm^{-2} min^{-1}) × 10^{-3}	%IE	C_R (mg cm^{-2} min^{-1}) × 10^{-3}	%IE
0	165.8	–	275.8	–	398.4	–
50	58.4	64.8	159.3	42.2	242.1	39.2
100	37.7	77.2	106.6	61.3	171.7	56.9
150	33.8	79.6	97.2	64.7	160.6	59.7
200	29.0	82.5	73.5	73.3	152.4	61.7
250	21.4	87.1	67.9	75.4	127.0	68.1
300	16.9	89.8	64.8	76.5	99.8	74.9

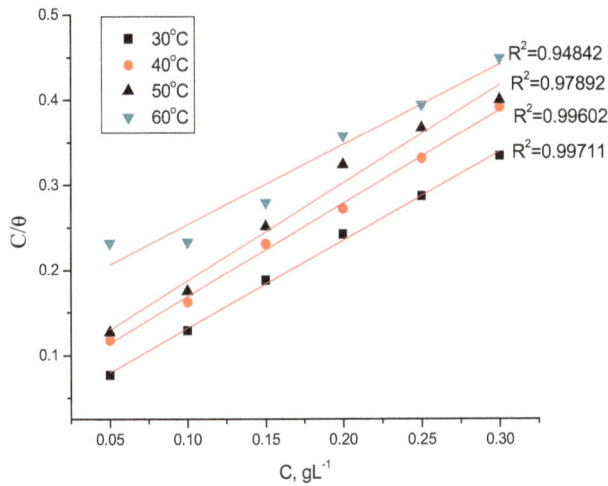

Fig. 2 Langmuir adsorption plots for carbon steel in 1 M HCl at different temperatures in the presence of various concentration of *Tilia cordata* extract

Effect of temperature

The influence of temperature, on the inhibition efficiency and nature of corrosion process, was explained by Arrhenius and transition state equations. Arrhenius equation:

$$\text{Log } k_{corr} = -E_a^*/2.303 \text{ RT } + \text{constant.} \quad (6)$$

Transition state equation [31]:

$$\text{Rate } (k_{corr}) = \text{RT/ Nhexp } (\Delta S^*/R)\exp(-\Delta H^*/RT). \quad (7)$$

The average value of corrosion rate (k_{corr}) obtained from weight loss experiment at different temperatures was used to study the effect of temperature. According to Arrhenius equation, the values of Log k_{corr} were plotted vs. 1/T. The resultant curves were straight lines with slopes ($-E_a^*/2.303$ R), from which the activation energy of the process can be calculated (E_a^*). Likewise, straight lines were obtained when plotting the values of Log (k_{corr}/T) vs. 1/T as shown in Fig. 5. According to the transition state equation, the slope of these curves equal ($-\Delta H^*/2.303R$) and the intercept is [(log (R/Nh)) + ($\Delta S^*/2.303R$)], from which the values of ΔH^* and ΔS^* were determined and listed in Table 4. The results in Table 4 indicate that the activation energy increased in presence of extract than in case of free acid solution. This indicates that the extract molecules adsorbed physically on carbon steel surface. By increasing the temperature the %IE decreased due to desorption of extract molecules from the metal surfaces takes place.

The study of the activation parameters reveals that the activation energy (E_a) increases when the concentration of plant extract increases. To illustrate, the energy barrier of corrosion reaction increases, as a result, the corrosion rate decreases. Moreover, the interpretation of the values and positive sign of the enthalpies (ΔH^*) reflect the endothermic nature of corrosion process of carbon steel in hydrochloric acid solutions. On the other hand, the values of entropy ΔS^* pertinent to uninhibited and inhibited acid solutions are negative. This implies that the activated complex, in the rate determining step, accompanied by dissociation rather than association, meaning that

Table 3 Thermodynamic parameters for the adsorption process of *Tilia cordata* on carbon steel surface in 1 M HCl at different temperatures

Temp. K	K_{ads} M^{-1}	$-\Delta G_{ads}^\circ$ kJ mol^{-1}	$-\Delta H_{ads}^\circ$ kJ mol^{-1}	$-\Delta S_{ads}^\circ$ J mol^{-1} K^{-1}
303	35	18.8	44.9	86.5
313	16.6	16.9	44.9	89.6
323	13.6	16.4	44.9	88.3
333	6.2	14.5	44.9	91.5

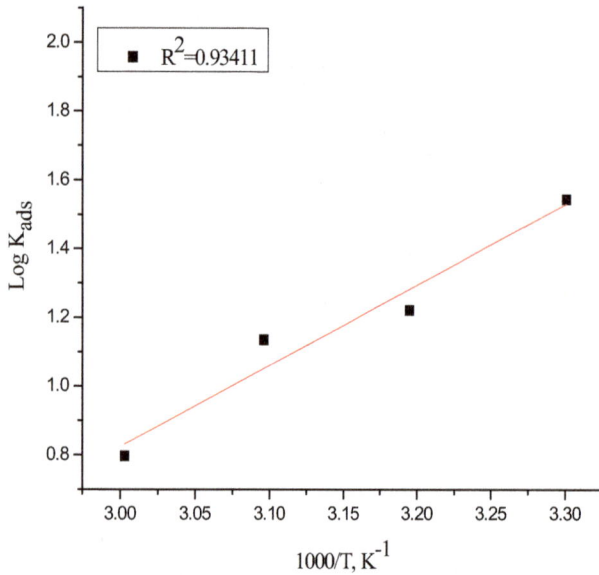

Fig. 3 (Log K_{ads}) versus ($1000/T$) curve for the dissolution of carbon steel in 1 M HCl in the presence of *Tilia cordata* extract

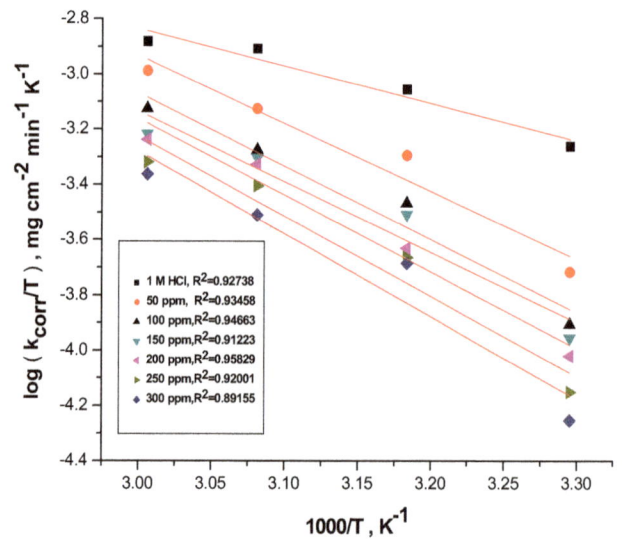

Fig. 4 (Log k_{corr} vs $1/T$) plots for corrosion of carbon steel in 1 M HCl in the absence and presence of different concentration of *Tilia cordata* extract

disordering increases on going from reactants to the activated complex [32] (Fig. 4).

Electrochemical measurements

Potentiodynamic polarization tests

The kinetics of anodic and cathodic reactions was studied by Tafel polarization measurements. Figure 6 indicates the effect of addition of *Tilia cordata* on anodic and cathodic polarization curves of carbon steel in 1 M HCl. Both cathodic and anodic reactions were noted to subsidize when the plant extract is added to the test solution, in turn, confirms that this additive reduced the metal dissolution and suppressed the hydrogen evolution reaction largely. Electrochemical corrosion kinetics parameters, i.e., corrosion potential (E_{corr}), cathodic and anodic Tafel slopes (β_a, β_c) and corrosion current density (i_{corr}) obtained from the extrapolation of the polarization curves, are given in Table 5. The region between linear part of cathodic and anodic branch of polarization curves becomes wider when

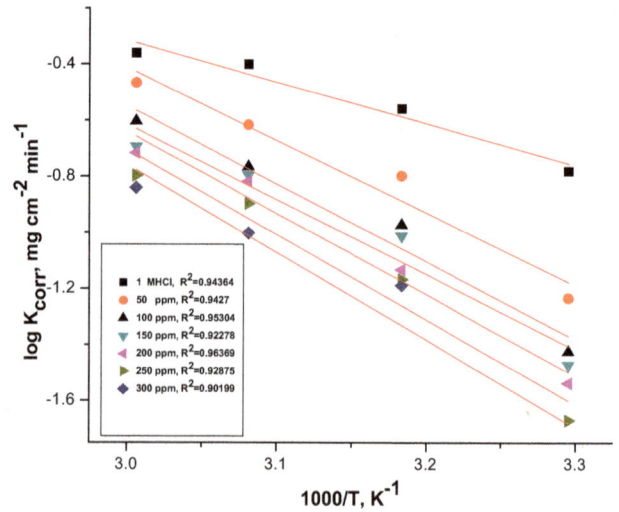

Fig. 5 Log (k_{corr}/T) vs ($1/T$) curves for corrosion of carbon steel in 1 M HCl in absence and presence of different concentration of *Tilia cordata* extract

Table 4 Activation parameters for corrosion of carbon steel in 1 M HCl in absence and presence of different concentration of *Tilia cordata* extract	$[C_{inh}]$ (mg L^{-1})	E_a^* kJ mol^{-1}	ΔH^* kJ mol^{-1}	$-\Delta S^*$ J mol^{-1} K^{-1}
	0	28.5	25.8	174.3
	50	49.7	47.0	112.6
	100	53.3	50.6	104.5
	150	51.3	48.6	111.7
	200	55.7	53.0	98.9
	250	58.1	55.4	92.9
	300	59.9	57.2	88.6

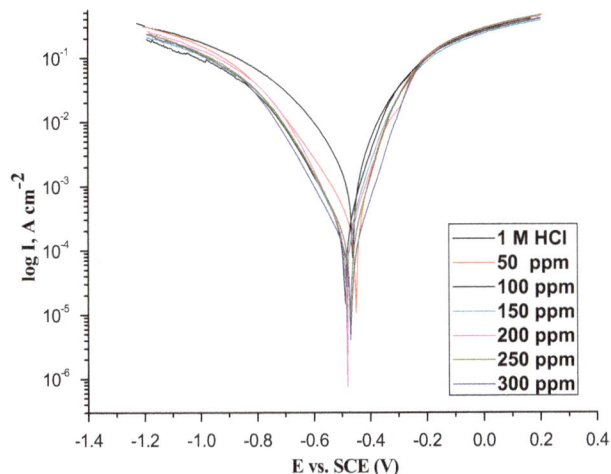

Fig. 6 Potentiodynamic polarization curves for the dissolution of carbon steel in 1 M HCl in the absence and presence of different concentrations of Tilia extract at 25 °C

Tilia cordata extract is added to the acid solution [33, 34]. The results recorded in Table 5 showed that the addition of the investigated extract lowers both cathodic and anodic currents without any significant change in corrosion potential. This indicates that this extract acts as mixed type inhibitors. The Tafel polarization results were consistent with those obtained from weight loss method. The surface coverage (θ) and inhibition efficiency (%IE) were calculated from the below Eq. (8).

$$\%IE = \theta \times 100 = \left[1 - \left(i^{\circ}/i\right)\right] \times 100, \qquad (8)$$

where i°, i are the values of corrosion current density in the presence and absence of Tilia cordata extracts, respectively.

Electrochemical impedance spectroscopy (EIS) tests

The two main parameters, R_{ct} and C_{dl}, deduced from electrochemical impedance techniques are listed in Table 6. Furthermore, Fig. 7 presents the Nyquist plots of carbon steel in uninhibited and inhibited test solution of 1 M hydrochloric acid solution. The semicircular nature of impedance diagrams indicates that a charge transfer

Table 6 Data from electrochemical impedance measurements for corrosion of carbon steel in 1 M HCl solutions at various concentrations of *Tilia cordata* extract

$[C_{inh}]$ (mg L^{-1})	R_{ct}, Ohm cm^2	C_{dl}, μF/cm^2	%IE
0	12.4	8.33	–
50	155.4	5.28	92.0
100	165.9	4.00	92.5
150	195.6	3.87	93.6
200	198.8	3.68	93.7
250	209.6	3.64	94.0
300	357.2	2.22	96.5

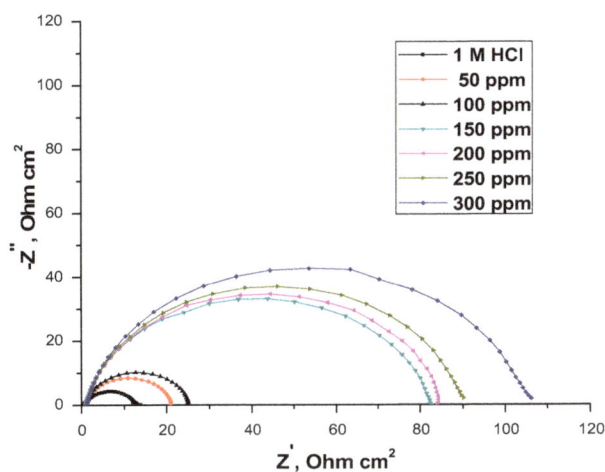

Fig. 7 Nyquist plots of carbon steel in 1 M HCl in absence and presence of different concentrations of Tilia extract at 25 °C

process mainly governs the corrosion of carbon steel, and the dissolution mechanism of carbon steel remains unaffected when the plant extract is added to the test solution [35] (Fig. 8). The experimental data were fitted to an equivalent electrical circuit, as shown in Fig. 9, so that the resultant spectra of the Nyquist plots impedance can be investigated. This simulation modeling assists to determine not only the solution resistance R_s, but also is useful for predicting other important parameters such as

Table 5 Potentiodynamic data of carbon steel in 1 M HCl in absence and presence of different concentrations of rosemary extract

$[C_{inh}]$ (mg L^{-1})	i_{corr}, μA cm^{-2}	$-E_{corr}$, mV	β_a, mV dec^{-1}	β_c, mV dec^{-1}	k_{corr} mpy	θ	%IE
0	1750	461	119	195	801.1	–	–
50	456	451	102	187	208.2	0.739	73.9
100	274	489	89	147	125.2	0.843	84.3
150	262	482	91	155	119.6	0.850	85.0
200	251	481	100	143	114.7	0.856	85.9
250	226	470	90	158	103.2	0.870	87.0
300	124	468	86	147	56.6	0.929	92.9

Fig. 8 Bode plot for corrosion of carbon steel in 1 M HCl in absence and presence of different concentrations of Tilia extract at 25 °C

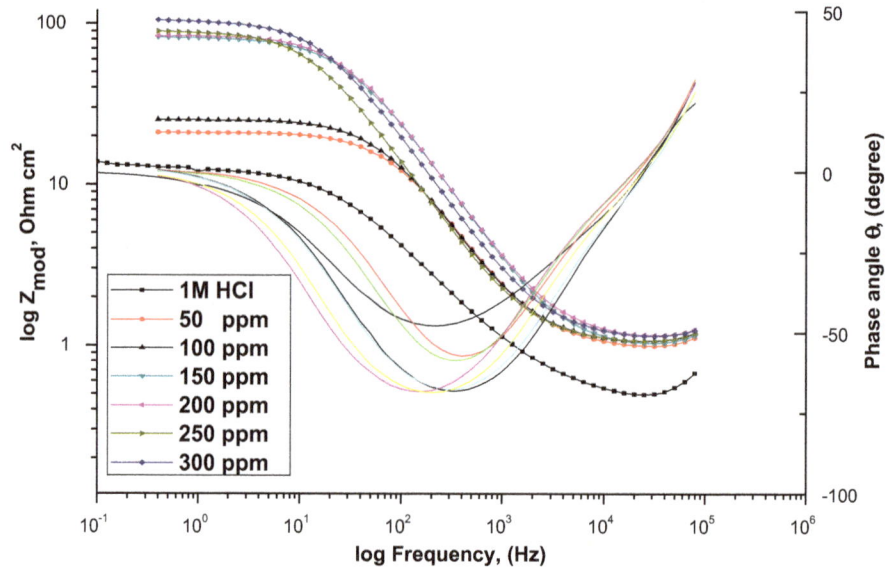

Fig. 9 Equivalent circuit model fitting impedance spectra

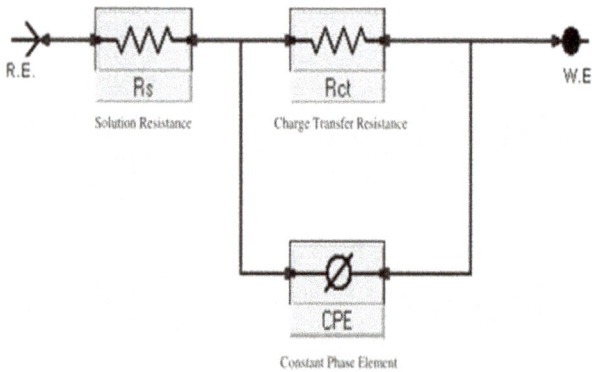

the double layer capacitance C_{dl} and the charge transfer resistance R_{ct}.

The values of charge transfer resistance R_{ct} were calculated from the difference in impedance at low and high frequencies obtained from Bode plots. The value of R_{ct} is considered a measure of electron transfer across the carbon steel surface and it is inversely proportional to the corrosion rate. The capacitance double layer C_{dl} was calculated at the frequency f_{max} at which the imaginary component of the impedance is maximal using the equation [36].

$$C_{dl} = \frac{1}{2\pi f_{max} R_{ct}} \quad (9)$$

The impedance results presented in Table 6 indicate that the magnitude of R_{ct} value increased while that of C_{dl} decreased with the addition of various concentrations of *Tilia cordata* to 1 M HCl. The decrease in C_{dl} values results from the adsorption of the plant extract compounds

at the metal surface. The double layer between the charged metal surface and the solution is considered as an electrical capacitor. The adsorption of plant extract on carbon steel surface reduces its electrical capacity as they replace the water molecules and other ions adsorbed at the interface surface leading to the formation of a protective adsorption layer on the metal (working electrode) surface which increases the thickness of the electrical double layer. The thickness of the protective layer (d) is related to C_{dl} in accordance with Helmholtz model, given by the following equation [37]:

$$C_{dl} = \frac{\varepsilon \varepsilon_o A}{d} \quad (10)$$

where ε is the dielectric constant of the medium and ε_0 is the permittivity of free space (8.854×10^{-14} F/cm) and A is the effective surface area of the electrode. From Table 6, it is clear that as the thickness of the protective layer formed by inhibitor molecules, increases, the C_{dl} should decrease. In the present work C_{dl} value was found to reach the maximum in case of the uninhibited solution. Addition of *Tilia cordata* extract to the aggressive medium is found to decrease the C_{dl} value and also the lowest value was obtained for the investigated extract with the highest inhibition efficiency. The decrease in C_{dl} values on increasing the extract inhibitor concentration is due to building an adsorbed layer on interface surface between the metal and acid solution. The inhibition efficiency of extract inhibitor for the corrosion of carbon steel in 1M HCl is calculated using R_{ct} values as follows:

$$\%IE = \theta \times 100 = \left[1 - \left(R_{ct}^{\circ}/R_{ct}\right)\right] \times 100. \quad (11)$$

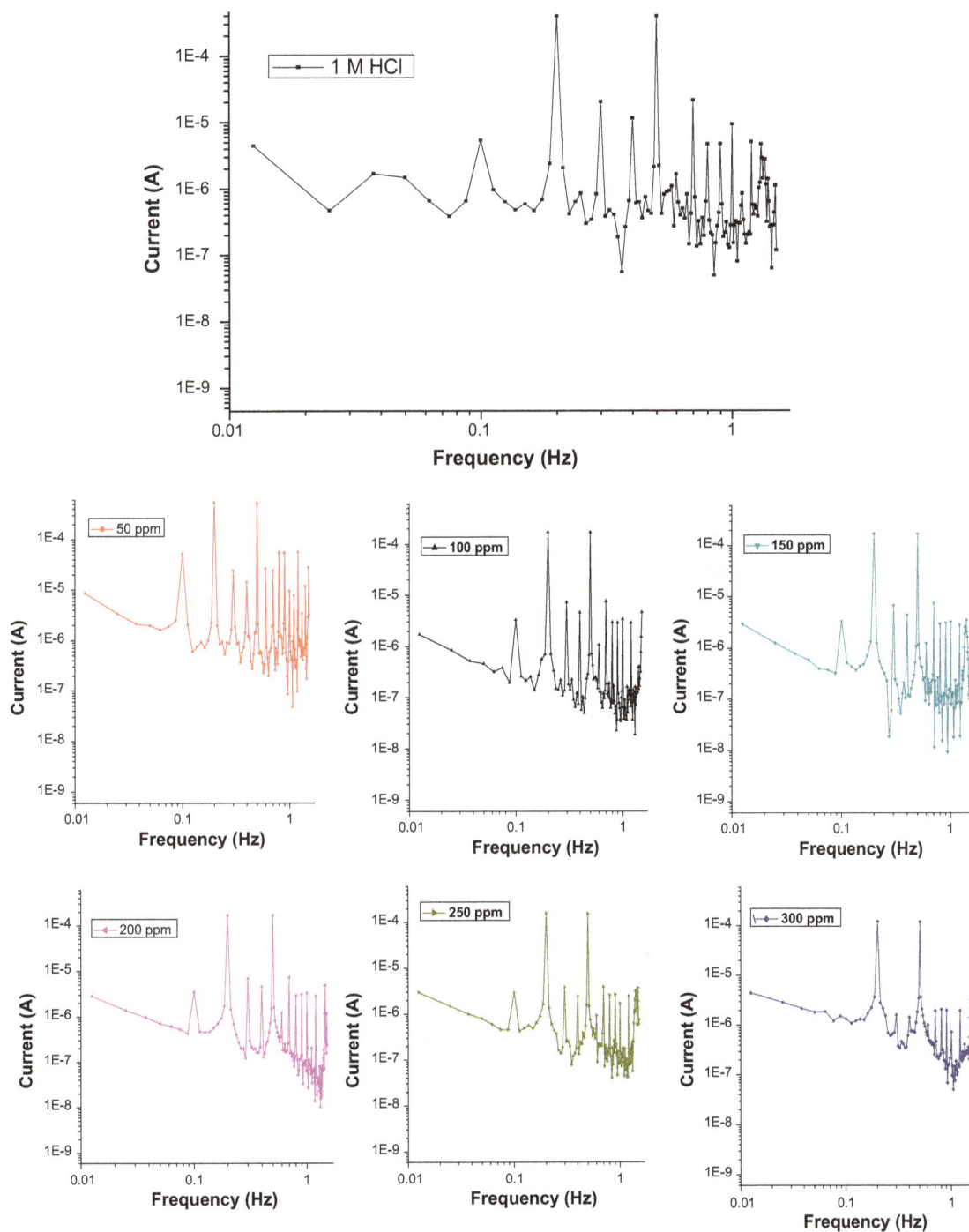

Fig. 10 EFM spectra for carbon steel in 1 M HCl in the absence and presence of 300 mg/l of Tilia extract at 25 °C

Electrochemical frequency modulation measurements

The current response, due to the potential perturbation by applying one or more sine waves, contains useful information about the corroding system. The larger peaks were used to calculate the corrosion current values without prior knowledge of Tafel slopes, and with only a small polarization signal [38, 39]. The causality factors CF-2 and CF-3 are calculated from the frequency spectrum of the current responses. Figure 10 shows the EFM spectra (current vs frequency) of carbon steel in HCl solution in absence and presence of different concentrations of *Tilia cordata* extract. The corrosion current density decreases on increasing the concentration of *Tilia cordata* extract, indicating that this extract inhibits the corrosion of carbon

Table 7 Electrochemical kinetic parameters obtained by EFM technique for carbon steel in 1 M HCl solution containing various concentrations of the *Tilia cordata* extract at 25 °C

$[C_{inh}]$ (mg L^{-1})	i_{corr} μA cm^{-2}	β_a mV dec^{-1}	β_c mV dec^{-1}	k_{corr} mpy	CF(2)	CF(3)	θ	%IE
0	655.7	92	131	299.6	2.06	3.56	–	–
50	300.1	34	38	137.1	2.00	2.00	0.542	54.2
100	233.8	79	100	106.8	1.86	2.29	0.643	64.3
150	229	78	99	104.6	1.86	2.73	0.650	65.0
200	226.4	78	97	103.5	1.81	2.58	0.654	65.4
250	211.5	83	95	96.67	1.86	3.51	0.677	67.7
300	176.3	89	97	80.55	1.63	3.78	0.731	73.1

Fig. 11 SEM images of carbon steel surface **a** polished surface, **b** exposed to free acid solution only (blank) and **c** exposed to test solution containing 300 mg/l of *Tilia cordata* extract

(a) polished surface

(b) exposed to free acid solution only (Blank)

(c) exposed to test solution containing 300 mg.L^{-1}of Tilia cordata extract

steel in 1 M HCl through forming a barrier film by adsorption onto the metal surface. The causality factors obtained under different experimental conditions are approximately equal to its theoretical values (2 and 3) indicating that the measured data are valid. The increase in *Tilia cordata* concentration accompanied by an increase in the inhibition efficiency and the inhibition efficiency was calculated using Eq. (12):

$$\%IE_{EFM} = \left[1 - \left(i_{corr}/i^{\circ}_{corr}\right)\right] \times 100 \tag{12}$$

where i_{corr} and i°_{corr} are the corrosion current densities in absence and presence of *Tilia cordata* extract (Table 7).

Scanning electron microscopy (SEM) studies

Figure 11 represents the micrographs obtained for carbon steel samples before and after immersed into 1 M HCl for 12 h, without, and with further treatment of plant extract additive. It can be shown that the carbon steel surface suffer from severe dissolution in case of exposure to the free acid. On the contrary, the surface of carbon steel is found to be less affected by the aggressiveness of the acid solution when exposed to 1 M HCl solution containing 300 mg L^{-1} of *Tilia cordata* extract. Besides, the morphology of carbon steel surface is apparently smooth

compared to the micrograph obtained in case of the free acid. Moreover, a thin protective layer is formed and evenly distributed on carbon steel surface. This corroborates the involvement of plant extract molecules in blocking the corrosion cells on carbon steel surface by decreasing the contact between carbon steel surface and the corrosive solution that sequentially inhance the inhibition characteristics and show better corrosion protection for the metal surface [40, 41].

EDX analysis

The elemental composition of the thin film formed on carbon steel surface was analyzed using energy EDX. The results listed in Table 8 provide a comparison between the elemental compositions of carbon steel surface before and after 12 h exposure to the uninhibited and inhibited 1 M hydrochloric acid solution. The EDX analysis indicates that the weight % of carbon is increased, that can be attributed to the carbon atoms of some adsorbed organic molecules, in case of the test solution containing 300 mg L^{-1} of *Tilia cordata* extract. This confirms the formation of protective layer, because of the adsorption of some organic constituents dissolved in the aqueous extract of *Tilia cordata* (Fig. 12).

Table 8 Elemental composition (wt%) of carbon steel surface after 12 h. of immersion in 1 M HCl in absence and presence of 300 mg L^{-1} of *Tilia cordata* extract

Wt%	Iron	Carbon	Silicone	Manganese
Carbon steel surface	87.72	11.59	0.38	0.76
Free acid solution (Blank)	82.06	16.66	0.26	0.74
300 mg L^{-1} of *Tilia cordata*	80.45	18.40	0.35	0.81

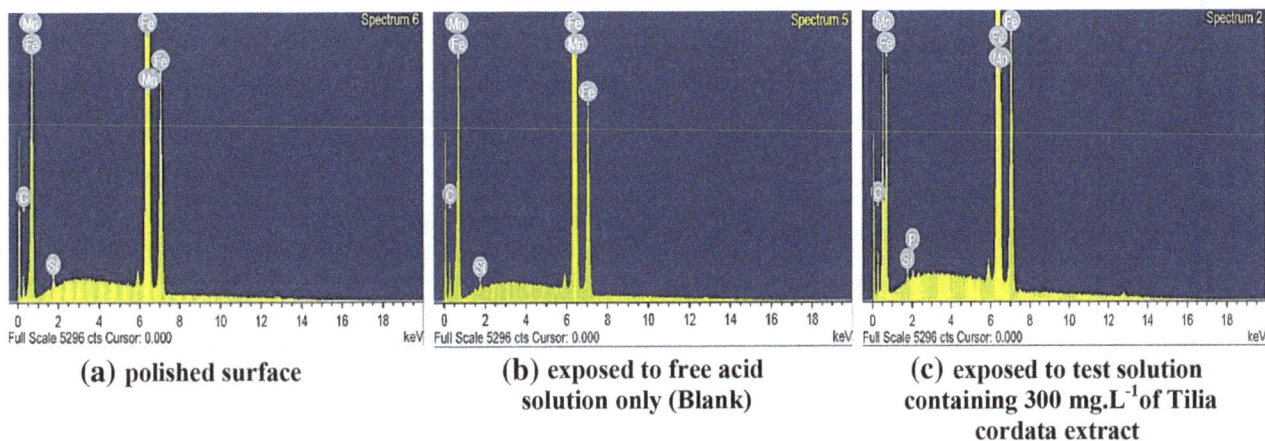

(a) polished surface   **(b) exposed to free acid solution only (Blank)**   **(c) exposed to test solution containing 300 mg.L⁻¹ of Tilia cordata extract**

Fig. 12 EDX analysis of carbon steel surface **a** polished surface, **b** exposed to free acid solution only (blank) and **c** exposed to test solution containing 300 mg/l of *Tilia cordata* extract

Mechanism of inhibition

However, it is not readily available at this point to determine the specific constituent or group of molecules of the plant extract that are adsorbed onto the metal surface, *Tilia cordata* exhibited a high content of flavonol *O*-glycosides (mono- and di-) quercetin and kaempferol derivatives and tiliroside. These constituents are adsorbed onto the metal surface, the adsorbed molecules constitute a physical barrier between the metal and the corrosion medium, thereby reducing the metal dissolution at anodic sites and hydrogen evolution at cathodic sites, also interfering the corrosive attack of acid solutions and protecting the metal surface. *Tilia cordata* extract provide greater inhibition due to the large degree of surface coverage resulting from the adsorption of particular molecules or a group of constituents from the crude extract [42].

efficiency percentages increased with increasing the concentration of plant extract, but decreased with raising the test solution temperature. The free energy change of adsorption, enthalpy of adsorption and entropy of adsorption indicated that the adsorption process is spontaneous and exothermic and the phytochemical constituents of *Tilia cordata* extract adsorbed at the metal-surface interface and produce a protective barrier by the process of electrostatic adsorption (physisorption). The activation parameters reveal that the activation energy increases as the concentration of *Tilia cordata* increases. Furthermore, the endothermic nature of the carbon steel dissolution process can be inferred from the values of activation enthalpy. In addition, the entropy of activation increased with increasing inhibitor concentration; hence, the system disorder is increased.

Conclusion

The good accordance between the results obtained from different experimental methods leads to conclude that an aqueous extract of *Tilia cordata* can be used as an effective green corrosion inhibitor for the corrosion of carbon steel in hydrochloric acid solutions. The inhibition

Compliance with ethical standards

Conflict of interest The authors declare that they have no competing interests.

References

1. Fekry AM, Mohamed RR (2010) Acetyl thiourea chitosan as an eco-friendly inhibitor for mild steel in sulphuric acid medium. Electrochim Acta 55(6):1933–1939
2. Mitchell KE (1998) US Patent 5,746,908. U.S. Patent and Trademark Office, Washington
3. Gale WF, Totemeier TC, (Eds.) (2003) Smithells metals reference book. Butterworth-Heinemann, UK
4. Lalitha A, Ramesh S, Rajeswari S (2005) Surface protection of copper in acid medium by azoles and surfactants. Electrochim Acta 51(1):47–55
5. Anthony N, Malarvizhi E, Maheshwari P et al (2004) Corrosion inhibition by caffeine—Mn. 2 + system. Indian. J Chem Technol 11(3):346–350
6. Benabdellah M, Hammouti B (2005) Corrosion behaviour of steel in concentrated phosphoric acid solutions. Appl Surf Sci 252(5):1657–1661
7. Elouafi M, Abed Y, Hammouti B et al (2001) Effect of acidity level on the corrosion of steel in concentrated HCl solutions. Ann Chim Sci Mater 26(5):79–84
8. Abed Y, Hammouti B (2000) Corrosion of steel in concentrated H_2SO_4 solutions. Bull Electrochem 16(7):296–298
9. Weina Su, Tian Yimei, Peng Sen (2014) The influence of sodium hypochlorite biocide on the corrosion of carbon steel in reclaimed water used as circulating cooling water. Appl Surf Sci 315(1):95–103
10. El-Etre AY (2006) Khillah extract as inhibitor for acid corrosion of SX 316 steel. Appl Surf Sci 252(24):8521–8525
11. Oguzie EE (2007) Corrosion inhibition of aluminium in acidic and alkaline media by Sansevieria trifasciata extract. Corros Sci 49:1527–1539
12. Fouda AS, Nofal AM, El-Ewady GY, Abousalem AS (2015) Eco-friendly impact of rosmarinus officinalis as corrosion inhibitor for carbon steel in hydrochloric acid solutions. Der Pharma Chemica 7(5):183–197
13. Fouda AS, El-Awady GY, Abousalem AS (2014) Corrosion inhibition and thermodynamic activation parameters of arcatium lappa extract on mild steel in acidic medium. Chem Sci Rev Lett 3(12):1277–1290
14. Li Yan, Zhao P, Liang Q et al (2005) Berberine as a natural source inhibitor for mild steel in 1 M H_2SO_4. Appl Surf Sci 252(5):1245–1253
15. Zucchi F, Omar IH (1985) Plant extracts as corrosion inhibitors of mild steel in HCl solutions. Surf Tech 24(4):391–399
16. Negri G, Santi D, Tabach R (2013) Flavonol glycosides found in hydroethanolic extracts from *Tilia cordata*, a species utilized as anxiolytics. Revista Brasileira de Plantas Medicinais 15(2):217–224
17. Karioti A, Chiarabini L, Alachkar A, Chehna MF, Vincieri FF, Bilia AR (2014) HPLC–DAD and HPLC–ESI-MS analyses of Tiliaeflos and its preparations. J Pharm Biomed Anal 100:205–214
18. Behrens A, Maie N, Knicker H, Kögel-Knabner I (2003) MALDI-TOF mass spectrometry and PSD fragmentation as means for the analysis of condensed tannins in plant leaves and needles. Phytochemistry 62(7):1159–1170
19. Oniszczuk A, Podgórski R (2015) Influence of different extraction methods on the quantification of selected flavonoids and phenolic acids from *Tilia cordata* inflorescence. Ind Crops Prod 76:509–514
20. Khaled KF (2008) Application of electrochemical frequency modulation for monitoring corrosion and corrosion inhibition of iron by some indole derivatives in molar hydrochloric acid. Mater Chem Phys 112(1):290–300
21. Khaled KF (2009) Evaluation of electrochemical frequency modulation as a new technique for monitoring corrosion and corrosion inhibition of carbon steel in perchloric acid using hydrazine carbodithioic acid derivatives. J Appl Electrochem 39(3):429–438
22. Bosch RW, Hubrecht J, Bogaerts WF et al (2001) Mobile hydrogen monitoring in the wall of hydrogenation reactors. Corrosion 57(1):60–71
23. Abdel-Rehim SS, Khaled KF, Abd-Elshafi NS (2006) Electrochemical frequency modulation as a new technique for monitoring corrosion inhibition of iron in acid media by new thiourea derivative. Electrochim Acta 51(16):3269–3277
24. Şahin M, Bilgiç S, Yilmaz H (2002) The inhibition effects of some cyclic nitrogen compounds on the corrosion of the steel in NaCl mediums. Appl Surf Sci 195(1–4):1–7
25. Schorr M, Yahalom J (1977) the significance of the energy of activation for the dissolution reaction of metal in acids. Corros Sci 12(11):867–868
26. Tang L, Lie X, Si Y et al (2006) the synergistic inhibition between 8-hydroxyquinoline and chloride ion for the corrosion of cold rolled steel in 0.5 M sulfuric acid. Mater Chem Phys 95(1):29–38
27. Maayta AK, Al-Rawashdeh NAF (2004) Inhibition of acidic corrosion of pure aluminum by some organic compounds. Corros Sci 46(5):1129–1140
28. Abboud Y, Abourriche A, Saffaj T et al (2009) A novel azo dye, 8-quinolinol-5-azoantipyrine as corrosion inhibitor for mild steel in acidic media. Desalination 237(1–3):175–189
29. Khamis E (1990) the effect of temperature on the acidic dissolution of steel in the presence of inhibitors. Corrosion 46(6):476–484
30. Benabdellah M, Aouniti A, Dafali A et al (2006) Investigation of the inhibitive effect of triphenyltin 2-thiophene carboxylate on corrosion of steel in 2 M H_3PO_4 solutions. Appl Surf Sci 252(23):8341–8347
31. Khaled KF, Samardzija KB, Hackerman N (2004) Piperidines as corrosion inhibitors for iron in hydrochloric acid. J Appl Electrochem 34(7):697–704
32. Singh AK, Singh AK, Ebenso EE (2014) Inhibition effect of cefradine on corrosion of mild steel in HCl solution. Int J Electrochem Sci 9:352–364
33. Morad MS, El-Dean AK (2006) 2, 2′-Dithiobis (3-cyano-4, 6-dimethylpyridine): a new class of acid corrosion inhibitors for mild steel. Corros Sci 48(11):3398–3412
34. Singh AK, Quraishi MA (2010) Piroxicam; a novel corrosion inhibitor for mild steel corrosion in HCl acid solution. J Mater Environ Sci 1(2):101–110
35. Arab ST, Al-Turkustani AM (2006) Corrosion inhibition of steel in phosphoric acid by phenacyldimethylsulfonium bromide and some of its *p*-substituted derivatives. Portugaliae Electrochimica Acta 24:53–69
36. Elewady GY (2008) Pyrimidine derivatives as corrosion inhibitors for carbon-steel in 2 M hydrochloric acid solution. Int J Electrochem Sci 3(9):1149–1161
37. Ahamad I, Prasad R, Quraishi MA (2010) Thermodynamic, electrochemical and quantum chemical investigation of some Schiff bases as corrosion inhibitors for mild steel in hydrochloric acid solutions. Corros Sci 52(3):933–942
38. Kus E, Mansfeld F (2006) An evaluation of the electrochemical frequency modulation (EFM) technique. Corros Sci 48(4):965–979

39. Caigman GA, Metcalf SK, Holt EM (2000) Thiophene substituted dihydropyridines. J Chem Cryst 30(6):415–422
40. Muralidharan S, Phani KLN Pitchumani S et al (1995) Poly-amino-benzoquinone polymers: a new class of corrosion inhibitors for mild steel. J Electrochem Soc 142(5):1478–1483
41. Prabhu RA, Venkatesha TV, Shanbhag AV et al (2008) Inhibition effects of some Schiff's bases on the corrosion of mild steel in hydrochloric acid solution. Corros Sci 50(12):3356–3362
42. Oguzie EE (2008) Corrosion inhibitive effect and adsorption behavior of Hibiscus sabdariffa extract on mild steel in acidic media. Port Electrochim Acta 26(3):303–314

6

Experimental and computational studies on the inhibition performances of benzimidazole and its derivatives for the corrosion of copper in nitric acid

Loutfy H. Madkour[1] ⓘ · I. H. Elshamy[2]

Abstract The inhibitive performance of seven synthesized 2-(2-benzimidazolyl)-4 (phenylazo) phenol (BPP_1–7) derivatives was investigated experimentally on the corrosion of copper in 2.0 M HNO₃ acid using mass loss, thermometric and DC potentiodynamic polarization techniques. Quantum chemical calculations was investigated to correlate the electronic structure parameters of the investigated benzimidazole derivatives with their inhibition efficiencies (IE%) values. Global reactivity parameters such as E_{HOMO}, E_{LUMO}, the energy gap between E_{LUMO} and E_{HOMO} (ΔE), chemical hardness, softness, electronegativity, proton affinity, electrophilicity and nucleophilicity have been calculated and discussed. Molecular dynamics simulation was applied on the compounds, to optimize the equilibrium configurations of the molecules on the copper surface. The areas containing N atoms are most possible sites for bonding Cu (111) surface by donating electrons. Binding constant (K_b), active sites (1/y), lateral interaction (f), equilibrium constant (K_{ads}) and standard free energy of adsorption ($\Delta G°$) values obtained from either kinetic model and/or Frumkin adsorption isotherm were compared and discussed. Thermodynamic functions and activation parameters such as: E_a, $\Delta H*$, $\Delta S*$ and $\Delta G*$ at temperatures 303, 313, 323 and 333 K were determined and explained. IE% values of the examined compounds on Cu (111) surface followed the order arrangement: BPP_1 > BPP_2 > BPP_3 > BPP_4 > BPP_5 > BPP_6 > BPP_7. The theoretical data obtained as compatible with experimental results showed that the studied benzimidazole derivatives (BPP_1–7) are effective inhibitors for the corrosion of copper in nitric acid solution.

Keywords Corrosion inhibition · Copper · Benzimidazole derivatives · Quantum chemical calculations · Molecular dynamics simulation · Kinetic · Thermodynamic parameters

Introduction

Copper and its alloys are widely used materials for their excellent electrical and thermal conductivities in many applications such as electronics, production of wires, sheets, tubes and recently in the manufacture of integrated circuits [1]. Copper is resistant toward the influence of atmosphere and many chemicals; however, it is known that in aggressive media, it is susceptible to corrosion. Copper is relatively noble metal, requiring strong oxidants for its corrosion or dissolution. Nitric acid is one of the most widely used corrosive media that attracted a great deal of research on copper corrosion [2, 3]. The use of copper corrosion inhibitors in such conditions is necessary since no protective passive layer can be expected. Experimental techniques are very convenient in the understanding of inhibition mechanism but they are generally expensive and time-consuming. With the improvement of computer hardware and software, density functional theory (DFT) [4, 5] and molecular dynamics (MD) simulation methods in recent times have become fast and powerful tools to predict the corrosion inhibition efficiencies of inhibitor molecules [6–11]. In the conceptual DFT, quantum chemical parameters such as chemical hardness [12, 13], softness [14],

✉ Loutfy H. Madkour
  lha.madkour@gmail.com; loutfy_madkour@yahoo.com

[1] Chemistry Department, Faculty of Science and Arts, Baljarashi, Al-Baha University, P.O. Box 1988, Al-Baha, Saudi Arabia

[2] Chemistry Department, Faculty of Science, Tanta University, Tanta 31527, Egypt

electronegativity [15], proton affinity [16], electrophilicity [17] and nucleophilicity are considered in the prediction of chemical reactivity or stability. In the calculation of these mentioned chemical properties, Koopmans Theorem [18] provides great facilities to computational and theoretical chemists. According to this theory, ionization energy and electron affinity values of chemical species are associated with their HOMO and LUMO energy values, respectively. Hard-soft acid and base (HSAB) [19] principle introduced by Pearson states that "hard acids prefer to coordinate to hard bases and soft acids prefer to coordinate to soft bases." Polarizable chemical species is defined with soft concept. As can be understood from this classification of Pearson, nitrogen containing structures give electrons easily to metals [20]. In recent years, there is a considerable amount of effort devoted to studying inhibition properties of benzimidazole and its derivatives for metallic corrosion [21]. Benzimidazole is a heterocyclic aromatic organic compound with a bicyclic structure consisting of the fusion of benzene and imidazole rings [10], and hydrogen atoms on the rings can be substituted by other groups or atoms. Some derivatives of benzimidazole have been demonstrated as excellent inhibitors [22–27] for copper and its alloys in acidic solution, and exhibit different inhibition performance with the difference in substituent groups and substituent positions on the imidazole ring [21, 28, 29]. Thus, benzimidazoles are N- and O-containing compounds and good electron donors; this will enhance their adsorption on the surface of the copper metal and increase binding capabilities. Chemical structures of 2-(2-benzimidazolyl)-4(phenylazo) phenol (BPP_1–7) molecules considered in this study are shown in Fig. 1. The corrosion characteristics of copper surface and effectiveness of these compounds as chemical inhibitors were described from a quantitative point of view.

Experimental details

Weight loss measurements

Weight loss experiments were done according to the standard methods as reported in literature [30].

The corrosion rate, W (expressed in mg cm^{-2} min^{-1}) as well as the inhibition efficiency ($E_W\%$) over the exposure time period were calculated. The corrosion rates of the copper coupons have been determined for 4-h immersion period at 30 ± 1 °C from mass loss, using Eq. (1) where Δm is the mass loss, S is the area, and t is the immersion period [31]. The standard deviation of the observed weight loss was ± 1 %. The percentage protection efficiency ($E_W\%$) was calculated according the relationship Eq. (2) where W and W_{inh} are the corrosion rates of copper without and with the inhibitor, respectively [32]:

$$W = \frac{\Delta m}{S.t.} \tag{1}$$

The corrosion rate W (expressed in mg cm^{-2} min^{-1}) as well as the inhibition efficiency ($E_W\%$) over the exposure time period were calculated according to the following equation:

$$E_W\% = \frac{W - W_{inh}}{W} \times 100. \tag{2}$$

Thermometric measurements

The reaction vessel used was basically the same as that described recently in the previous article [30]. The reaction number (RN) and the reduction in reaction number (% red RN) were calculated using the following Eqs. (3) and (4):

$$RN = \frac{(T_{max} - T_i)}{t} \tag{3}$$

$$\% \text{ reduction in RN} = \frac{(RN_{free} - RN_{inh})}{RN_{free}} \times 100 \tag{4}$$

where T_{max} and T_i are the maximum and initial temperatures, respectively; t is the immersion time (in minutes) required to reach T_{max}; RN_{free} and RN_{inh} are the reaction number in the absence and presence of BPP inhibitors, respectively.

Electrochemical measurements

The copper metal (99.98 % Copper Egyptian Company) was used as foils with a surface area of 2.54 cm^2,

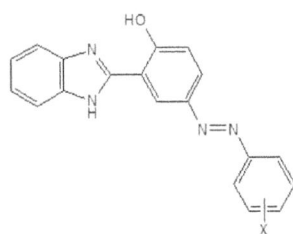

X = o-OCH$_3$ (BPP_1)

   = p-OCH$_3$ (BPP_2)

   = o-CH$_3$ (BPP_3)

   = m-CH$_3$ (BPP_4)

   = m-OCH$_3$ (BPP_5)

   = p-NO$_2$ (BPP_6)

   = m-NO$_2$ (BPP_7)

Fig. 1 Molecular structure of 2-(2-benzimidazolyl)-4 (phenylazo) phenol (BPP_1–7) derivatives

containing <0.001 % Ag, 0.001 % Sn as impurities. Description details of the electrochemical corrosion cell were explained in my published papers [33, 34].

The corrosion rate R_{corr} in milli-inches per year (MPY) can be obtained from the following equation:

$$R_{corr} = 0.13 I_{corr} E \cdot W / d \tag{5}$$

where I_{corr} is the corrosion current density ($\mu A\ cm^{-2}$), $E \cdot W$ is the equivalent weight of copper and d is its density.

Computational details

Quantum chemical calculations

Density functional theory (DFT) is certainly most widely used for the prediction of chemical reactivity of molecules, clusters and solids. It is important to note that DFT methods have become very popular in recent times [30, 35–37]. In the present study, quantum chemical calculations were performed with complete geometry optimizations using Gaussian-09 software package [38]. Geometry optimization were carried out by B3LYP functional at the 6–31G (d,p) basis set [39–42] and at the density functional theory (DFT) level. Frequency calculations were carried at the same levels of the theory to characterize the stationary points as local minima. The molecular orbital structures and energies were also calculated at the B3LYP method with 6-31+G** basis sets for both HOMO and LUMO levels.

In the conceptual density functional theory, chemical reactivity indices such as chemical hardness, electronegativity, chemical potential are defined as derivatives of the electronic energy (E) with respect to number of electrons (N) at a constant external potential, $\upsilon(r)$. Within the framework of this theory, mentioned chemical properties are given as [43, 44]:

$$\chi = -\mu = -\left(\frac{\partial E}{\partial N}\right)_{v(r)} \tag{6}$$

$$\eta = \frac{1}{2}\left(\frac{\partial \mu}{\partial N}\right)_{v(r)} = \frac{1}{2}\left(\frac{\partial^2 E}{\partial N^2}\right)_{v(r)} \tag{7}$$

With the help of finite differences method, for chemical hardness, electronegativity and chemical hardness, the following expressions based on first vertical ionization energy and electron affinity values of chemical species are given.

$$\eta = \frac{I - A}{2} \tag{8}$$

$$\chi = -\mu = \frac{I + A}{2} \tag{9}$$

According to Koopmans's Theorem [18], the negative of the highest occupied molecular orbital energy and the negative of the lowest unoccupied molecular orbital energy corresponds to ionization energy and electron affinity, respectively ($-E_{HOMO} = I$ and $-E_{LUMO} = A$). As a result of this theorem, chemical hardness and chemical potential can be expressed as [45]:

$$\mu = \frac{E_{LUMO} + E_{HOMO}}{2} \tag{10}$$

$$\eta = \frac{E_{LUMO} - E_{HOMO}}{2} \tag{11}$$

The global softness is defined as the inverse of the global hardness and this quantity is given as [14]:

$$S = \frac{1}{\eta} = -\left(\frac{\partial N}{\partial \mu}\right)_{v(r)} \tag{12}$$

The global electrophilicity index (ω) introduced by Parr [46] is the inverse of nucleophilicity and is given below in Eq. 8. From the light of this index, electrophilic power of a chemical compounds is associated with its electronegativity and chemical hardness. Nucleophilicity (ε) is the inverse of the electrophilicity as is given below in Eq. (14) [17]:

$$\varepsilon = 1/\omega \tag{13}$$

$$\omega = \frac{\mu^2}{2\eta} = \frac{\chi^2}{2\eta}. \tag{14}$$

Molecular dynamics simulations

Molecular dynamics (MD) simulation is very popular for the investigation regarding the interaction between the inhibitor molecule and the concerned metal surface. The interaction between inhibitors and the copper surface was simulated using Forcite module of Materials Studio 6.0 program developed by Accelrys Inc. [47]. Herein, we had chosen the Cu (111) surface to simulate the adsorption process. Five layers of copper atoms were used to ensure that the depth of the surface was greater than the non-bond cutoff radius used in the calculation. The MD simulation was performed at 298 K controlled by the Andersen thermostat, NVT ensemble, with a time step of 1.0 fs and simulation time of 1000 ps, using the COMPASS forcefield [48]. Non-bond Interactions, Van der Waals and electrostatic, were set as atom-based summation method and Ewald summation method, respectively, with a cutoff radius of 1.55 nm. Details of simulation process can be referred to some previous literature [49, 50].

The interaction energy between the inhibitor molecules and the Cu (111) surface is calculated by Eq. (15).

$$E_{interaction} = E_{total} - (E_{surface} + E_{inhibitor}) \tag{15}$$

Herein, the total energy of the surface and inhibitor molecule is designated as E_{total}, $E_{surface}$ is the surface energy without the inhibitor and $E_{inhibitor}$ is the energy of the adsorbed inhibitor on the surface. The binding energy of the inhibitor molecule is expressed as $E_{binding} = -E_{interaction}$.

Obviously, a larger $E_{binding}$ implies that the corrosion inhibitor combines with the copper surface more easily and tightly, and a higher and spontaneous inhibitive performance of the benzimidazole (BPP_1–7) derivatives.

Results and discussion

Chemical measurements

Weight loss measurements

The value of percentage inhibition efficiency (E_W %) and corrosion rate (W_{corr}) obtained from weight loss method at different concentrations of (BPP_1) in 2.0 M HNO$_3$ at 303 K are summarized in Table 1 and shown in Fig. 2. It is evident from Table 1 that the corrosion rate is decreased from 1.87 mg/cm^2 h to 0.11 mg/cm^2 h on the addition of 10^{-3}M of BPP_1. The corrosion data for benzimidazole (BPP_1–7) inhibitors are reported in Table 2. According to this data, it is clear that the addition of BPP compounds reduces the corrosion rate of copper in nitric acid solution. The variation of inhibition efficiency with increase in inhibitor concentrations is shown in Fig. 3. It was observed that BPP_1 inhibits the corrosion of copper in HNO$_3$ solution, at all concentrations, i.e., from 10^{-9} to 10^{-3} M. Maximum inhibition efficiency was shown at 10^{-3} M concentration of the inhibitor in 2.0 M HNO$_3$ to depend on the substituents.

Table 2 Corrosion rate of copper in 2.0 M HNO$_3$ with and without benzimidazole (BPP_1–7) additives at 10^{-4} M, and the corresponding inhibition efficiency from weight loss measurements at 303 K for 4 h

Solution	W_{corr} (mg/cm^2 h)	E_W (%)
Blank	1.87	–
BPP_1	0.24	87
BPP_2	0.52	72
BPP_3	0.56	70
BPP_4	0.60	68
BPP_5	0.77	59
BPP_6	0.84	55
BPP_7	0.88	53

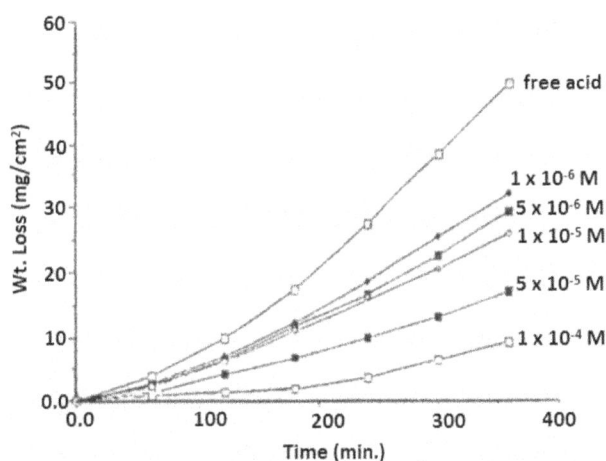

Fig. 2 Variation of weight loss against time for copper corrosion in 2.0 M HNO$_3$ in the presence of different concentrations of BPP_1 at 303 K

Thermometric measurements

In this method, the temperature change of the system involving copper dissolution in 2.0 M HNO$_3$ was followed without and with different concentrations of the investigated BPP_1 derivative, as tabulated in Table 3 and shown in Fig. 4 for BPP_1 as an example for the additives. An incubation period is first recognized along which the temperature rises gradually with time. The temperature–time curves provide a mean of differentiating between weak and strong adsorption. The thermometric data for all benzimidazole (BPP_1–7) derivatives are depicted in Table 4. It is evident that the dissolution of copper in 2.0 M HNO$_3$ starts from the moment of immersion. On increasing the concentration of the inhibitor from 10^{-6} to 10^{-4} M the value of T_{max} decreases, whereas the time (t) required for reaching T_{max} increases, and both factors cause a large

Table 1 Corrosion parameters for copper in aqueous solution of 2.0 M HNO$_3$ in the absence and presence of different concentrations of BPP_1 from weight loss measurements at 303 K for 4 h

Inhibitor	Conc (M)	W_{corr} (mg/cm^2 h)	E_W (%)	θ
Blank	2.0 M HNO$_3$	1.87	–	–
BPP_1	10^{-9}	1.46	22	0.22
	10^{-8}	1.38	26	0.26
	10^{-7}	1.33	29	0.29
	10^{-6}	1.27	32	0.32
	5×10^{-6}	1.14	39	0.39
	10^{-5}	1.08	42	0.42
	5×10^{-5}	0.67	64	0.64
	10^{-4}	0.24	87	0.87
	10^{-3}	0.11	94	0.94

Fig. 3 Variation of inhibition efficiency E_W (%) on copper with different concentrations of BPP_**1** inhibitor in 2.0 M HNO$_3$ at 303 K for 4 h immersion time

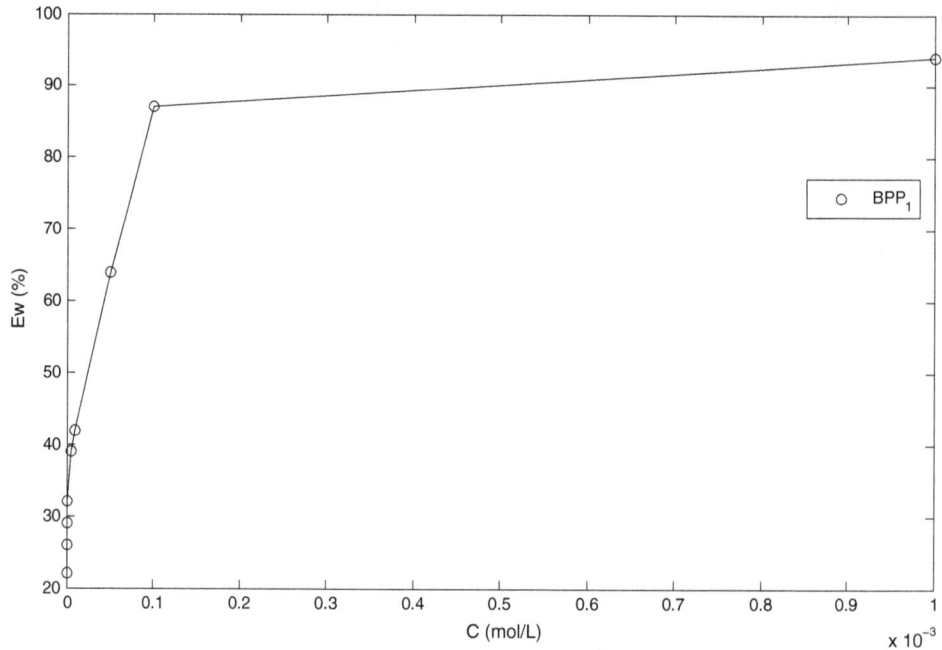

Table 3 Effect of different concentrations of BPP_**1** on the thermometric parameters of copper in 2.0 M HNO$_3$

Conc. (M)	LogC	T_i (°C)	T_{max} (°C)	t (min)	Δt (min)	LogΔt	θ	R.N.	% red in R.N.
2.0 M HNO$_3$		26.0	35.5	120				0.079	
10^{-6}	−6.0	25.9	34.6	165	45	1.65	0.342	0.052	34.17
5×10^{-6}	−5.3	26.0	33.5	180	60	1.77	0.481	0.041	48.10
10^{-5}	−5.0	26.0	32.1	210	90	1.95	0.658	0.027	65.82
5×10^{-5}	−4.3	26.1	31.4	225	105	2.02	0.708	0.023	70.88
10^{-4}	−4.0	26.2	29.6	240	120	2.07	0.823	0.014	82.27

decrease in *RN* and increasing of % red *RN* of the system, as shown in Tables 3 and 4. This indicates that the synthesized BPP_**1**–**7** additives retard the dissolution presumably by strong adsorption onto the copper surface. The extent of inhibition depends on the degree of the surface coverage (θ) of the copper surface with the adsorbate. Strehblow and Titze [51] showed that the Cu passive layer consists of a duplex structure of oxides, with an inner cuprous oxide and an outer cupric hydroxide. Copper will not corrode in non-oxidizing acidic environments, since hydrogen evolution is not a part of its corrosion process. However, when oxygen or other oxidants such as Fe^{3+} and NO_3^- ions are present, corrosion becomes important. Copper oxides are stable only in the pH range of 8–12 [52], but not in acidic solutions [Eqs. (16) and (17)] in which surface roughening can occur:

$$Cu_2O + 2H^+ \rightleftarrows 2Cu^+ + H_2O \tag{16}$$

$$CuO + 2H^+ \rightleftarrows Cu^{2+} + H_2O \tag{17}$$

Moreover, Cu^+ ions can undergo disproportionation according to Eq. (18).

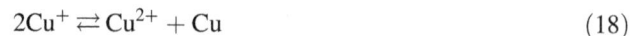

$$2Cu^+ \rightleftarrows Cu^{2+} + Cu \tag{18}$$

These reactions take place along the incubation period. The heat evolved from the above reactions accelerates further dissolution of the oxide and activates the dissolution of the copper metal exposed to the aggressive medium. The thermometric parameters measured experimentally such as *RN*, time delay (Δt) and/or log (Δt) versus molar concentration of the investigated benzimidazole (BPP_**1**-**7**) derivatives confirms and support a two-step adsorption process, at first a monolayer of the adsorbed is formed on the copper electrode surface, and then it is followed by the adsorption of a second adsorbed layer or a chemical reaction leading to the deposition of the BPP-Cu complex on the metal surface. The plot of Δt and/or log (Δt) as a function of log C_{In} yields a linear relation shape for the first region of the curve then a region of constancy; this reveals the completion of the adsorbed monolayer of the inhibitor. In thermometric measurements % red *RN* values are taken as the measure for the corrosion inhibition efficiency (IE%). Plots of % red RN versus molar concentration (C_{In}) of the additives for copper corrosion in 2.0 M HNO$_3$ are invariably sigmoidal in nature

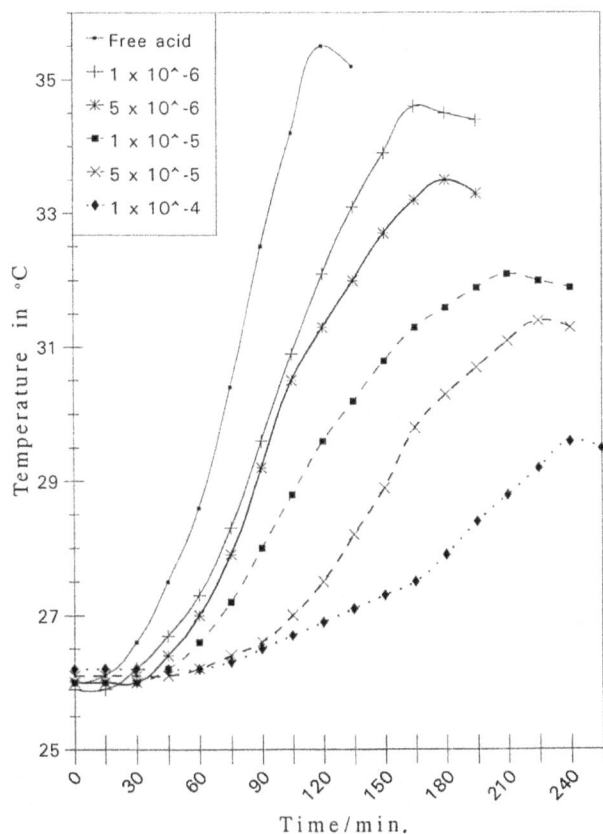

Fig. 4 Temperature versus time curves of copper corrosion in 2.0 M HNO₃ solution in the absence and presence of different concentrations of BPP_1

as shown in Fig. 5. The inhibition efficiency of the benzimidazole (BPP_1–7) derivatives depends on many factors, including the molecular size, heat of hydrogenation, mode of interaction with copper electrode surface, formation of metallic complexes and the charge density on the adsorption sites. Adsorption is expected to take place primarily through functional groups; essentially OH and OCH₃ would depend on its charge density. The thermometric technique cannot be applied for the copper corrosion in alkaline media because of the formation of Cu_2O, CuO and Cu $(OH)_2$ oxide films on the copper electrode surface, which formed only in near neutral and slightly alkaline solutions (pH = 8–12) [52].

Electrochemical measurements

Potentiodynamic polarization measurement

Potentiodynamic measurements were performed in the presence of different concentrations of 2-(2-benzimidazolyl)-4(phenylazo) phenol (BPP_1–7) derivatives. The results are tabulated in Tables 5, 6 and plotted in Figs. 6, 7, which show that the anodic and cathodic polarization curves of copper in 2.0 M HNO₃ solution, without and with BPP_1 derivative at 303 K. It is observed that both the cathodic curves and anodic curves show lower current density in the presence of the investigated inhibitors (BPP) than that of the blank. This indicates that the benzimidazoles inhibit the corrosion rate. Potentiokinetic polarization curves were plotted for copper corrosion in 2.0 M HNO₃ in the presence of the benzimidazole (BPP_1–7) derivatives as given in Fig. 7. A linear region with apparent Tafel was observed. The cathodic reaction was activation controlled and the addition of the compounds tested decreased the current densities in large anodic and cathodic domains of potential. The results indicated that the BPP_1–7 derivatives acted as mixed-type inhibitors. Generally, the addition of mixed inhibitors in solution does not change corrosion potential significantly because they inhibit both the anodic and cathodic reactions. Small changes in potentials can be a result of the competition of the anodic and the cathodic inhibiting reactions, and of the copper surface condition. The results of Fig. 7 illustrate that all BPP_1–7 derivatives bring down the corrosion current without causing any considerable change in the corrosion mechanism. According to corrosion theory, the rightward shift of the cathodic curves reveal that corrosion is mainly accelerated by cathode reactions. HNO₃ is a strong copper oxidizer capable of rapidly attacking copper. The potentiodynamic behavior of pure Cu in near neutral and slightly alkaline solutions exhibits three anodic peaks associated with the formation of Cu_2O, CuO and Cu $(OH)_2$. Cu_2O is first formed [Eq. (19)], which subsequently oxidizes to CuO [Eq. (20)] or, at more positive potentials, to Cu $(OH)_2$ [Eq. (21)].

Table 4 Effect of different concentrations of benzimidazole (BPP_1–7) derivatives on the inhibition efficiency IE% of copper in 2.0 M HNO₃ solution as determined by thermometric technique	Conc. (M)	% red in R.N.						
		BPP_1	BPP_2	BPP_3	BPP_4	BPP_5	BPP_6	BPP_7
	1×10^{-6}	34.17	32.91	30.37	35.44	13.92	34.17	15.18
	5×10^{-6}	48.10	41.77	43.03	44.30	26.58	45.56	29.11
	1×10^{-5}	65.82	63.29	48.10	51.89	37.97	49.36	39.24
	5×10^{-5}	70.88	67.08	54.49	56.96	49.36	53.16	46.83
	1×10^{-4}	82.27	74.68	68.30	63.29	62.02	58.22	55.65

Fig. 5 Effect of concentration
of BPP_1 on % reduction in
RN for copper in 2.0 M HNO₃

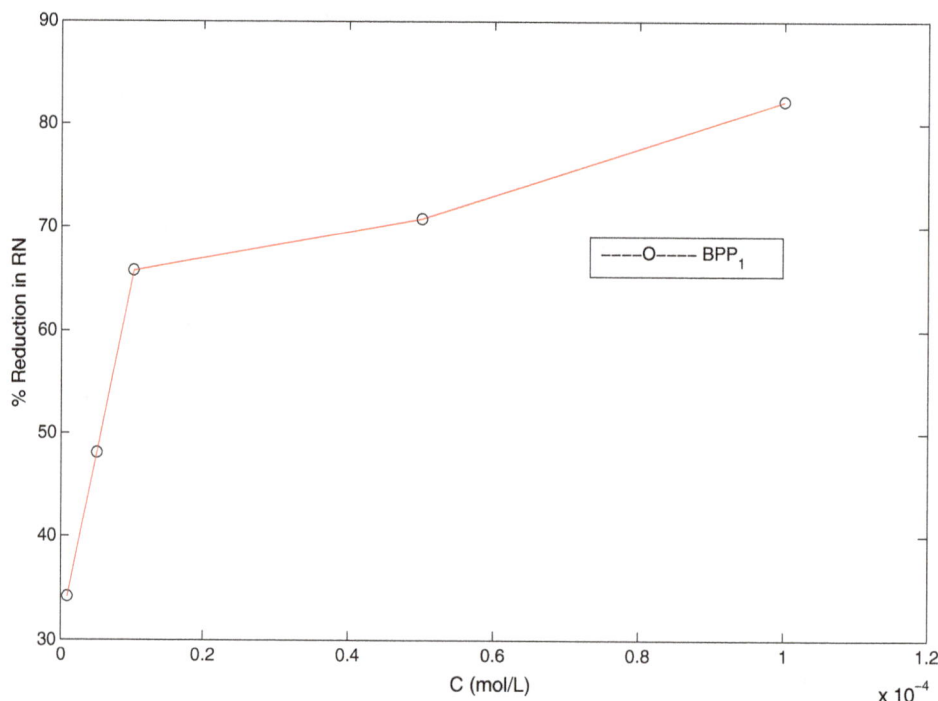

Table 5 Polarization corrosion parameters of Cu in 2.0 M HNO₃ solution containing BPP_1 at 295 K

Inhibitor type	Conc. (M)	R_p (Ω cm^2)	I_{corr} (μA cm^{-2})	R_{corr} (MPY)	$-E_{corr}$ (mV(SCE))	β_c (V dec^{-1})	β_a (V dec^{-1})	$E_{R_p}\%$ (Ω cm^2)	IE%
2.0 M HNO₃	–	7.053	7236	6671	61	0.233	0.237	–	–
2.0 M HNO₃ + (BPP_1)	10^{-7}	13.52	3468	3197	68	0.196	0.241	47.83	52.07
	10^{-6}	17.74	2317	2137	47	0.176	0.204	60.24	67.96
	5×10^{-6}	221.7	104.8	96.63	10	0.160	0.081	96.82	98.55

Table 6 Polarization corrosion parameters of Cu in 2.0 M HNO₃ solution containing 1×10^{-6} M of benzimidazole (BPP_1–7) derivatives at 303 K

Inhibitor type	R_p (Ω cm^2)	I_{corr} (μA cm^{-2})	R_{corr} (MPY)	E_{corr} (mV(SCE))	β_c (V dec^{-1})	β_a (V dec^{-1})	$E_{R_p}\%$ (Ω cm^2)	IE%
2.0 M HNO₃	6.360	7287	6719	29	0.302	0.165	–	–
BPP_1	45.08	235	731.3	23	0.177	0.154	85.9	96.8
BPP_2	75.34	406.7	374.9	12	0.162	0.125	91.5	94.4
BPP_3	63.36	507	467.4	7	0.145	0.151	89.9	93.0
BPP_4	30.93	1180	1206	26	0.185	0.188	79.4	83.8
BPP_5	76.40	397.2	366.2	7	0.202	0.107	91.7	94.5
BPP_6	147.9	320	295.1	6	0.204	0.234	95.7	95.6
BPP_7	113.0	242.5	290	−3	0.124	0.129	94.4	96.6

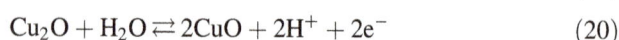

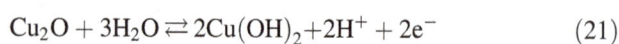

$$2Cu + H_2O \rightleftharpoons Cu_2O + 2H^+ + 2e^- \quad (19)$$

$$Cu_2O + H_2O \rightleftharpoons 2CuO + 2H^+ + 2e^- \quad (20)$$

$$Cu_2O + 3H_2O \rightleftharpoons 2Cu(OH)_2 + 2H^+ + 2e^- \quad (21)$$

Additionally, the potentiodynamic polarization curves in Figs. 6, 7 exhibit no steep slope in the anodic range, meaning that no passive films are formed on the copper surface. Consequently, copper may directly dissolve in

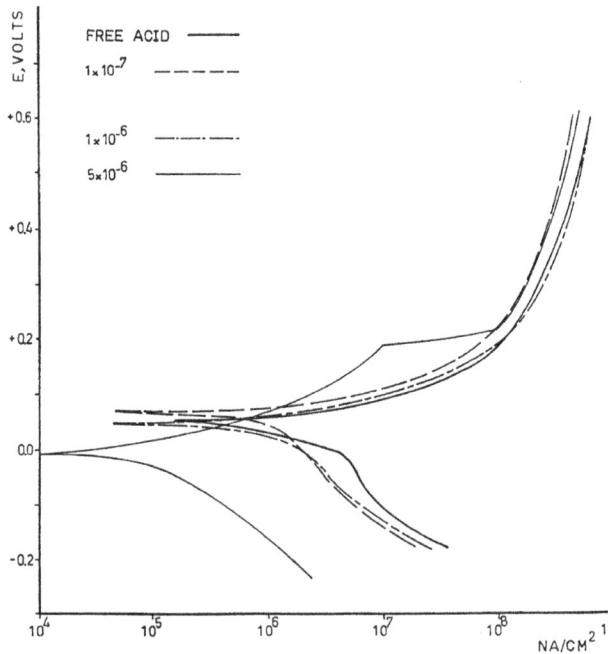

Fig. 6 Cathodic and anodic polarization curves of copper in 2.0 M HNO$_3$ solution with the BPP_1 in different concentrations at 303 K

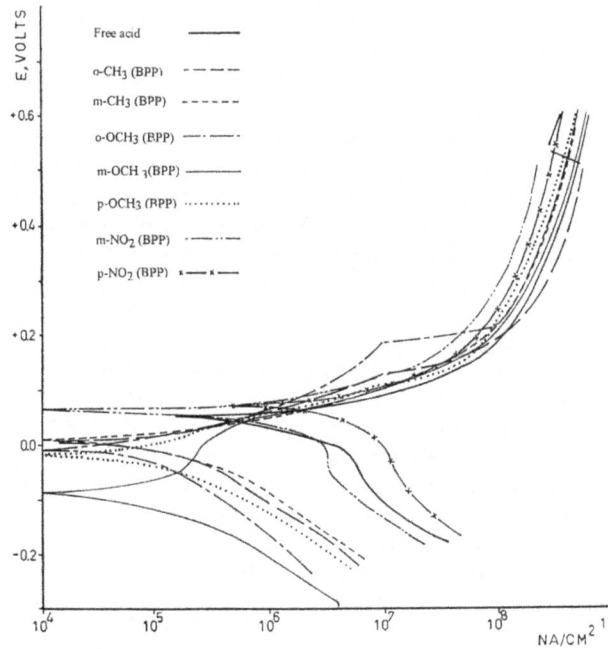

Fig. 7 Potentiodynamic polarization curves of copper in 2.0 M HNO$_3$ solution with the 5 × 10^{-6} M of different BPP_1–7 inhibitors at 303 K

Fig. 8 Potential–pH equilibrium diagram for the system, copper–water, at 25 ± 1 °C [53]. The dashed regions and the equilibrium components corresponding to the Pourbaix diagram in HNO$_3$

2.0 M HNO$_3$ solutions. The Pourbaix diagram for copper–water system is shown in Fig. 8 [53]. It indicates that copper is corroded to Cu^{2+} in HNO$_3$ solutions, and no oxide film is formed to protect the surface from corrosion. Copper dissolution is, thus, expected to be the dominant

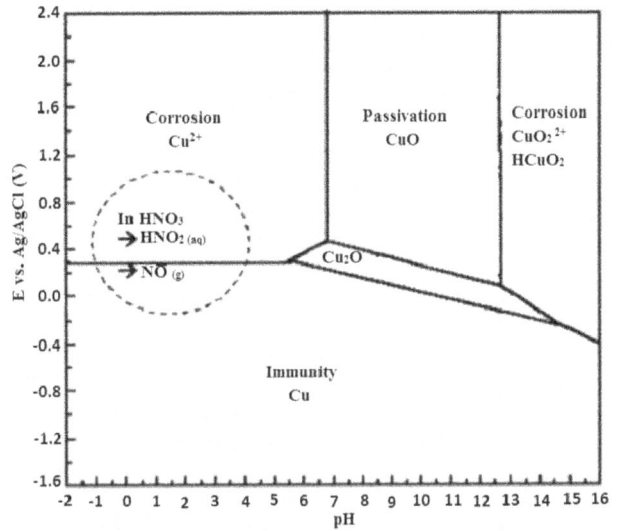

reaction in HNO$_3$ solutions. The electrochemical reactions for copper in HNO$_3$ solution can be described as follows:

Anodic reaction:

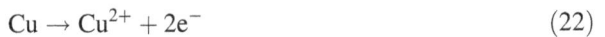

$$Cu \rightarrow Cu^{2+} + 2e^- \qquad (22)$$

Cathodic reactions [54]:

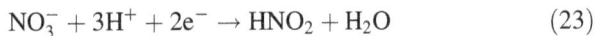

$$NO_3^- + 3H^+ + 2e^- \rightarrow HNO_2 + H_2O \qquad (23)$$

$$NO_3^- + 4H^+ + 3e^- \rightarrow NO + 2H_2O \qquad (24)$$

$$O_2 + 4H^+ + 4e^- \rightarrow 2H_2O \qquad (25)$$

Figure 7 shows the anodic and cathodic Tafel plots of copper in the presence of all seven studied BPP_1–7 inhibitors, for comparison polarization curves measured in 2.0 M HNO$_3$ acid solution are plotted. These compounds induce an increase in both the cathodic and anodic over voltages and cause a parallel displacement of both the cathodic and the anodic Tafel curves. The results indicate that the presence of the BPP compounds in the solution inhibits both the hydrogen evolution and the anodic dissolution processes. The data suggest that all tested BPP act as mixed-type inhibitors. They also indicate that the BPP molecules are mainly adsorbed at the copper surface via the oxygen and nitrogen centers of adsorption. The corrosion current density, I_{corr}, for copper in free 2.0 M HNO$_3$ acid is 7236 (μ A cm^{-2}). Addition of the BPP derivatives tested induces a decrease in both I_{corr} and R_{corr} whereas, the R_p values at the same time increases; thus BPP derivatives act as inhibitors for copper dissolution in nitric acid. Table 5 gives the effect of BPP_1 concentrations on corrosion parameters. In general, the inhibition efficiency depends on the concentration of the inhibitors. When an external emf was applied, both the cathode and the anode were polarized due to the presence of the nitrogen atom which is responsible for the cathodic behavior and the oxygen atom which is responsible for the anodic behavior. Similar behavior was observed in the presence of different BPP additives, from Table 5, the slopes of cathodic and anodic Tafel lines (β_c and β_a) were more changed on increasing the concentration of the tested inhibitor and have the values $\beta_c = 0.233 - 0.160$ V/decade and $\beta_a = 0.237 - 0.081$ V/decade. This behavior indicates that the adsorbed BPP derivatives mechanically screen the coated part of the copper electrode and therefore, protect it from the action of the corrosive medium. BPP inhibitor molecules have the effect on the mechanism of dissolution of copper and cause activation of a part of the surface respect to the corrosive medium. The high Tafel constant values support the probability of the presence of a complicated surface processes involving Cu ions and BPP molecules. Moreover, Tafel slopes of about 75 mV decade^{-1} can be attributed to a surface kinetic process rather than a diffusion controlled process. Inspection of polarization curves in Figs. 6, 7 show that, it was not possible to evaluate the cathodic Tafel slope as there is no visible linear region that prevents linear extrapolation to E_{corr} of the cathodic polarization curves. It has been shown that in the Tafel extrapolation method, use of both the anodic and cathodic Tafel regions is undoubtedly preferred over the use of only one Tafel region [55]. However, the corrosion rate can also be determined by Tafel extrapolation of either the cathodic or anodic polarization curve alone. If only one polarization curve alone is used, it is generally the cathodic curve which usually produces a longer and better defined Tafel region. Anodic polarization may sometimes produce concentration effects, due to passivation and dissolution, as noted above, as well as roughening of the surface which can lead to deviations from Tafel behavior. The situation is quite different here; the anodic dissolution of copper in aerated 2.0 M HNO$_3$ solutions obeys Tafel's law. The anodic curve is, therefore, preferred over the cathodic one for the Tafel extrapolation method. However, the cathodic polarization curve displays a limiting diffusion current due to the reduction of dissolved oxygen. Thus, the cathodic process is controlled, as will be seen latter, by concentration polarization rather than activation polarization, which prevented linear extrapolation of the cathodic curves. The corrosion current densities were obtained here by extrapolation of the anodic polarization curves to E_{corr}. Similar results were reported previously [54, 56, 57]. This irregularity was confirmed by other researchers and can be explained as the superposition of at least two cathodic current contributions: one arises from oxygen reduction and the second one consequential of copper ion re-deposition [58]. It is common practice and it was possible in this case to evaluate I_{corr} by extrapolation of the anodic polarization curves only to E_{corr}. At more cathodic potential with respect to E_{corr}, the characteristic horizontal line resulting from limiting current density for oxygen reduction can be observed. Because of uncertainty and source of error in the numerical values of the cathodic Tafel slope (β_c) calculated by the software; We did not introduce here β_c values recorded by the software. Table 5 gives the values of the associated electrochemical parameters of corrosion such as corrosion current density I_{corr} which calculated with the extrapolation of the linear parts of Tafel lines, corrosion potential E_{corr}, anodic Tafel constants β_a. The inhibition efficiency (IE%) was determined from the polarization curves using the Eq. (26):

$$IE\% = \frac{I_{corr} - I'_{corr}}{I_{corr}} \times 100 \qquad (26)$$

I_{corr} and I'_{corr} are the uninhibited and inhibited corrosion current densities, respectively, determined by extrapolation of the cathodic Tafel lines to corrosion potential (E_{corr}). It was clearly seen that cathodic slope was found equal indicating that the reduction of hydrogen did not modified in the presence of the azoles (BPP_1–7) inhibitors. Thus, the presence of the inhibitors at 10^{-6} M leads to decrease in the values of I_{corr}, which was particularly significant in the case of BPP_1 as shown in Table 5. The inhibition properties of BPP derivatives were evaluated also by polarization resistance method. Tables 5 and 6 gathered the corresponding values of polarization resistance R_p and the inhibition efficiency (E_{R_p}%) which is calculated using Eq. (27).

$$E_{R_p}\% = \frac{R'_p - R_p}{R'_p} \times 100 \qquad (27)$$

R_p and R'_p are the polarization resistance values without and with inhibitor, respectively. From polarization resistance measurements, the weakest value of R_p is found for copper in 2.0 M HNO$_3$ without inhibitor. However, the addition of the BPP derivatives in solution leads to an increase in the polarization resistance values. In fact, the value of R_p is 45.08 Ω cm^2 in the case of BPP_1 and increases to (63.36 Ω cm^2) for BPP_3, then, increases to (76.40 Ω cm^2) for BPP_5 and reaches to (147.9 Ω cm^2) for BPP_7 as given in Table 6. This in turn leads to a decrease in corrosion current density I_{corr} values because this later is inversely proportional to R_p. Table 5 show that the increasing of the concentration from 10^{-7} to 5×10^{-6} M (BPP_1) reduces significantly the dissolution rate of copper and consequently increasing IE%. Anodic Tafel constants change with inhibitor concentration, which is an indication of its effect on the copper dissolution reaction. No definite trend was observed in the shift of E_{corr} values, in the presence of various concentrations of BPP derivatives in 2.0 M HNO$_3$ solutions. BPP_1–7 derivatives may be classified as mixed-type inhibitors. Figure 7 shows that the polarization curves of copper in the presence of 5×10^{-6} M of all BPP_ 1–7 inhibitors. The arrangement of IE% follows the order: BPP_1 > BPP_2 > BPP_3 > BPP_4 > BPP_5 > BPP_6 > BPP_7. In the case of p-nitro derivative BPP_6, the inhibition efficiency (IE%) firstly increases, then sharply decreases by increasing the inhibitor concentration. This can be explained on the basis that, the p-NO$_2$ group characterized by critical inhibitor concentration (CIC) as well as the critical micelle concentration (CMC). The inhibition efficiency (IE%) of BPP derivatives depend on many factors, including the number of adsorption sites, their charge density, molecular size and mode of interaction with the copper metal surface. Retardation of copper dissolution by the BPP substituents is expected to be due to their adsorption on the copper surface via oxygen and nitrogen atoms. In general o-OCH$_3$ 2-(2(benzimidazolyl)-4-(phenylazo)-phenol (BPP_1) inhibitor is found to be the most efficient inhibitor; this may be due to the position of –OCH$_3$ group at ortho position in the phenylazo and its nature to the nearest on the phenol ring. This aids the more flat of the molecules and coverage the copper substrate to be more possible. This flat formation and coating effect decreases in the case of m-OCH$_3$ derivative.

Adsorption isotherm and thermodynamic activation parameters

The adsorption isotherm can be determined by assuming that inhibition effect is due mainly to the adsorption at metal/solution interface. Basic information on the adsorption of inhibitors on the metal surface can be provided by adsorption isotherm. In order to obtain the isotherm, the fractional surface coverage values (θ) as a function of inhibitor concentration must be obtained. The values of θ can be easily determined from the weight loss measurements by the ratio $E_W\%/100$, where $E_W\%$ is inhibition efficiency obtained by weight loss method. So it is necessary to determine empirically which isotherm fits best to the adsorption of inhibitors on the copper surface. Several adsorption isotherms (viz., Frumkin, Langmuir, Temkin, and Freundlich) were tested. Data were tested graphically by fitting to various isotherms. Figure 9 shows the dependence of the fraction (θ) as function of the logarithm of the concentration of BPP_1. The obtained plot is consistent with an S-shape adsorbed isotherm for BPP_1 showing an adsorption on the copper surface according to the Frumkin isotherm.

$$\frac{\theta}{1-\theta}\exp(-f\theta) = K_{ads}C \qquad (28)$$

with

$$K_{ads} = \frac{1}{55.5}\exp\left(\frac{-\Delta G^{\circ}_{ads}}{RT}\right) \quad \text{or} \atop \Delta G^{\circ}_{ads} = -RT(\ln 55.5 K_{ads}) \qquad (29)$$

where C is the concentration of the adsorbed substance in the bulk of the solution and θ is the degree of surface coverage of the metal surface by the inhibitor. K_{ads} is the modified equilibrium constant of the adsorption process, f is a constant depending on intermolecular interactions in the adsorption layer and on the heterogeneity of the surface, R is the universal gas constant and T is the absolute temperature. The value 55.55 in the above equation is the concentration of water in solution in mol L^{-1}. The values of K_{ads} and ΔG°_{ads} were calculated at 303 K and are listed in Table 7. ΔG°_{ads} is the free energy of adsorption and f is a function of adsorption energy. The average value of K, f and ΔG°_{ads} calculated from $\theta = \text{Log}\,([BPP\_1])$ curve are: $f = -7.59$; $K = 4.66 \times 10^6$ and $\Delta G^{\circ}_{ads} = -48.75$ kJ·mol^{-1}.

It is well known that values of $\Delta G^{\circ}_{ads} > -40$ kJ mol^{-1} (Table 7), indicate a chemisorption mechanism. In addition to electrostatic interaction, there may be some other interactions [59]. The high K_{ads} and ΔG°_{ads} values may be attributed to higher adsorption of the inhibitor molecules at the metal–solution interface [60]. Moreover, the inhibition of copper by BPP compounds is often explained by the formation of Cu (II)-BPP through its heteroatoms [61]. A plot of IE% versus Log ([BPP]) substituents has the character of an S-shape adsorbed isotherm as shown in Fig. 10. In physisorption process, it is assumed that acid anions such as NO$_3^-$ ions are specifically adsorbed on the metal surface, donating an excess negative charge to the metal surface. In this way, potential of zero charge becomes less negative which promotes the adsorption of inhibitors in cationic form [62]; those of order of 40 kJ mol^{-1} or higher involve charge sharing or transfer from the inhibitor molecules to the metal surface to form a coordinate type of bond (chemisorption) [63]. The strong correlation coefficients of the fitted curves, reveals that the inhibition tendency of the inhibitors is due to the adsorption of these synthesized molecules on the metal surface [64], as given in Table 7. The slopes of ln $(\theta/1-\theta)C$ versus θ plots are close to $\equiv 7$–18 in case of nitric acid solution which indicates the ideal simulating and expected from Frumkin adsorption isotherm [64]. K_{ads} values were calculated from the intercepts of the straight lines on the ln $(\theta/1-\theta)$ C [65]. Generally, the relatively high values of the adsorption equilibrium constant (K_{ads}) as shown in Table 7, reflect the

Fig. 9 Frumkin isotherm adsorption of (BPP_**1**) on the copper surface in 2.0 M HNO$_3$ solution

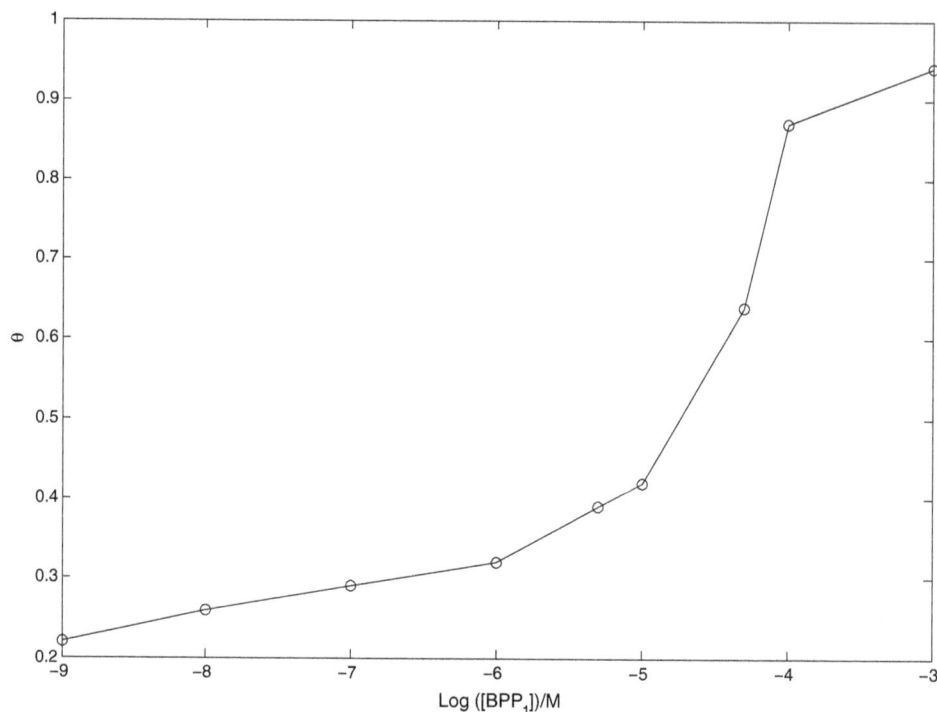

Table 7 Binding constant (K_b), active sites ($1/y$), lateral interaction (f), equilibrium constant (K_{ads}) and standard free energy of adsorption ($\Delta G°$) for copper in 2.0 M HNO$_3$ for benzimidazole (BPP_**1–7**) molecules at 303 K

BPP_1–7 inhibitors	Kinetic model			Frumkin adsorption isotherm		
	$1/y$	K_b	$-\Delta G°$ (kJ/mol)	$-f$	K_{ads}	$-\Delta G_{ads}°$ (kJ/mol)
BPP_1	1.88	499.39	25.75	7.59	4.66×10^6	48.75
BPP_2	2.62	92.73	21.51	7.08	4.05×10^6	48.40
BPP_3	1.77	570.29	26.08	6.83	2.20×10^6	46.87
BPP_4	3.14	16.43	17.16	3.37	1.16×10^5	39.46
BPP_5	2.12	104.04	21.80	5.13	2.53×10^5	41.42
BPP_6	5.78	6.23	14.72	18.50	4.28×10^8	60.13
BPP_7	5.88	4.99	14.16	18.58	1.56×10^8	57.60

high adsorption ability [66] of the BPP_**1–7** molecules on the copper surface.

The corrosion inhibition mechanism for the BPP inhibitors on Cu surface in HNO$_3$ is confirmed by finding the correlation between the kinetic parameters or the generalized mechanistic scheme of the kinetic–thermodynamic model proposed by El-Awady et al. [67, 68] (Eq. 30) with the experimental Frumkin adsorption isotherm.

$$\theta/(1 - \theta) = K'[I]^y \quad \text{or} \quad \log(\theta/1 - \theta) = \log K' + y \log[I]$$
(30)

where y is the number of inhibitors molecules [I] occupying one active site, and K' is a constant, the relationship in Eq. (30) gives a satisfactory linear relation. Hence, the suggested model fits the obtained experimental data. The

slope of such lines is the number of inhibitor molecules occupying a single active site, (y) and the intercept is the binding constant ($\log K'$). As mentioned, $1/y$ gives the number of active sites occupied by a single organic molecule and K'^y is the equilibrium constant for the adsorption process. The binding constant (K_b) corresponding to that obtained from the known adsorption isotherms curve fitting is given by the following equation:

$$K_b = K'^{(1/y)}$$
(31)

Table 7 comprises the values of $1/y$ and K_b for the studied BPP. This table show that the number of active sites occupied by one molecule in the case of BPP ($1/y \equiv 2$–6). Values of $1/y$ greater than unity implies the formation of multilayer of the inhibitor molecules on the

Fig. 10 Variation of inhibition efficiency IE% on copper with different concentrations of BPP_1–7 inhibitors in 2.0 M HNO$_3$ at 303 K for 4 h immersion time

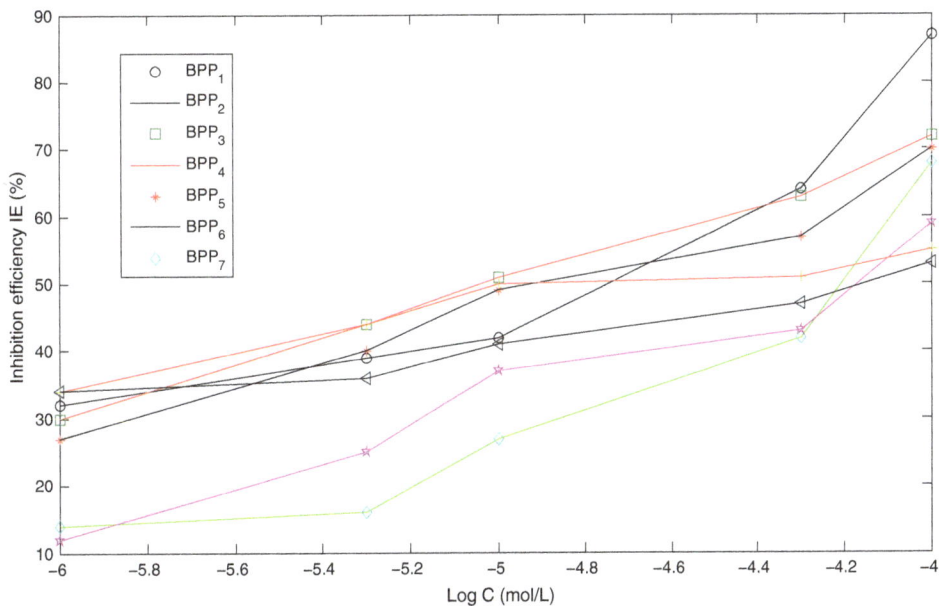

metal surface, whereas, values of $1/y$ less than unity indicates that a given inhibitor molecule will occupy more than one active site [68]. According to the proposed kinetic–thermodynamic model, the adsorption takes place via formation of multilayer of the inhibitor molecules on the copper electrode surface. The slope values do not equal unity (gradient slopes <1); hence, the adsorption of these synthesized BPP on copper surface does not obey a Langmuir adsorption isotherm. Frumkin adsorption isotherm (Eq. 28) represents best fit for experimental data obtained from applying BPP as chemical inhibitors on copper in 2.0 M HNO$_3$ acid solution. The values of K_{ads} (equilibrium constant of the inhibitor adsorption process) and (f) are tabulated in Table 7. The lateral interaction parameter (f) has negative values; this parameter is a measure of the degree of steepness of the adsorption isotherm. The adsorption equilibrium constant (K_{ads}) calculated from Frumkin equation acquires lower values than those binding constant (K_b) obtained and calculated from the kinetic–thermodynamic model. The lack of compatibility of the calculated (K_b) and experimental (K_{ads}) values may be attributed to the fact that Temkin adsorption isotherm is only applicable to cases where one active site per inhibitor molecule is occupied. The lateral interaction parameter was introduced to treat of deviations from Langmuir ideal behavior, whereas the kinetic–thermodynamic model uses the size parameter. The values of the lateral interaction parameter $(-f)$ were found to be negative and increase from $\equiv 6.83$ to 18.58. This denotes that an increase in the adsorption energy takes place with the increase in the surface coverage (θ). Adsorption process is a displacement reaction involving removal of adsorbed

water molecules from the electrode metal surface and their substitution by inhibitor molecules. Thus, during adsorption, the adsorption equilibrium forms an important part in the overall free energy changes in the process of adsorption. It has been shown that, the free energy change (ΔG°_{ads}) increases with increase of the solvating energy of adsorbing species, which in turn increases with the size of hydrocarbon portion in the organic molecule and the number of active sites. Hence, the increase of the molecular size leads to decreased solubility, and increased absorbability. The large negative values of the standard free energy changes of adsorption (ΔG°_{ads}), obtained for BPP, indicate that the reaction is proceeding spontaneously and accompanied with a high efficient adsorption. Although, the obtained values of the binding constant (K_b) from the kinetic model and the modified equilibrium constant (K_{ads}) from Temkin equation are incompatible, generally have large values (Table 7), mean better inhibition efficiency of the investigated synthesized BPP_1–7, i.e., stronger electrical interaction between the double layer existing at the phase boundary and the adsorbing molecules. In general, the equilibrium constant of adsorption (K_{ads}) was found to become higher with increasing the inhibition efficiency of the inhibitor studied as given in Table 7. Application of kinetic–thermodynamic model on benzimidazole (BPP_1–7) inhibitors of copper in 2.0 M HNO$_3$ at 303 K is shown in Fig. 11.

Effect of temperature

Temperature has a great effect on the rate of metal electrochemical corrosion. In case of corrosion in a neutral

(a)

(b)

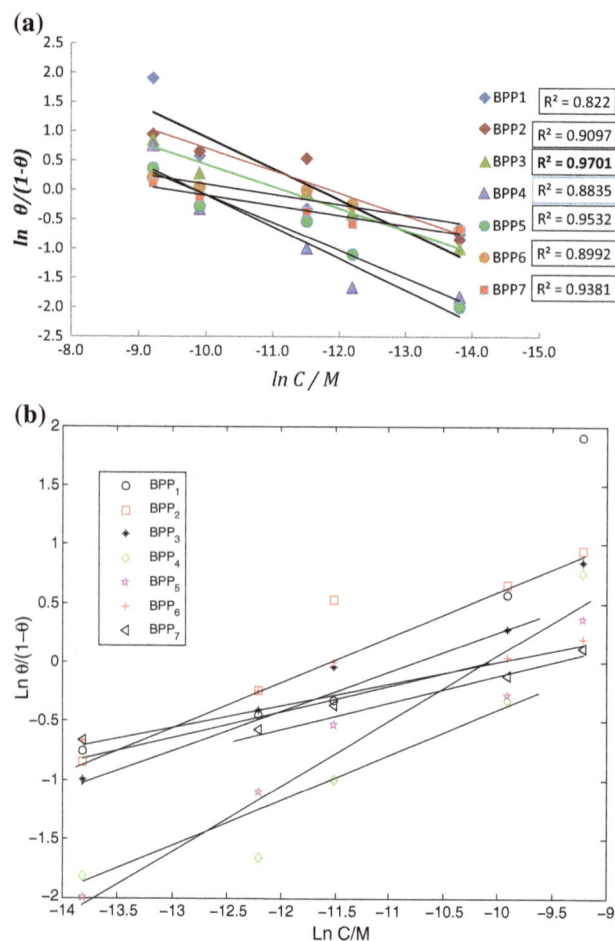

Fig. 11 Application of kinetic–thermodynamic model on benzimidazole (BPP_**1**–**7**) inhibitors of copper in 2.0 M HNO₃ at 303 K

Fig. 12 Potentiodynamic polarization curves of copper in 2.0 M HNO₃ solution with the 1×10^{-6} M of BPP_**1** at different temperatures

solution (oxygen depolarization), the increase in temperature has a favourable effect on the overpotential of oxygen depolarization and the rate of oxygen diffusion but it leads to a decrease of oxygen solubility. In case of corrosion in acidic medium (hydrogen depolarization), the corrosion rate increases exponentially with temperature increase because the hydrogen evolution over potential decreases [69]. The potentiodynamic polarization curves for copper corrosion in 2.0 M HNO₃ acid in the presence of 1×10^{-6} M BBP_**1** at temperatures 30, 40, 50 and 60 °C is given in Fig. 12.

The relationship between the corrosion rate (R_{corr}) of copper in acidic media and temperature (T) is often expressed by the Arrhenius equation [70, 71]:

$$\ln R_{corr} = \ln A - \frac{E_a}{RT} \qquad (32)$$

where E_a is the apparent activation energy, R is the molar gas constant (8.314 J K⁻¹ mol⁻¹), T is the absolute temperature, and A is the frequency factor. The plot of Ln R_{corr}

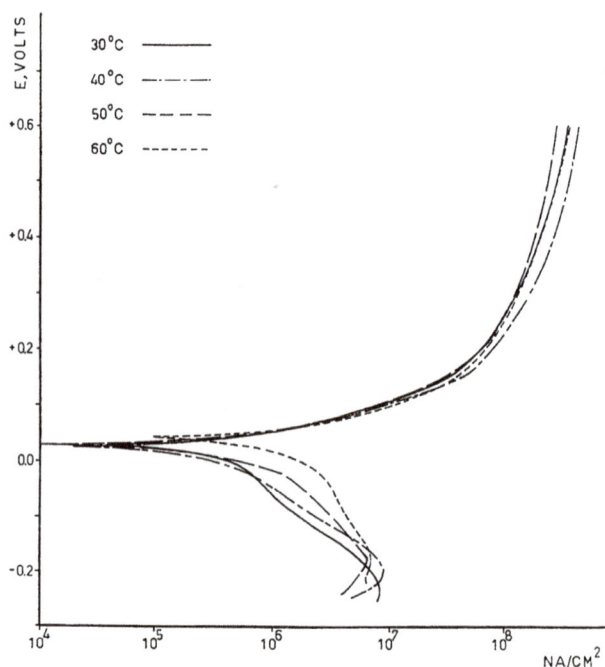

against $1/T$ for copper corrosion in 2.0 M HNO₃ in the absence and presence of different concentrations of BPP is shown in Fig. 13. All parameters were given in Table 8. The activation energy increased in the presence of BPP, which indicated physical (electrostatic) adsorption. Furthermore, the activation energy rose with increasing inhibitor concentration, suggesting strong adsorption of inhibitor molecules at the metal surface [60]. The increase in activation energy was due to the corrosion reaction mechanism in which charge transfer was blocked by the adsorption of BPP molecules on the copper surface [72]. It also revealed that the whole process was controlled by the surface reaction since the energy of the activation corrosion process in both the absence and presence of BPP was greater than 20 kJ mol⁻¹ [73].

Experimental corrosion rate values obtained from Potentiodynamic polarization measurements for copper in 2.0 M HNO₃ in the absence and presence of BPP was used to further gain insight on the change of enthalpy (ΔH^*) and entropy (ΔS^*) of activation for the formation of the activation complex in the transition state using transition equation [63]:

$$R_{corr} = \frac{RT}{Nh} \exp\frac{(\Delta S^*)}{R} \exp\left(-\frac{\Delta H^*}{RT}\right) \qquad (33)$$

where h is the Plank's constant (6.626176×10^{-34} J s), N is the Avogadro's number (6.02252×10^{-23} mol⁻¹), R is the universal gas constant and T is absolute temperature.

Fig. 13 Arrhenius plots of Ln R_{corr} versus 1/T for copper in 2.0 M HNO_3 in the absence and the presence of BPP_1 at optimum concentration

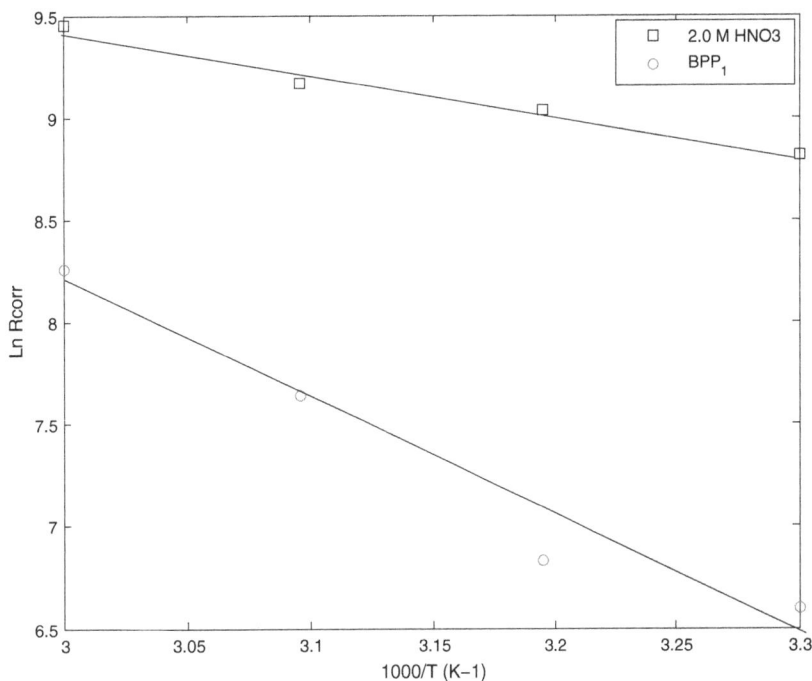

The free energy change of activation (ΔG^*) is obtained from the Eyring equation: (Eq. (34).

$$k = \frac{k_B T}{h} e^{-\Delta G^*/RT} \qquad (34)$$

Another convenient form of Eq. (34) is Eq. (35).

$$\Delta G^* = \Delta H^* - T\Delta S^* \qquad (35)$$

The relation of Ln R_{corr}/T versus $1/T$ for copper corrosion in 2.0 M HNO_3 in the absence and presence of different concentrations of BPP is given in Fig. 14. Straight lines were obtained with slope of ($\Delta H^*/R$) and an intercept of [Ln (R/Nh) + ($\Delta S^*/R$)] from which the values of ΔH^* and ΔS^*, respectively, were computed.

The activation parameters were computed and listed in Table 9. Inspection of these data reveal that the activation parameters (E_a, ΔH^*, ΔS^* and ΔG^*) of dissolution reaction of copper in 2.0 M HNO_3 in the presence of BPP are higher than in the absence of inhibitor. The positive sign of the enthalpy of activation reflect the endothermic nature of copper dissolution process meaning that dissolution of copper is difficult [74]. The entropy of activation was negative in absence of BPP, whereas ΔS^* was changed between negative and positive values in presence of BPP implying that the rate-determining step for the activated complex is dissociation step rather than association. In other words, the adsorption process is accompanied by an increase in entropy, which is the driving force for the adsorption of inhibitor onto the copper surface [75]. From the thermodynamic parameters in Table 9, it can be seen that E_a increases as the inhibition efficiency of the additives increases. This suggests that the

process is controlled by a surface reaction, since the energy of activation for the corrosion process is above 20 kJ mol^{-1}. The higher entropy values (ΔS^*) as shown in Table 9 in the case of different inhibitors compared to a high negative values of (-121.16 J. K^{-1} mol^{-1}) in the case of free nitric acid solution indicates a slower reaction [76]. The negative entropy (ΔS^*) values only in the case of BPP_1 and BPP_4 inhibitors, with their lower activation energy (E_a) values indicate that the copper corrosion inhibition process in these derivatives only is entropy controlled rather than activation energy controlled (Table 9). The copper corrosion inhibition process in the case of all other BPP inhibitors is activation energy controlled, since the activation energy (E_a) values are relatively high, as shown in Table 9. The activation energy (E_a) values support the sequence arrangement of different inhibitors according to their increasing inhibition efficiency (IE%) values. The activation energy for copper was found between 2, 4.6 and11.7 kcal mol^{-1} in 0.1 M HCl, 4.0 M HNO_3 and 3.0 M HNO_3 solutions, respectively. Generally, one can say that the nature and concentration of the electrolyte greatly affect the activation energy for the corrosion process.

Theoretical study

Quantum chemical calculations

For the purpose of determining the active sites of the inhibitor molecule, three influence factors: natural atomic

Table 8 Various corrosion parameters for copper in 2.0 M HNO_3 in absence and presence of optimum concentration 1×10^{-6} M of benzimidazole (BPP_ **1–7**) at different temperatures, $303 \leq T \leq 333$ K

System	Temperature (K)	R_p (Ω cm^2)	I_{corr} (μA cm^{-2})	R_{corr} (MPY)	E_{corr} mV(SCE)	β_c (V dec^{-1})	β_a (V dec^{-1})	IE%
2.0 M HNO$_3$	303	6.360	7287	6719	29	0.302	0.165	–
	313	5.433	10,180	8384	49	0.366	0.195	–
	323	4.577	10,380	9570	43	0.250	0.195	–
	333	3.001	13,860	12,770	45	0.170	0.219	–
2.0 M HNO$_3$ + BPP_**1**	303	45.08	235	731.3	23	0.177	0.154	96.8
	313	37.14	1004	926.1	26	0.178	0.166	90.1
	323	30.30	2257	2080	25	0.244	0.186	78.2
	333	14.50	4194	3866	39	0.381	0.222	69.7
2.0 M HNO$_3$ + BPP_**2**	303	75.34	406.7	374.9	12	0.162	0.125	94.4
	313	32.96	1399	1289	21	0.132	0.186	86.2
	323	74.83	2807	8119	41	0.335	0.212	72.9
2.0 M HNO$_3$ + BPP_**3**	303	63.36	507	467.4	7	0.145	0.151	93.0
	313	26.35	1647	1518	21	0.220	0.183	83.8
	323	25.80	2100	3567	23	0.215	0.190	79.7
	333	50.67	4190	11,240	44	0.333	0.248	69.7
2.0 M HNO$_3$ + BPP_**4**	303	30.93	1180	1206	26	0.185	0.188	83.8
	313	24.13	1766	1546	17	0.187	0.186	82.6
	323	13.83	2916	2688	−34	0.162	0.219	71.9
	333	7.026	7287	6718	35	0.252	0.222	47.4
2.0 M HNO$_3$ + BPP_**5**	303	76.40	397.2	366.2	7	0.202	0.107	94.5
	313	50.24	700.6	645.9	15	0.185	0.144	93.1
	323	14.05	3067	2827	26	0.193	0.204	70.4
2.0 M HNO$_3$ + BPP_**6**	303	147.9	320	295.1	6	0.204	0.234	95.6
	313	20.81	2440	344	25	0.402	0.165	76.0
	323	15.72	3590	2249	30	0.302	0.228	65.4
2.0 M HNO$_3$ + BPP_**7**	303	113.0	242.5	290	−3	0.124	0.129	96.6
	313	46.8	724.7	668.1	8	0.139	0.179	92.9
	323	9.743	4087	4690	30	0.213	0.245	60.6
	333	9.410	6575	5140	30	0.236	0.229	52.5

charge, distribution of frontier orbital, and Fukui indices are considered. According to classical chemical theory, all chemical interactions are by either electrostatic or orbital. Electrical charges in the molecule were obviously the driving force of electrostatic interactions it has been proven that local electron densities or charges are important in many chemical reactions and physico-chemical properties of compounds [30]. The inhibition efficiencies of seven benzimidazole (BPP_**1–7**) derivatives on the corrosion of copper were investigated by quantum chemical and molecular dynamics simulation studies.

The molecular reactivity of the studied molecules was investigated and compared via analysis of frontier molecular orbitals. The energy of HOMO is associated with the electron donating ability of a molecule. High values of energy of HOMO state that the molecule is prone to donate electrons to appropriate acceptor molecules with low-energy and empty molecular orbital. On the other hand, LUMO energy level is an indicator of electron accepting abilities of molecules. It is important to note that the molecules that have lower LUMO energy value have more electron accepting ability. The reactive ability of the inhibitor is considered to be closely related to their frontier molecular orbitals, the HOMO and LUMO. The frontier molecule orbital density distributions involving the optimized structures, HOMOs, LUMOs with Mulliken orbital charges population's analyses of the investigated seven (BPP_**1–7**) molecules are presented in Figs. 15 and 16. The optimized structures, HOMOs, LUMOs of non-protonated inhibitor molecules (BPP_**1–7**) using DFT/B3LYP/6-31++G (d,p) are shown in Fig. 15. As seen from the figure, the populations of the HOMO focused around the carbon chain containing imidazole and azo nitrogen. But the LUMO densities were mainly around the benzene

Fig. 14 Arrhenius plots of Ln R_{corr}/T versus $1/T$ for copper in 2.0 M HNO$_3$ in the absence and the presence of BPP_**1** at optimum concentration

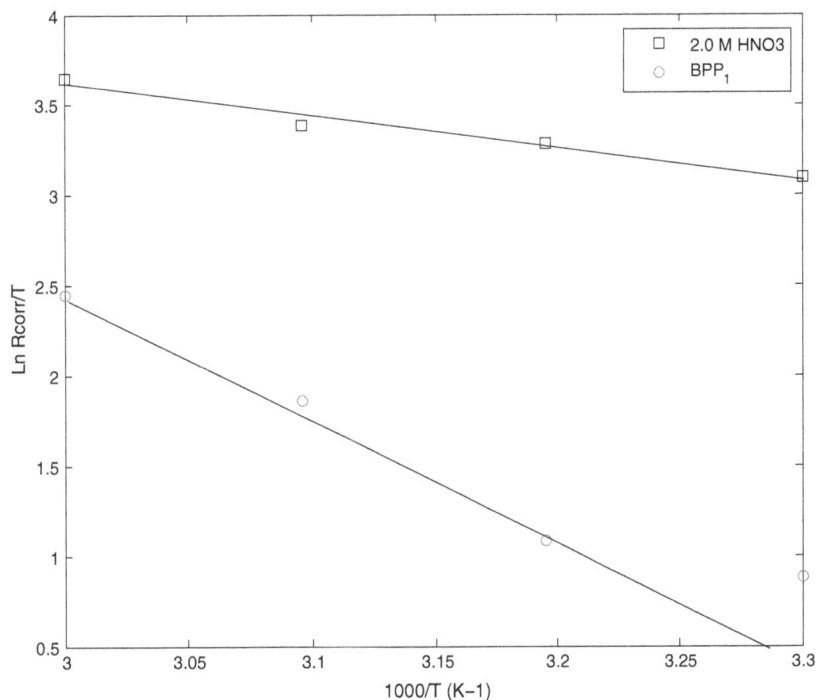

Table 9 Activation energy (E_a), enthalpy change (ΔH^*), free energy change (ΔG^*), and entropy change (ΔS^*) for the corrosion of copper in 2.0 M HNO$_3$ in absence and presence of 1×10^{-6} M of benzimidazole (BPP_**1–7**) inhibitors at different temperatures, $303 \leq T \leq 333$ K

System	A (MPY)	E_a (kJ mol^{-1})	ΔH^* (kJ mol^{-1})	ΔG^* (kJ mol^{-1})	ΔS^* (J K^{-1} mol^{-1})
2.0 M HNO$_3$	6.953×10^6	17.481	15.403	53.932	−121.16
2.0 M HNO$_3$ + BPP_**1**	4.883×10^{11}	51.632	46.473	58.608	−38.161
2.0 M HNO$_3$ + BPP_**2**	2.201×10^{24}	126.453	123.100	57.546	209.438
2.0 M HNO$_3$ + BPP_**3**	1.225×10^{18}	89.407	85.280	57.482	87.413
2.0 M HNO$_3$ + BPP_**4**	8.419×10^4	9.351	44.853	57.408	−39.483
2.0 M HNO$_3$ + BPP_**5**	1.683×10^{17}	85.419	80.922	59.107	69.696
2.0 M HNO$_3$ + BPP_**6**	3.470×10^{16}	82.253	80.287	60.036	64.700
2.0 M HNO$_3$ + BPP_**7**	9.336×10^{17}	89.883	86.928	58.67	88.860

cyclic. Higher HOMO energy (E_{HOMO}) of the molecule means a higher electron-donating ability to appropriate acceptor molecules with low-energy empty molecular orbital and thus explains the adsorption on metallic surfaces by way of delocalized pairs of π-electrons. E_{LUMO}, the energy of the lowest unoccupied molecular orbital, signifies the electron receiving tendency of a molecule. From Fig. 15, it could be seen that BPP_**1–7** have similar HOMO and LUMO distributions, which were all located on the entire BPP moiety. This is due to the presence of nitrogen and oxygen atoms together with several π-electrons on the entire molecule. Thus, unoccupied d orbital of Cu atom can accept electrons from inhibitor molecule to form coordinate bond. Also the inhibitor molecule can accept electrons from Cu atom with its anti-bonding orbitals to form back donating bond. Figure 16 shows B3LYP/

6-311G** selected bond length for the optimized geometry of the studied compounds calculated for BPP. It has been reported that the more negative the atomic charges of the adsorbed center, the more easily the atom donates its electron to the unoccupied orbital of the metal [14, 30]. Mulliken orbital charges populations analysis of BPP_ **1–7** derivatives using DFT method at the B3LYP method with 6–31+G** basis sets level are presented in Fig. 16. It can be seen that the area of carbon bone chain containing imidazole and azo nitrogen, hydroxyl, and methyl charged a large electron density and might form adsorption active centers. It is clear from Fig. 16 that nitrogen and oxygen as well as some carbon atoms carry negative charge centers which could offer electrons to the copper surface to form a coordinate bond. This shows that the N and O atoms are the probable reactive sites for the adsorption of copper. Higher

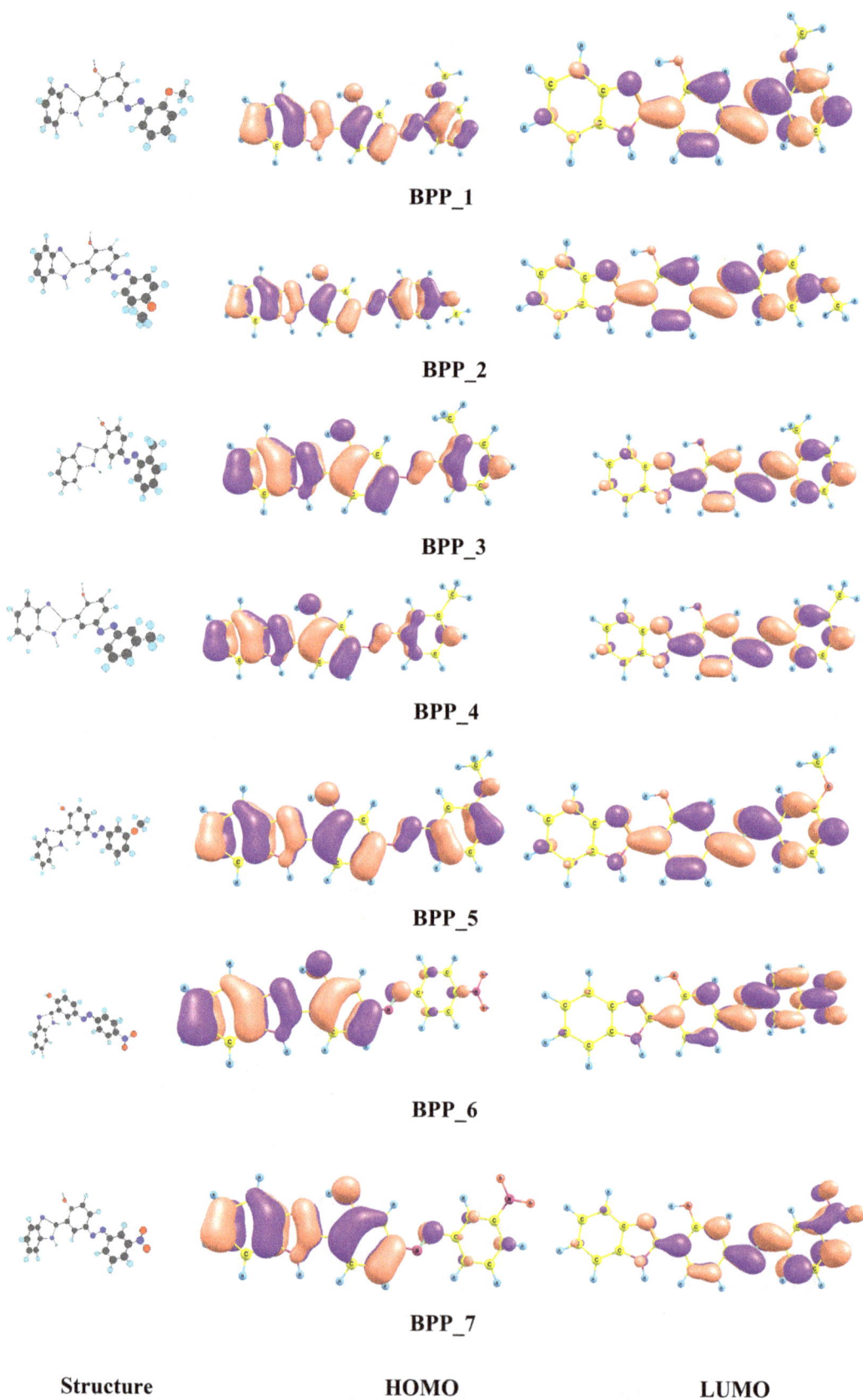

BPP_1

BPP_2

BPP_3

BPP_4

BPP_5

BPP_6

BPP_7

Structure   **HOMO**   **LUMO**

Fig. 15 The optimized structures, HOMOs, LUMOs of non-protonated inhibitor molecules (BPP_**1–7**) using DFT/B3LYP/6-31++G (*d,p*)

Fig. 16 Mulliken orbital charges populations analysis of BPP_**1–7** derivatives using DFT method at the B3LYP method with 6-31+G** basis sets level

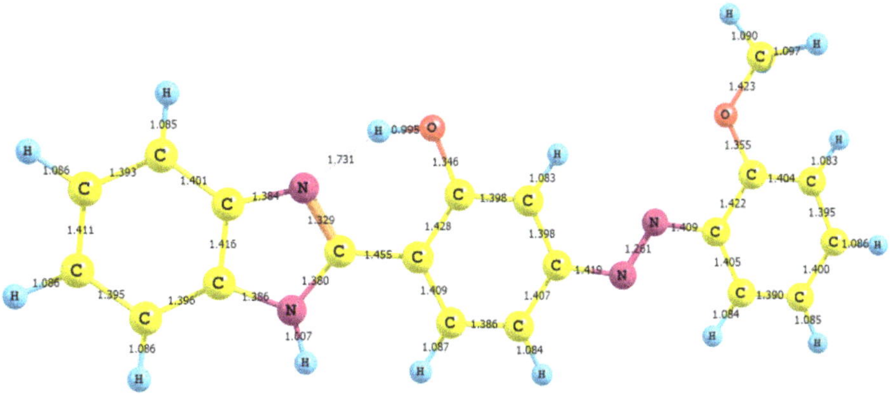

BPP_1

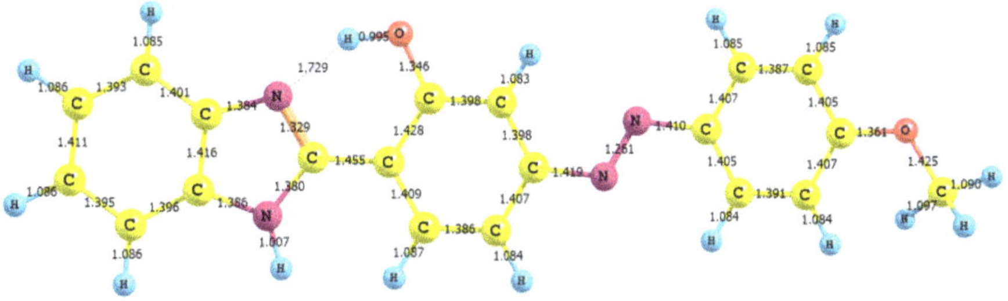

BPP_2

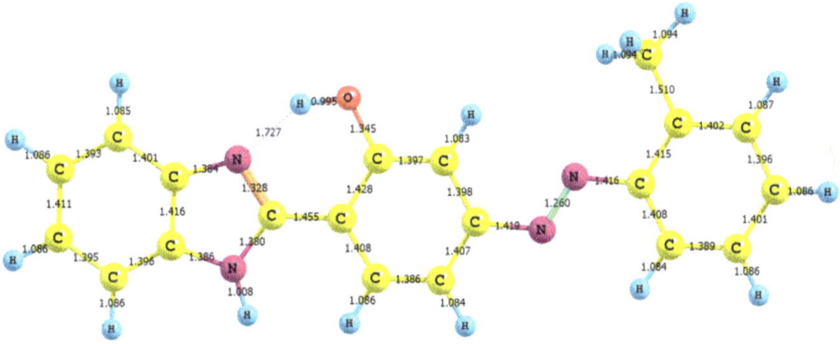

BPP_3

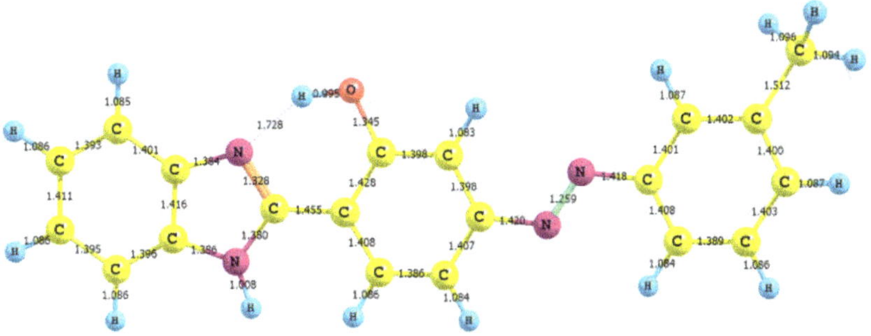

BPP_4

Fig. 16 continued

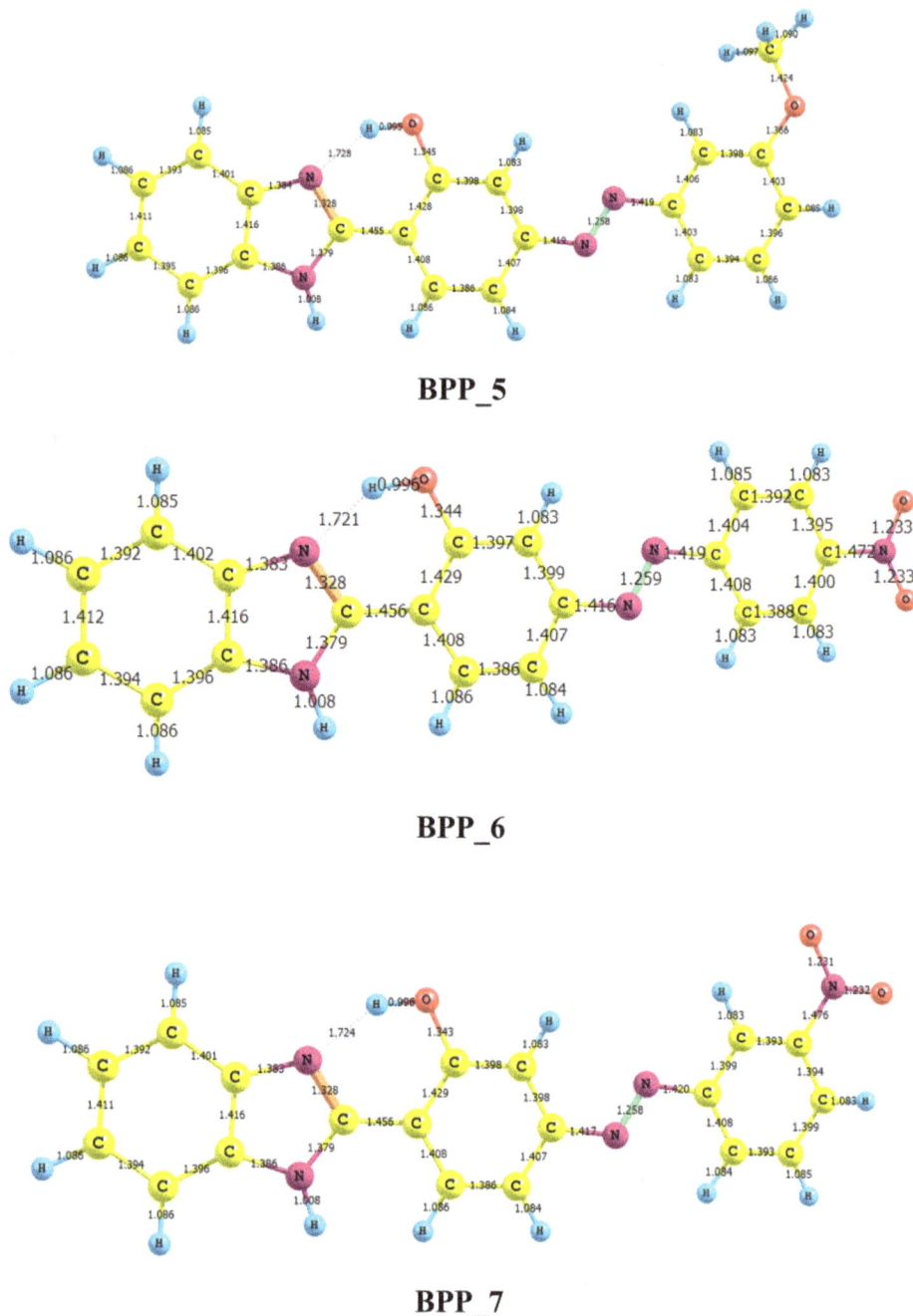

BPP_5

BPP_6

BPP_7

values of E_{HOMO} are likely to indicate a tendency of the molecule to donate electrons to appropriate acceptor molecules with low-energy or empty electron orbital. Quantum chemical parameters such as E_{HOMO}, E_{LUMO}, ΔE (HOMO–LUMO energy gap), chemical hardness, softness, electronegativity, proton affinity, electrophilicity and nucleophilicity are important and useful tools to compare the corrosion inhibition performances of molecules. Calculated quantum chemical parameters of studied molecules in aqueous phase are presented in Table 10. It is evident from Table 10 that BPP has the highest E_{HOMO} and a lower

E_{HOMO} in the protonated form. This means that the electron-donating ability of BPP is weaker in the protonated form. This confirms the experimental results that interaction between BPP and copper is electrostatic in nature (physisorption). The energy of the LUMO is directly related to the electron affinity and characterizes the susceptibility of the molecule towards attack by nucleophiles. The lower the values of E_{LUMO}, the stronger the electron accepting abilities of molecules. It is clear that the protonated form of BPP exhibits the lowest E_{HOMO}; thus, the protonated form is the most likely form for the interaction

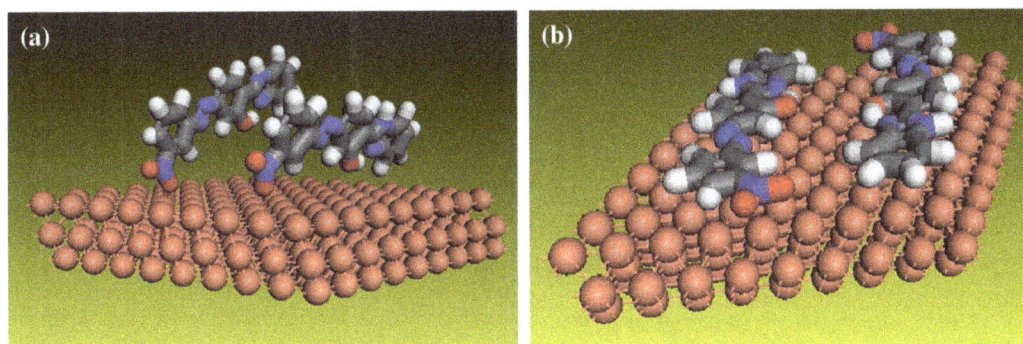

Fig. 17 The schematic proposed representation structure models of the adsorption behavior of the single BPP_1 inhibitor molecule on Cu (111) with different orientations in 2.0 M HNO$_3$ solution

of copper with BPP molecule. Low values of the energy gap (ΔE) will provide good inhibition efficiencies, because the excitation energy to remove an electron from the last occupied orbital will be low [54]. Accordingly, the difference between E_{LUMO} and E_{HOMO} energy levels ($\Delta E = E_{LUMO} - E_{HOMO}$) and the dipole moment (μ) was also determined. The global hardness (η) is approximated as $\Delta E/2$, and can be defined under the principle of chemical hardness and softness [14]. These parameters also provide information about the reactive behavior of molecules and are presented in Table 10. A molecule with a low-energy gap is more polarizable and is generally associated with a high chemical reactivity and low kinetic stability, and is termed soft molecule [14, 30]. According to Wang et al. [46], adsorption of inhibitor onto a metallic surface occurs at the part of the molecule which has the greatest softness and lowest hardness. The results show that BPP in the protonated form have the lowest energy gap and lowest hardness; this agrees with the experimental results that BPP could have better inhibitive performance on copper surface, i.e., through electrostatic interaction between the cation form of BPP and the vacant d orbital of copper physisorption. BPP had the highest inhibition efficiency because it had the highest HOMO energy values, and it had the greatest ability of offering electrons. This also agrees well with the value of (ΔG_{ads}°) obtained experimentally (Table 7). The dipole moment of (BPP) is highest in the protonated form [($\mu = 9.5177$ Debye (28.818×10^{-30} cm)], which is higher than that of H$_2$O ($\mu = 6.23 \times 10^{-30}$ cm). The high value of dipole moment probably increases the adsorption between chemical compound and metal surface [17, 19, 30]. Accordingly, the adsorption of BPP molecules can be regarded as a quasi-substitution process between the benzimidazole compound and water molecules at the electrode surface. Frontier orbital energy level indicates the tendency of bonding to the metal surface. Further study on the formation of chelating centers in an inhibitor requires the information of

spatial distribution of electronic density of the compound molecules [30]. The structure of the molecules can affect the adsorption by influencing the electron density at the functional group. Generally, electrophiles attack the molecules at negative charged sites. As seen from Fig. 16, the electron density focused on N atoms, O atoms, and C atoms in methyl. The regions of highest electron density are generally the sites to which electrophiles attacked. So, N, O, and C atoms were the active center, which had the strongest ability of bonding to the metal surface. On the other side, HOMO (Fig. 15) was mainly distributed on the areas containing imidazole; azo nitrogen. Thus, the areas containing N atoms were probably the primary sites of the bonding. As showed in Table 10, the values of HOMO energy increases with increasing length of carbon bone chain containing imidazole nitrogen. Similar situation can be also seen in Figs. 15 and 16 the configuration changes led to the increase in electron density; and inhibition efficiency was enhanced by increase in HOMO energy and electron density. The region of active centers transforming electrons from N atoms to copper surface. The electron configuration of copper is [Ar] $4s^2 3d^9$, the $3d$ orbitals are not fully filled with electrons. N heteroatom's has lonely electron pairs that is important for bonding unfilled $3d$ orbitals of Cu atom and determining the adsorption of the molecules on the metal surface. BPP_1 had the highest inhibition efficiency among the BPP_1–7, which was resulted from the geometry change that led to HOMO energy increase and electron density distribution in the molecule. Based on the discussion above, it can be concluded that the benzimidazole (BPP) molecules have many active centers of negative charge. In addition, the areas containing N and O atoms are the most possible sites of bonding metal surface by donating electrons to the copper surface.

According to HOMO and LUMO orbital energies given in Table 10, we can write the corrosion inhibition efficiency order as: BPP_1 > BPP_2 > BPP_3 >

BPP_4 > BPP_5 > BPP_6 > BPP_7 (in terms of HOMO and LUMO energies). Chemical hardness is the resistance against electron cloud polarization or deformation of chemical species. As can be understood from this definition, chemical hardness of a molecule and its inhibition efficiency are inversely proportional to each other because a hard molecule is reluctant to give electrons. Chemical hardness, softness and ΔE are quantum chemical parameters closely associated with each other [77]. As is known, both softness [78] and hardness are given based on HOMO and LUMO orbital energies as a result of Koopmans's theorem [18]. Hard molecules which have high HOMO–LUMO energy gap cannot act as good corrosion inhibitor. However, soft molecules which have low HOMO–LUMO energy gap are good corrosion inhibitors because they can easily give to metals. It is clear that we can write the same corrosion inhibition ranking considering these three chemical properties. From the light of the results given in the Table 10, one can write the corrosion inhibition ranking of studied molecules based on their hardness, softness and HOMO–LUMO energy gap values as: BPP_1 > BPP_2 > BPP_3 > BPP_4 > BPP_5 > BPP_6 > BPP_7.

Electronegativity is an important parameter in terms of the prediction and comparison of corrosion inhibition efficiencies of molecules. The number of electrons transferred between metal and corrosion inhibitor can be calculated using Eq. (36). It is seen from the equation given below that the electron transfer value metal and inhibitor decreases as the electronegativity of inhibitor increases. According to Sanderson's electronegativity equalization principle [79], the electron transfer between metal and inhibitor continues until their electronegativity values become equal with each other. As a matter of fact, Eq. (36) has been derived taking advantage from hardness equalization principle and electronegativity equalization principle.

$$\Delta N = \frac{\chi_{Cu} - \chi_{inh}}{2(\eta_{Cu} + \eta_{inh})} \tag{36}$$

where ΔN is electron transfer between metal and inhibitor. X_{Cu} and χ_{inh} are electronegativity of metal and electronegativity of inhibitor, respectively. η_{Cu} and η_{inh} represent chemical hardness value of metal and chemical hardness value of inhibitor, respectively. Two systems, copper and inhibitor, are brought together, electrons will flow from lower χ (inhibitor) to higher χ (Cu), until the chemical potentials become equal. Copper surface is the Lewis acid according to HSAB theory [80]. The difference in electronegativity drives the electron transfer, and the sum of the hardness parameters acts as a resistance [81]. To calculate the fraction of electrons transferred, a theoretical value for the absolute electronegativity of copper according to Pearson was used $\chi_{Cu} = 463.1$ kJ mol^{-1} [82], and a

Table 10 Quantum chemical and molecular dynamics parameters derived for benzimidazole (BPP_1–7) derivatives calculated with DFT at the B3LYP method with 6-31+G** basis sets level

BPP_1–7 inhibitors	Total energy (a.u.)	E_{HOMO} (eV)	E_{LUMO} (eV)	$\Delta E = E_{LUMO} - E_{HOMO}$ (eV)	μ (Debye)	$I = -E_H$	$A = -E_L$	$\chi = \frac{(I+A)}{2}$	$\eta = \frac{(I-A)}{2}$	ω	ε	S	PA	$\Delta N = \frac{\chi_{Cu} - \chi_{inh}}{2(\eta_{Cu} + \eta_{inh})}$
BPP_1	−1141.4939	−5.884	−2.695	3.189	2.2951	5.884	2.695	4.2895	1.5945	5.769	0.173	0.627	−4.02936	143.873
BPP_2	−1141.5003	−5.866	−2.656	3.210	5.0769	5.866	2.656	4.2610	1.6050	5.656	0.177	0.623	−4.12228	142.940
BPP_3	−1066.2689	−6.027	−2.813	3.214	3.4745	6.027	2.813	4.4200	1.6070	6.078	0.164	0.622	−4.09453-	142.713
BPP_4	−1066.2700	−6.029	−2.765	3.264	3.3291	6.029	2.765	4.3970	1.6320	5.923	0.169	0.613	−4.04310	140.534
BPP_5	−1141.4987	−6.024	−2.802	3.222	3.0989	6.024	2.802	4.4130	1.6110	6.044	0.165	0.621	−4.11181	142.361
BPP_6	−1231.5044	−6.376	−3.562	2.814	8.9287	6.376	3.562	4.9690	1.4070	8.774	0.114	0.711	−2.65518	162.804
BPP_7	−1231.5037	−6.313	−3.304	3.009	9.5177	6.313	3.304	4.8085	1.5045	7.684	0.130	0.665	−3.24566	152.307

global hardness of $\eta_{Cu} = 0$, by assuming that for a metallic bulk $I = A$ [81] because they are softer than the neutral metallic atoms. From Table 10, it is possible to observe that molecule BPP_**6** has a lower value of global hardness. The fraction of transferred electrons (ΔN) is also the largest for molecule BPP_**6** then BPP_**7** and, in turn, is BPP_**1**, $\approx$ BPP_**2** $\approx$ BPP_**3** $\approx$ BPP_**5**, then BPP_**4**. Quantum chemical and molecular dynamics parameters derived for BPP_**1–7** derivatives calculated with DFT at the B3LYP method with 6-31+G** basis sets level are listed in Table 10. The values of the interaction energy and the total energy of the BPP_**1–7** derivatives on copper (1 1 1) surface are listed in Table 10. It is clear from Table 10 that the total energy has a negative value, whereas the binding energy has a positive value (Table 7). As the value of the binding energy increases, the more easily the inhibitor adsorbs on the metal surface, the higher the inhibition efficiency [83]. BPP_**1** has the highest binding energy compared to the other BPP derivatives to the copper surface that are found during the molecular dynamics simulation process described elsewhere [83]. High values of binding energy obtained with BPP_**1** molecules explain its highest inhibition efficiency from the theoretical point of view. Therefore, according to a series of properties calculated for each molecule shown in Table 10 the reactivity order, that is, the inhibitive effectiveness order for the BPP molecules are: $-OCH_3 > -CH_3 > -NO_2$ substituents. The calculated theoretical results are in agreement with experimental results. The use of Mulliken population analysis for the prediction of the adsorption center of inhibitors is widely used. The partial atomic charges on atoms of the inhibitor molecules provide important clues about the identifying of reactive center. The atoms with the highest negative charge represent the high tendency on the metal surface [78]. The inhibitors can easily interact with the metal surface through such atoms. Proton affinity can be defined as the enthalpy of the reaction with H^+ ion of a chemical species in gas phase and this parameter for inhibitors is one of the useful tools to compare their electron donating abilities [84]. The presence of the heteroatoms such as oxygen and nitrogen in the molecules of azole BPP_ **1–7** derivatives leads to high tendency for protonation in acidic medium. Thus, analysis of the protonated forms of BPP_**1–7** derivatives is important in terms of the calculation of the proton affinities of neutral inhibitors. It should be stated that proton affinity is a measure of the basicity. In this sense, corrosion inhibitors act as Lewis bases. The basicity of a molecule will increase with increasing of its proton affinity. We calculated the proton affinities of the studied benzimidazole compounds considering Eqs. (37) and (38). According to proton affinity values given in the Table 10 for studied compounds, the

inhibition efficiencies of mentioned compounds follow the same previous order.

$$PA = E_{(pro)} - \left(E_{(non-pro)} + E_{H+} \right) \tag{37}$$

where $E_{non-pro}$ and E_{pro} are the energies of the non-protonated and protonated inhibitors, respectively. E_{H^+} is the energy of H^+ ion and was calculated as:

$$E_{H^+} = E_{(H_3O^+)} - E_{(H_2O)} \tag{38}$$

The electrophilicity index (ω) is an important parameter that indicates the tendency of the inhibitor molecule to accept the electrons. Nucleophilicity (ε) is physically the inverse of electrophilicity ($1/\omega$). For this reason, it should be stated that a molecule that have large electrophilicity value is ineffective against corrosion while a molecule that have large nucleophilicity value is a good corrosion inhibitor. Thus, for studied molecules, we can write the inhibition efficiency ranking as: BPP_**1** > BPP_**2** > BPP_**3** > BPP_**4** > BPP_**5** > BPP_**6** > BPP_**7**.

Molecular dynamics (MD) simulation

The use of the molecular dynamics simulation is a useful and modern tool to investigate the interaction between inhibitors and metal surface. Thus, in this study, molecular dynamics simulation studies were performed to predict the binding energies of these azole derivatives on copper surface and to show whether there is a remarkable correlation between experimental inhibition efficiencies and binding energies for molecules considered in this study. The binding energies between Cu (111) surface and the benzimidazole (BPP) derivatives were obtained using Eq. (15). The schematic proposed representation structure models of the adsorption behavior of the single BPP_**1** inhibitor molecule on Cu (111) substrate with different orientations in 2.0 M HNO₃ solution has been presented in Fig. 17, i.e. representative snapshots of BPP_**1** on Cu (111) surface (inset images show the on-top views). This indicates that the adsorption density of BPP_**1** is higher and equilibrium adsorption configurations of the studied synthesized benzimidazole derivatives on Cu (111). Thus, Fig. 17 indicates the close contacts between the benzimidazole derivatives and Cu (111) metal surface as well as the best equilibrium adsorption configuration for the compounds considered. The obtained results given in Tables 7 and 10 show that the binding energies calculated for the interactions between inhibitors and metal surface are very high. It is important to note that high binding energy leads to a more stable inhibitor/surface interaction [85]. The binding energies obtained are observed to increase in the order: BPP_**1** > BPP_**2** > BPP_**3** > BPP_**4** > BPP_**5** > BPP_**6** > BPP_**7**. It is understood from the results, the benzimidazole

(BPP_1–7) derivatives can act as good corrosion inhibitors against copper corrosion. It can also be seen from Fig. 11 and Table 7, there is a good linear correlation between experimental inhibition efficiencies and calculated binding energies in this study. Therefore, the studied BPP_1–7 molecules are likely to adsorb on the copper surface to form stable adsorbed layers and protect copper from corrosion.

The results obtained in the study showed that these compounds considered are good inhibitors against corrosion of copper. Quantum chemical calculations and molecular dynamics simulations carried out and the calculated binding energies of the studied molecules on copper surface demonstrated that these molecules are very effective against the corrosion of copper.

Inhibition mechanism

In acid solutions, organic inhibitors may interact with the corroding metal and hence affect the corrosion reaction in more than one way [86], sometimes simultaneously. It is, therefore, often difficult to assign a single general inhibition mechanism, since the mechanism may change with experimental conditions.

Both molecular and protonated species can adsorb on the copper surface. The adsorption of BPP derivatives through the lone pairs in the groups (–N = N–, –OCH$_3$, –OH, –NO$_2$, –NH) can occur on the positive copper surface. Adsorption of the protonated BPP derivatives on the cathodic sites on copper surface will retard the oxygen evolution reaction. Adsorption on the anodic sites of copper surface can occur via N and O atoms to retard copper dissolution process. Adsorption of BPP derivatives on copper surface is assisted by hydrogen bond formation between BPP derivatives and the Cu$_2$O and/or CuO formed on the copper surface. This type of adsorption should be more prevalent for protonated inhibitors, because the positive charge on the N-atom is conductive to the formation of hydrogen bonds. Unprotonated N atoms may adsorb by direct chemisorption or by hydrogen bonding [87]. BPP derivatives, thus, have the ability to influence both the cathodic and anodic partial reactions, giving rise to the mixed-inhibition mechanism observed.

The well-known Pourbaix diagram (Fig. 8) for copper–water system, indicates that copper is corroded to Cu^{2+} in HNO$_3$ solutions, and no oxide film is formed to protect the surface from corrosion. Copper dissolution is thus expected to be the dominant reaction in HNO$_3$ solutions. The pure nitric acid and inhibitor-containing nitric acid solutions used in our experiments were all aerated where dissolved oxygen may be reduced on copper surface and this will allow some copper corrosion to occur [88]. It is a good

approximation to ignore the hydrogen evolution reaction and only consider oxygen reduction in the aerated nitric acid solutions at potentials near the corrosion potential. Cathodic reduction of oxygen can be expressed either by a direct 4e$^-$ transfer, Eq. (25). Or by two consecutive 2e$^-$ steps involving a reduction to hydrogen peroxide first, Eq. (39), followed by a further reduction, according to Eq. (40):

$$O_2 + 2H^+ + 2e^- \rightarrow H_2O_2 \tag{39}$$

$$H_2O_2 + 2H^+ + 2e^- \rightarrow 2H_2O \tag{40}$$

The transfer of oxygen from the bulk solution to the copper/solution interface will strongly affect the rate of oxygen reduction reaction, despite how oxygen reduction takes place, either in 4e$^-$ transfer or two consecutive 2e$^-$ transfer steps. Dissolution of copper in nitric acid is described by the following two continuous steps:

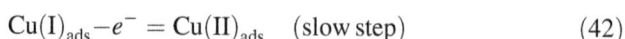

$$Cu - e^- = Cu(I)_{ads} \quad \text{(fast step)} \tag{41}$$

$$Cu(I)_{ads} - e^- = Cu(II)_{ads} \quad \text{(slow step)} \tag{42}$$

Where Cu(I)$_{ads}$ is an adsorbed species at the copper surface and does not diffuse into the bulk solution. The dissolution of copper is controlled by the diffusion of soluble Cu(II) species from the outer Helmholtz plane to the bulk solution. Upon addition of BPP, it is obvious that the slopes of the anodic (β_a) and cathodic (β_c) Tafel lines remain almost unchanged, giving rise to a nearly parallel set of anodic lines, and an almost parallel cathodic plots results too. Thus, the adsorbed BPP inhibitors act by simple blocking of the active sites for both anodic and cathodic processes. In other words, the adsorbed inhibitor decreases the surface area for corrosion without affecting the corrosion mechanism of copper in these solutions, and only causes inactivation of a part of the surface with respect to the corrosive medium [89]. From the experimental results obtained, we note that a plausible mechanism of corrosion inhibition of copper in 2.0 M HNO$_3$ by BPP may be deduced on the basis of adsorption. In acidic solutions, the inhibitor can exist as cationic spices' (Eq. (43)) which may be adsorbed on the cathodic sites of copper and reduce the evolution of hydrogen:

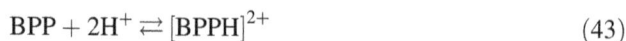

$$BPP + 2H^+ \rightleftarrows [BPPH]^{2+} \tag{43}$$

The protonated BPP, however, could be attached to the copper surface by means of electrostatic interaction between NO$_3^-$ and protonated BPP since the copper surface has positive charges in the acid medium [90]. This could further be explained based on the assumption that in the presence of NO$_3^-$ the negatively charged NO$_3^-$ would attach to positively charged surface. When BPP molecule adsorbs on the copper surface, electrostatic interaction takes place by partial transference of electrons from the polar atoms (N and O atoms and the delocalized π-electrons around the heterocyclic ring) of azole molecules to the metal surface. In addition to electrostatic interaction (physisorption) of BPP molecules on the copper surface, molecular adsorption may also play a role in the adsorption process. The inhibition of copper corrosion in acid solution by the investigated 2-(2-benzimidazolyl)-4(phenylazo) phenol derivatives was found to depend on the concentration and nature of the inhibitor. Variation in structure of the inhibitor molecules takes place through the arylazo side chain, the benzimidazolyl moiety being the same. The electron charge density of the adsorption centers would depend on substituents in the arylazo conjugated with the benzimidazolyl. Also, substituents on the phenyl ring may participate in adsorption. Depending on polar substituent constant (σ), the electron-donating character of the substituents used is increased in the order: $OCH_3 > CH_3 > -NO_2$. This order is almost concordant with the observed order of inhibition efficiency IE% for our BPP compounds. The o-OCH_3 compound is the most efficient inhibitor. Owing to the higher electron donating character of $-OCH_3$ compared to the other substituents. On the other hand, p-NO_2 and m-NO_2 compounds were the least effective inhibitors in this series. This may be attributed to the planarity of the p-NO_2 and m-NO_2 groups with the phenyl ring brings about maximum electron withdrawal. Also, the NO_2 group is easily reduced in acid medium and this process is exothermic, heat evolved aids desorption of the compound on copper surface. o-CH3 and m-CH_3 compounds comes after p-OCH_3 and o-OCH_3 compound in inhibition efficiency. This is due to the lower electron-donating character of $-CH_3$ group compared to $-OCH_3$ group. Thus, the overall mechanism constituents electron release which maintains sufficient electron charge density on the molecule.

Conclusions

The following conclusions may be drawn from the study:

1. 2-(2-benzimidazolyl)-4 (phenylazo) phenol (BPP_1–7) derivatives were found to act as safe effective corrosion inhibitors for copper surface in 2.0 M HNO_3. Inhibition efficiency (IE%) values increase with the inhibitor concentration but decrease with rise in temperature suggesting physical adsorption, with some chemisorption ($\Delta G°_{ads}$ more negative than -40 kJ/mol). Moreover, the inhibition of copper by BPP_1–7 substituents is often explained by the formation of Cu (II)-BPP through its heteroatoms.

2. Polarization measurements show that BPP_1–7 derivatives act essentially as a mixed-type inhibitors and their inhibition mechanism is adsorption assisted by hydrogen bond formation.

3. The adsorption of BPP_1–7 inhibitors on copper surface was found to accord with Frumkin adsorption isotherm model. The adsorption process is strongly, spontaneously, exothermic and accompanied with an increase in entropy of the system from thermodynamic point of view.

4. Phenomenon of adsorption is proposed from the values of kinetic/thermodynamics parameters ($\Delta G°_{ads}$, $1/y$, K_b, f, K_{ads}, E_a, ΔH^*, ΔS^* and ΔG^*) obtained.

5. The theoretical quantum study demonstrated that the inhibition efficiency increases with increase in E_{HOMO} and decrease in E_{LUMO}, dipole moment (μ) and energy gap ΔE. The quantum mechanical approach may well be able to foretell molecule structures that are better for corrosion inhibition.

6. Equilibrium adsorption configurations of the studied synthesized BPP_1 derivative on Cu (111) has been presented.

7. The efficiency order of the studied benzimidazole BPP_1–7 inhibitors obtained by experimental results was verified by theoretical calculations.

Acknowledgments I gratefully acknowledge Tanta University, Chemistry Department, at Tanta, Egypt for the financial assistance and Department of Chemistry, Al-Baha University, Baljarashi, Saudi Arabia for facilitation of our study.

References

1. Krishnamoorthy A, Chanda K, Murarka SP, Ramanath G, Ryan JG (2001) Self-assembled near-zero-thickness molecular layers as diffusion barriers for Cu metallization. Appl Phys Lett 78:2467–2469
2. Khaled KF (2010) Corrosion control of copper in nitric acid solutions using some amino acids: a combined experimental and theoretical study. Corros Sci 52:3225–3234
3. Mihit M, Laarej K, Abou El Makarim H, Bazzi L, Salghi R, Hammouti B (2010) Study of the inhibition of the corrosion of copper and zinc in HNO3 solution by electrochemical technique and quantum chemical calculations. Arab J Chem 3:55–60
4. Parr RG, Yang W (1989) Density functional theory of atoms and molecules. Oxford University Press, Oxford UK

5. Dreizler RM, Gross EKU (1990) Density functional theory. Springer, Berlin

6. Obot IB, Ebenso EE, Obi-Egbedi NO, Afolabi AS, Gasem ZM (2012) Experimental and theoretical investigations of adsorption characteristics of itraconazole as green corrosion inhibitor at a mild steel/hydrochloric acid interface. Res Chem Intermed 38:1761–1779

7. Khaled KF (2008) Molecular simulation, quantum chemical calculations and electrochemical studies for inhibition of mild steel by triazoles. Electrochim Acta 53:3484–3492

8. Obot IB, Ebenso EE, Kabanda MM (2013) Metronidazole as environmentally safe corrosion inhibitor for mild steel in 0.5 M HCl: experimental and theoretical investigation. J Environ Chem Eng 1:431–439

9. Kabanda MM, Obot IB, Ebenso EE (2013) Computational study of some amino acid derivatives as potential corrosion inhibitors for different metal surfaces and in different media. Int J Electrochem Sci 8:10839–10850

10. Obot IB, Obi-Egbedi NO (2010) Theoretical study of benzimidazole and its derivatives and their potential activity as corrosion inhibitors. Corros Sci 52:657–660

11. Obi-Egbedi NO, Obot IB, El-Khaiary MI, Umoren SA, Ebenso EE (2012) Computational simulation and statistical analysis on the relationship between corrosion inhibition efficiency and molecular structure of some phenanthroline derivatives on mild steel surface. Int J Electrochem Sci 7:5649–5675

12. Kaya S, Kaya C (2015) A new equation for calculation of chemical hardness of groups and molecules. Mol Phys 113:1311–1319

13. Kaya S, Kaya C (1060) A new method for calculation of molecular hardness: a theoretical study. Comput Theor Chem 2015:66–70

14. Yang W, Parr RG (1985) Hardness, softness and the Fukui function in the electronic theory of metals and catalysis. Proc Natl Acad Sci 82:6723–6726

15. Kaya S, Kaya C (1052) A new equation based on ionization energies and electron affinities of atoms for calculating of group electronegativity. Comput Theor Chem 2015:42–46

16. Safi ZS, Omar S (2014) Proton affinity and molecular basicity of *m*- and *p*-substituted benzamides in gas phase and in solution: a theoretical study. Chem Phys Lett 610–611:321–330

17. Chattaraj PK, Sarkar U, Roy DR (2006) Electrophilicity Index. Chem. Rev. 106(6):2065–2091

18. Koopmans T (1933) Ordering of wave functions and eigen-energies to the individual electrons of an atom. Physica 1:104–113

19. Chattaraj PK, Lee H, Parr RG, Principle HSAB (1991) J Am Chem Soc 113(5):1855–1856

20. Pearson RG (1997) Chemical hardness: applications from molecules to solids. Wiley-VCH, Weinheim

21. Zhang F, Tang Y, Cao Z, Jing W, Wu Z, Chen Y (2012) Performance and theoretical study on corrosion inhibition of 2-(4-pyridyl)-benzimidazole for mild steel in hydrochloric acid. Corros Sci 61:1–9

22. Jiang L, Lan Y, He Y, Li Y, Li Y, Luo J (2014) 1,2,4-Triazole as a corrosion inhibitor in copper chemical mechanical polishing. Thin Solid Films 556:395–404

23. Ramezanzadeh B, Arman SY, Mehdipour M, Markhali BP (2014) Analysis of electrochemical noise (ECN) data in time and frequency domain for comparison corrosion inhibition of some azole compounds on Cu in 1.0 M H_2SO_4 solution. Appl Surf Sci 289:129–140

24. Kuznetsov YI, Agafonkina MO, Andreeva NP (2014) Mercaptobenzimidazole (MBIMD) physical and chemical adsorption with ΔG of −31.1 kJ mol. Russ J Phys Chem A+ 88:702

25. Liu S, Duan JM, Jiang RY, Feng ZP, Xiao R (2011) Corrosion inhibition of copper in tetra-*n*-butylammonium bromide aqueous solution by benzotriazole. Mater Corros 62:47–52

26. Marija B, Mihajlović P, Antonijević MM (2015) Copper corrosion inhibitors period 2008–2014: a review. Int J Electrochem Sci 10:1027–1053

27. Antonijevic MM, Petrovic MB (2008) Copper corrosion inhibitors: a review. Int J Electrochem Sci 3:1–28

28. Ahamad I, Quraishi MA (2009) Bis (benzimidazol-2-yl) disulphide: an efficient water soluble inhibitor for corrosion of mild steel in acid media. Corros Sci 51(9):2006–2013

29. Ahamad I, Quraishi MA (2010) Mebendazole: new and efficient corrosion inhibitor for mild steel in acid medium. Corros Sci 52(2):651–656

30. Madkour LH, Elroby SK (2015) Inhibitive properties, thermodynamic, kinetics and quantum chemical calculations of polydentate Schiff base compounds as corrosion inhibitors for iron in acidic and alkaline media. Int J Indus Chem 6(3):165–184

31. Maayta AK, Al-Rawashded NAF (2004) Inhibition of acidic corrosion of pure aluminum by some organic compounds. Corros Sci 46(5):1129–1140

32. Emregül KC, Akay AA, Atakol O (2005) The corrosion inhibition of steel with Schiff base compounds in 2 M HCl. Mater Chem Phys 93:325–329

33. Madkour LH, Elmorsi MA, Ghoneim MM (1995) Inhibition of copper corrosion by arylazotriazoles in nitric acid solution. Monatshefte fiir Chemie 126:1087–1095

34. Madkou LH, Ghoneim MM (1997) Inhibition of the corrosion of 16/14 austenitic stainless steel by oxygen and nitrogen containing compounds. Bull Electrochem 13(1):1–7

35. Gece Gökhan (2008) The use of quantum chemical methods in corrosion inhibitor studies. Corros Sci 50(11):2981–2992

36. JiaJun Fu, Zang HaiShan, Wang Ying, Li SuNing, Chen Tao, Liu XiaoDong (2012) Experimental and theoretical study on the inhibition performances of quinoxaline and its derivatives for the corrosion of mild steel in hydrochloric acid. Ind Eng Chem Res 51(18):6377–6386

37. FuSu-ning Jia-jun, Li Su-Ning, Wang Ying, Liu Xiao-Dong, Lu-De Lu (2011) Computational and electrochemical studies on the inhibition of corrosion of mild steel by 1 Cysteine and its derivatives. J Mater Sci 46(10):3550–3559

38. Frisch MJ, Trucks GW, Schlegel HB, Scuseria GE, Robb MA, Cheeseman JR, Scalmani G, Barone V, Mennucci B, Petersson GA, Nakatsuji H, Caricato M, Li X, Hratchian HP, Izmaylov AF, Bloino J, Zheng G, Sonnenberg JL, Hada M, Ehara M, Toyota K, Fukuda R, Hasegawa J, Ishida M, Nakajima T, Honda Y, Kitao O, Nakai H, Vreven T, Montgomery JA Jr, Peralta JE, Ogliaro F, Bearpark M, Heyd JJ, Brothers E, Kudin KN, Staroverov VN, Kobayashi R, Normand J, Raghavachari K, Rendell A, Burant JC, Iyengar SS, Tomasi J, Cossi M, Rega N, Millam JM, Klene M, Knox JE, Cross JB, Bakken V, Adamo C, Jaramillo J, Gomperts R, Stratmann RE, Yazyev O, Austin AJ, Cammi R, Pomelli C, Ochterski JW, Martin RL, Morokuma K, Zakrzewski VG, Voth GA, Salvador P, Dannenberg JJ, Dapprich S, Daniels AD, Farkas O, Foresman JB, Ortiz JV, Cioslowski J, Fox DJ, Inc Gaussian, Wallingford CT (2009) Gaussian 09. Gaussian Inc, Pittsburgh PA, p 2009

39. Lee C, Yang W, Parr RG (1988) Development of the Colle–Salvetti correlation-energy formula into a functional of the electron density. Phys Rev B 37:785–789

40. Becke AD (1993) Density-functional thermochemistry. III. The role of exact exchange. J Chem Phys 98:5648–5652

41. Perdew JP, Burke K, Ernzerhof M (1997) Generalized gradient approximation made simple. Phys Rev Lett 78:1396

42. Perdew JP, Burke K, Ernzerhof M (1996) Generalized gradient approximation made simple. Phys Rev Lett 77:3865–3868
43. Chermette H (1999) Chemical reactivity indexes in density functional theory. J Comput Chem 20:129–154
44. Parr RG, Chattaraj PK (1991) Principle of maximum hardness. J Am Chem Soc 113:1854–1855
45. Islam N, Ghosh DC (2011) A new algorithm for the evaluation of the global hardness of polyatomic molecules. Int J Quant Chem 109:917–931
46. Parr RG, Sventpaly L, Liu S (1999) Electrophilicity index. J Am Chem Soc 121(9):1922–1924
47. Kirkpatrick S, Gelatt CD, Vecchi MP (1983) Optimization by simulated annealing. Science 220(4598):671–680
48. Sun H (1998) COMPASS: an ab initio force-field optimized for condensed-phase applications: overview with details on alkane and benzene compounds. J Phys Chem B 102(38):7338–7364
49. Guoa Lei, Zhub Shanhong, Zhang Shengtao (2015) Experimental and theoretical studies of benzalkonium chloride as an inhibitor for carbon steel corrosion in sulfuric acid. J Ind Eng Chem 24:174–180
50. Zhou Y, Xu S, Guo L, Zhang S, Lu H, Gong Y, Gao G (2015) Evaluating two new Schiff bases synthesized on the inhibition of corrosion of copper in NaCl solutions. RSC Adv 5:14804–14813
51. Strehblow HH, Titze B (1980) The investigation of the passive behaviour of copper in weakly acid and alkaline solutions and the examination of the passive film by Esca and ISS. Electrochim Acta 25(6):839–850
52. Brusic V, Frisch MA, Eldridge BN, Novak FP, Kaufman FB, Rush BM, Frankel GS (1991) Copper corrosion with and without inhibitors. J Electrochem Soc 138:2253–2259
53. Pourbaix M (1975) Atlas of electrochemical equilibria in aqueous solutions. NACE, Houston
54. Khaled KF, Fadl-Allah SA, Hammouti B (2009) Some benzo-triazole derivatives as corrosion inhibitors for copper in acidic medium: experimental and quantum chemical molecular dynamics approach. Mater Chem Phys 117(1):148–155
55. McCafferty E (2005) Validation of corrosion rates measured by the Tafel extrapolation method. Corros Sci 47(12):3202–3215
56. Quartarone G, Battilana M, Bonaldo L, Tortato T (2008) Investigation of the inhibition effect of indole-3-carboxylic acid on the copper corrosion in 0.5 M H₂SO₄. Corros Sci 50(12):3467–3474
57. Amin MA, Khaled KF (2010) Copper corrosion inhibition in O₂-saturated H₂SO₄ solutions. Corros Sci 52(4):1194–1204
58. Quartarone G, Bellomi T, Zingales A (2003) Inhibition of copper corrosion by isatin in aerated o.5 M H₂SO₄. Corros Sci 45(4):715–733
59. Behpour M, Ghoreishi SM, Soltani N, Salavati-Niasari M, Hamadanian M, Gandomi A (2008) Electrochemical and theoretical investigation on the corrosion inhibition of mild steel by thiosalicylaldehyde derivatives in hydrochloric acid solution. Corros Sci 50(8):2172–2181
60. Musa AY, Kadhum AAH, Mohamad AB, Daud AR, Takriff MS, Kamarudin SK (2009) A comparative study of the corrosion inhibition of mild steel in sulphuric acid by4,4-dimethyloxazo-lidine-2-thione. Corros Sci 51(10):2393–2399
61. Ye XR, Xin XQ, Zhu JJ, Xue ZL (1998) Coordination compound films of 1-phenyl-5-mercaptotetrazole on copper surface. Appl Surf Sci 135:307–317
62. Benali O, Larabi L, Traisnel M, Gengembra L, Harek Y (2007) Electrochemical, theoretical and XPS studies of 2-mercapto-1-methylimidazole adsorption on carbon steel in 1 M HClO₄. Appl Surf Sci 253(14):6130–6139
63. Noor EA, Al-Moubaraki AH (2008) Thermodynamic study of metal corrosion and inhibitor adsorption processes in mild steel/1-methyl-4[4′(-X)-styryl pyridinium iodides/hydrochloric acid systems. Mater Chem Phys 110(1):145–154

64. Yadav DK, Maiti B, Quraishi MA (2010) Electrochemical and quantum chemical studies of 3,4-dihydropyrimidin-2(1H)-ones as corrosion inhibitors for mild steel in hydrochloric acid solution. Corros Sci 52(11):3586–3598
65. Badawy WA, Ismail KM, Fathi AM (2006) Corrosion control of Cu-Ni alloys in neutral chloride solutions by amino acids. Electrochim Acta 51(20):4182–4189
66. Tang L, Li X, Si Y, Mu G, Liu G (2006) The synergistic inhibition between 8-hydroxyquinoline and chloride ion for the corrosion of cold rolled steel in 0.5 M sulfuric acid. Mater Chem Phys 95(1):29–38
67. Abdallah M (2002) Rhodanine azosulpha drugs as corrosion inhibitors for corrosion of 304 stainless steel in hydrochloric acid solution. Corros Sci 44(4):717–728
68. El-Awady AA, Abd-El-Nabey BA, Aziz SG (1992) Kinetic-thermodynamic and adsorption isotherms analyses for the inhibition of the acid corrosion of steel by cyclic and open-chain amines. J Electrochem Soc 139(8):2149–2154
69. Popova A, Sokolova E, Raicheva S, Christov M (2003) AC and DC study of the temperature effect on mild steel corrosion in acid media in the presence of benzimidazole derivatives. Corros Sci 45(1):33–58
70. Shukla SK, Quraishi MA (2009) Cefotaxime sodium: a new and efficient corrosion inhibitor for mild steel in hydrochloric acid solution. Corros Sci 51(5):1007–1011
71. Singh AK, Quraishi MA (2010) Effect of Cefazolin on the corrosion of mild steel in HCl solution. Corros Sci 52(1):152–160
72. Umoren SA, Ekanem UF (2010) Inhibition of mild steel corrosion in H2SO4 using exudate gum from Pachylobus edulis and synergistic potassium halide additives. Chem Eng Commun 197(10):1339–1356
73. Fouda AS, Al-Sarawy AA, El-Katori EE (2006) Pyrazolone derivatives as corrosion inhibitors for C-steel in hydrochloric acid solution. Desalination 201(1–3):1–13
74. Guan NM, Xueming L, Fei L (2004) Mater Chem Phys 86:59–68
75. Li Xianghong, Deng Shuduan (2009) Hui Fu., Synergism between red tetrazolium and uracil on the corrosion of cold rolled steel in H₂SO₄ solution. Corros Sci 51(6):1344–1355
76. Taqui Khan MM, Shukla RS (1991) Kinetic and spectroscopic study of the formation of an intermediate ruthenium(III) ascor-bate complex in the oxidation of L-ascorbic acid. Polyhedron 10(23–24):2711–2715
77. Obot IB, Obi-Egbedi NO, Eseola AO (2011) Anticorrosion potential of 2-mesityl-1H-imidazo[4,5-f][1,10]-phenanthroline on mild steel in sulfuric acid solution: experimental and theoretical study. Ind Eng Chem Res 50:2098–2110
78. Obi-Egbedi NO, Obot IB, El-Khaiary MI (1002) Quantum chemical investigation and statistical analysis of the relationship between corrosion inhibition efficiency and molecular structure of xanthene and its derivatives on mild steel in sulphuric acid. J Mol Struct 2011:86–96
79. Sanderson RT (1976) Chemical bond and bond energy. Academic Press, New York
80. Pearson RG (1988) Absolute electronegativity and hardness: application to inorganic chemistry. Inorg Chem 27(4):734–740
81. Rodríguez-Valdez LM, Martínez-Villafañe A, Glossman-Mitnik D (2005) CHIH-DFT theoretical study of isomeric thiatriazoles and their potential activity as corrosion inhibitors. J Mol Struct Thechem 716(1–3):61–65
82. Sastri VS, Perumareddi JR (1997) Molecular orbital theoretical studies of some organic corrosion inhibitors. Corros Sci 53(8):617–622
83. Khaled KF (2009) Monte Carlo simulations of corrosion inhibition of mild steel in 0.5 M sulphuric acid by some green corrosion inhibitors. J Solid State Electrochem 13:1743–1756

84. Kaya C (2011) Inorganic chemistry 1 and 2. Palme Publishing, Ankara
85. John S, Joy J, Prajila M, Joseph A (2011) Electrochemical, quantum chemical and molecular dynamics studies on the interaction of 4-amino-4H,3,5- di(methoxy)-1,2,4-triazole (ATD), BATD, and DBATD on copper metal in 1 N H_2SO_4. Mater Corros 62:1031–1041
86. Oguzie EE, Onuoha GN, Onuchukwu AI (2005) Inhibitory mechanism of mild steel corrosion in 2 M sulphuric acid solution by methylene blue dye. Mater Chem Phys 89:305–311
87. Khaled KF, Amin MA (2008) Computational and electrochemical investigation for corrosion inhibition of nickel in molar nitric acid by piperidines. J Appl Electrochem 38:1609–1621
88. Quartarone G, Moretti G, Bellomi T, Capobianco G, Zingales A (1998) Using indole to inhibit copper corrosion in aerated 0.5 M sulfuric acid. Corrosion 54(8):606–618
89. Ashassi-Sorkhabi H, Ghalebsaz-Jeddi N, Hashemzadeh F, Jahani H (2006) Corrosion inhibition of carbon steel in hydrochloric acid by some polyethylene glycols. Electrochim Acta 51(18):3848–3854
90. Li Y, Zhao P, Liang Q, Hou B (2005) Berberine as a natural source inhibitor for mild steel in 1 M H_2SO_4. Appl Surf Sci 252(5):1245–1253

7

Degradation of ortho-toluidine from aqueous solution by the TiO$_2$/O$_3$ process

Aref Shokri[1] · Kazem Mahanpoor[2]

Abstract In this work, the degradation and mineralization of ortho-toluidine (OT) that is one of the constituents of petrochemical wastewater was investigated by the TiO$_2$/O$_3$ process. The influence of some operational parameters such as concentration of pollutant (30–90 mg L^{-1}), initial pH and amounts of TiO$_2$ was investigated. A radical mechanism with the formation of an anion radical superoxide radical prior to hydroxyl radical is suggested for describing the interaction between ozone and TiO$_2$. These results were not similar to the ozonation process alone, in which higher pH had a positive effect on the removal of OT because of the generation of hydroxyl radicals. In optimum pH for the ozonation and O$_3$/TiO$_2$ processes, the degradation efficiency of the OT was 89.5 and 96%, respectively, at 60 min of reaction. Furthermore, it was made clear that in catalytic ozonation, the degradation efficiency of the OT was higher at neutral pH conditions (pH = 7). The removal of chemical oxygen demand (COD) was increased from 47.5% (only ozonation) to 73% (O$_3$/TiO$_2$) after 90 min of reaction. The kinetics of degradation was pseudo-first order; the degradation and relative mineralization of the OT were calculated by HPLC and COD tests, respectively.

Keywords TiO$_2$ nanocatalyst · Degradation · Ortho-toluidine (OT) · Chemical oxygen demand (COD) · Catalytic ozonation

Introduction

The wastewater generated from the Karoon Petrochemical Company in Iran contains ortho-toluidine (OT), 2-nitrophenol and other aromatic derivatives. Certain amounts of aromatic components are wasted during a process which contains a wide range of non-biodegradable pollutants that cause environmental problems [1]. *O*-toluidine is probably considered carcinogenic to humans, based on the international agency for research on cancer [2]. The conventional treatment methods have high operational costs, longer reaction time and secondary pollution [3], so the use of advanced oxidation processes (AOPs) is essential.

AOPs such as UV/H$_2$O$_2$, UV/O$_3$, TiO$_2$/UV, Fenton's reagents and catalytic ozonation include the production of non-selective oxidizing agents such as hydroxyl radicals, for the degradation of toxic and refractory pollutants in different wastewaters [4]. Ozone is a powerful oxidant and is used greatly in water treatment process [5, 6]. But in most cases, it has been reported that ozone cannot degrade organic pollutants completely and sometimes generates toxic intermediates [7].

In these conditions, catalytic ozonation has been attracting increasing attention as a result of its higher efficiency in the degradation and mineralization of organic pollutants and lower negative effect on the nature of water [8, 9]. Supported, unsupported metals and metal oxides are the most commonly used catalysts for the ozonation of organic pollutants in water. Among various

✉ Aref Shokri
aref.shokri3@gmail.com

Kazem Mahanpoor
k-mahanpoor@iau-arak.ac.ir

1 Young Researchers and Elite Club, Arak Branch, Islamic Azad University, Arak, Iran

2 Department of Chemistry, Faculty of Science, Arak Branch, Islamic Azad University, Arak, Iran

semiconductors, titanium dioxide has been recurrently reported as active, inexpensive and nontoxic. It can accelerate the ozonation process for degradation of a wide range of different pollutants [10, 11]. The most common applications of titania are in photocatalysis systems; however, it has been proposed as an active catalyst in catalytic ozonation of organic pollutants. TiO_2-catalyzed ozonation was more effective for the removal of some pollutant than ozone alone. The detailed and complete mechanism of catalytic ozonation is not clear up to this time and many studies have been done so far, but it is not completely obvious and is one of the gaps of catalytic ozonation. The effect of TiO_2 on ozone decomposition to produce hydroxyl radicals was not obvious. Some researchers proposed mechanisms based on non-radical pathway to form hydroxyl radical. A number of mechanisms are offered, but the direct formation of hydroxyl radicals from ozone decomposition on the surface of TiO_2 or indirect formation as a result of secondary reactions is still unknown [12]. Rosal et al. [13, 14] were investigated the degradation of clofibric acid by catalytic ozonation on titania. It was proposed that the adsorption and the following reaction of pollutants on catalyst sites are responsible for the improvement of catalytic ozonation.

According to the studies of many researchers, there is still a significant shortage of information concerning the role of TiO_2 in ozonation reactions, especially about the ozone decomposition reaction [12].

No study has been done on degradation of ortho-toluidine by catalytic ozonation up to this time. In this work, degradation of OT as an aromatic pollutant was studied by the O_3/TiO_2 process and the effects of pH, initial concentration of OT, amount of TiO_2 and the kinetics of the reaction for higher degradation of OT wre investigated.

Experimental

Materials

Ortho-toluidine (OT), HCl and NaOH, potassium iodide and sodium thiosulphate were of reagent grades and supplied from Merck. Titanium dioxide (P–25, 30% Rutile and 70% Anatase) purchased from Degussa, Germany, has a BET surface area of 55 m^2 g^{-1} and an average particle size of 20 nm.

Ozone was produced in an ozone generator fed by dry oxygen and all reagents were used as received without further purification. Distilled water was used throughout this study.

Apparatus

Experiments were carried out in a semi batch (batch for TiO_2 and OT and continuous for ozone) reactor. The pure oxygen, from a pressurized capsule, was entered into an ozone generator (214 V and 0.39 A) from the ARDA companies of Iran. The reactor was equipped with a water-flow jacket connected to a thermostat (BW20G model from Korean Company) for adjusting the temperature at 25 °C in all experiments as shown in Fig. 1. The pH was measured by the pH meter PT-10P Sartorius instrument, Germany. The progress in the degradation of the OT was recorded by a high-performance liquid chromatography (Knauer, Germany) equipped with a Spectrophotometer (Platm blue Germany). A reverse phase column was filled with 3 μm Separon C_{18} with 150 mm length and 4.6 mm diameter. The isocratic method was used with pH adjusted to 2.5, using orthophosphoric acid and a solvent mixture of acetonitrile and deionized water (60:40% v/v) at a flow rate of 1 mL min^{-1} at room temperature. In all tests, the

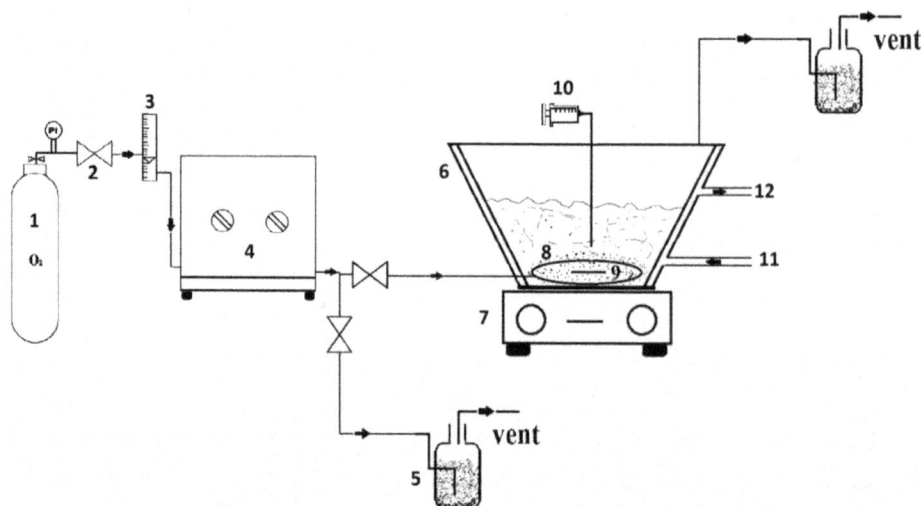

Fig. 1 Schematic diagram of the laboratory-scale installation. Notes: *1* pure oxygen, *2* cut off valve, *3* gas flow meter, *4* ozone generator, *5* washing bottle, *6* reactor, *7* magnetic stirrer, *8* ozone diffuser, *9* magnetic bar, *10* sampling port, *11* cooling water supply from thermostat, *12* cooling water return to thermostat

suspension was centrifuged and filtered to collect the catalyst particles.

Catalytic ozonation tests

About 2 L of aqueous solution containing a known concentration of the OT and nano TiO_2 was mixed completely in the reactor. A mixture of O_3/O_2 was produced by the ozone generator and entered from the bottom of the reactor by a porous diffuser for mixing well, saturating the solution with O_3, better mass transfer and reaction between ozone, TiO_2 and the pollutant. The concentration of gaseous ozone was measured by the iodometric method using 2% neutral buffered potassium iodide for ozone trapping and sodium thiosulfate as a titrant [15]. The flow rate of the O_3/O_2 mixture was kept constant at 0.4 L min^{-1}, based on the previous studies and initial tests, having an ozone concentration of 16.6 mg L^{-1}. To know the amount of ozone consumed, the reactor outlet gas was bubbled through a KI (2%w) tampshed solution for determining the ozone not reacted, while the potassium iodide solution reacted with the excess ozone based on the following equation (Eq. 1):

$$O_3 + 2KI + H_2O \rightarrow I_2 + 2KOH + O_2. \quad (1)$$

The produced iodine was titrated by standard sodium thiosulphate in the presence of starch as an indicator. The amounts of not reacted and reacted ozone, and the value of ozone in the tail gas were determined correspondingly. The residual of ozone in aqueous solution was estimated by a spectrophotometer using the indigo method [16].

TiO_2 particles were dispersed and suspended in the solution as the ozone gas entered into the reactor. Samples were withdrawn at different intervals and filtered to remove the TiO_2 particles. The concentration of OT was determined by spectrophotometry at two peaks, 230 and 280 nm. The HPLC and spectrophotometry methods (especially in 280 nm) gave similar results and the difference between them was little, which was corrected. But in acidic pH (lower than 4), the results of spectrophotometry cannot be related to HPLC, because the UV peaks of OT were destroyed in very acidic pH. In this condition, only HPLC should be employed. The experiments were carried out in the pH range of 4–10. The pH was adjusted only at the beginning of the reaction by adding NaOH (0.1 M) or HCl (0.1 M). A slight decrease in pH occurred based on the production of mineral acids.

COD was measured by the standard closed reflux and colorimetric method [17] and the absorbance of samples for COD was measured by a spectrophotometer at 600 nm.

The removal percent of OT and COD as a function of reaction time is given by the following equations (Eqs. 2–3):

$$Removal\ of\ OT\ (\%) = \left(\frac{[OT]_0 - [OT]_t}{[OT]_0} \right) \times 100, \quad (2)$$

$$Removal\ of\ COD\ (\%) = \left(\frac{[COD]_0 - [COD]}{[COD]_0} \right) \times 100, \quad (3)$$

where $[OT]_0$ and $[COD]_0$ are the initial concentrations of OT and COD at the start of the reaction, and [OT] and [COD] are the concentrations of the OT and amount of COD at time t, respectively.

Results and discussion

Effect of pH on ozonation

The degradation of ozone in aqueous solution depends on pH strongly, and it enhances with the increase in pH [18]. The experiments were carried out at pH values of 4, 6, 8, 9 and 10 and, as seen from Fig. 2, the results showed that the removal efficiency increased from 43 to 89.5% in 60 min of reaction as the pH increased from 4 to 9 and then decreased to 86%. So, the optimum pH during this process was 9.

At pH = 4, the formation of hydroxyl radicals was very low, so only slight radical reactions occurred. However, direct molecular ozonolysis was highly predominant and ozone reacted with the pollutant directly and the double bonds of the OT ring seemed to be destroyed by ozone. Hydroxyl radicals were strong oxidants that originated from the reaction of hydroxide ions with ozone at high pH and initiated the chain oxidation reaction of ozone, which was non-selective and very fast. At pH = 10, radical scavenging occurred because of a greater number of

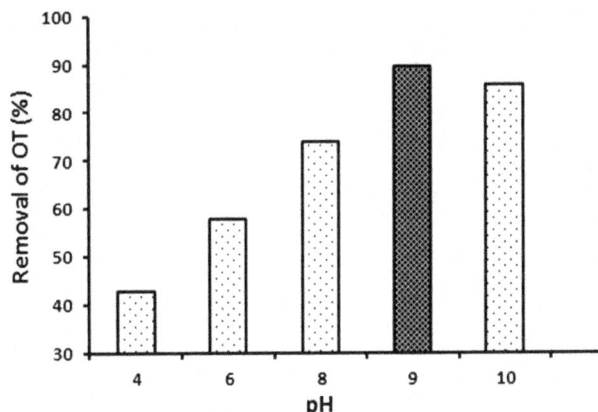

Fig. 2 Effect of pH on ozonation of OT ($[O_3]_0$ = 16.6 mg L^{-1}, $[OT]_o$ = 50 mg L^{-1})

hydroxide ions and consequently there was more hydroxyl radical production [19].

Effect of catalyst amount on the TiO$_2$/O$_3$ process

The effect of catalyst extent on the degradation of OT is presented in Fig. 3. In the non-catalytic ozonation process, the ordinary conversion of the OT was achieved because the action of free radicals created from the self-decomposition of ozone was poor. It is acceptable that ozonation of water also leads to the production of hydroxyl radicals through ozone disintegration [20]. The degradation efficiency of the OT was significantly increased in the presence of both ozone and catalyst. An exceptional feature of nano TiO$_2$ is its extremely high surface area. It is obvious that the degradation of the OT was based on the action of some ozone-absorbed species or free radicals produced perhaps on the catalyst surface or in the aqueous solution. Catalyst amounts employed a positive effect on OT removal in TiO$_2$/O$_3$ process. But as depicted in Fig. 3, the removal efficiency was increased from 80.5 to 93.5% by the increase in catalyst dose from 0.6 to 1.2 g L^{-1}. A further increase in catalyst amounts to 1.8 g L^{-1} did not yield any significant increase in the degradation rate. The difference between the removal efficiency of OT at 1.2 and 1.5 g L^{-1} of TiO$_2$ was very low, So, 1.2 g L^{-1} of catalyst was obtained as an optimum concentration of TiO$_2$ from an economic point of view and other experiments were performed in this concentration of catalyst.

Effect of pH on the TiO$_2$/O$_3$ process

In the TiO$_2$/O$_3$ system, pH had two direct effects on the process: one is ozone decomposition and the other is the surface charge and characteristics of the TiO$_2$ nanocatalyst, which has a direct influence on the adsorption of pollutant molecules [21, 22]. The point of zero charge (PZC) of TiO$_2$

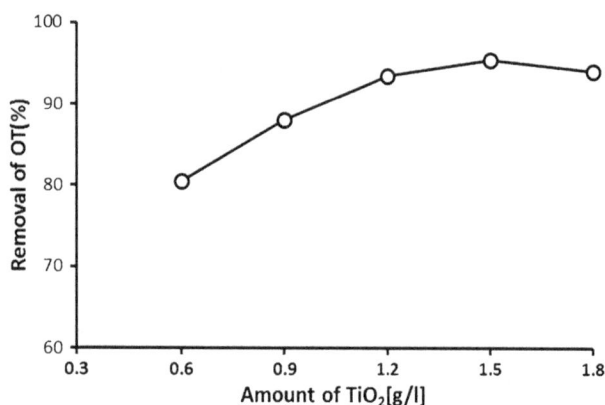

Fig. 3 Catalytic ozonation of OT with various catalyst doses ([O$_3$]$_o$ = 16.6 mg L^{-1}, [OT]$_o$ = 50 mg L^{-1}, pH = 6, time = 60 min)

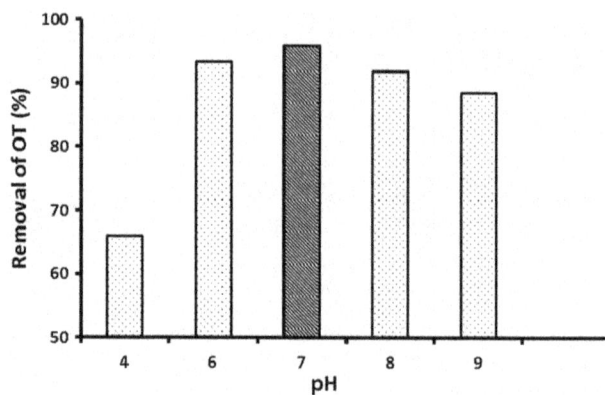

Fig. 4 Effect of pH on the TiO$_2$/O$_3$ process ([TiO$_2$] = 1.2 g L^{-1}, [O$_3$]$_o$ = 16.6 mg L^{-1}, [OT]$_o$ = 50 mg L^{-1}, time = 60 min)

was reported as 6.6 [23] and this factor affected the absorption of the pollutant on the catalyst and was determined by potentiometric titration as explained by Halter [24]. Organic pollutants in neutral state may be adsorbed on the surface of the catalyst if the surface is not charged near the PZC of the catalyst [25].

The effect of pH on the TiO$_2$/O$_3$ process is shown in Fig. 4. From the experimental results, it was clear that during treatment, the best results were obtained at a neutral pH. After 60 min, the degradation efficiency of OT was 96% in neutral condition (pH = 7), while in the solutions with pH 4, 6, 8 and 9 it was about 66, 93, 92, and 88.5%, respectively. The surface properties and the electrostatic interactions between TiO$_2$ and hydroxide ions in the solution were the main factors affecting the degradation of OT.

The effect of the initial concentration of OT on the removal efficiency

The effect of the initial concentration of OT on the efficiency of degradation in TiO$_2$/O$_3$ was investigated over the concentration range from 30 to 90 mg L^{-1} and the results are shown in Fig. 5. The results revealed that the rate of removal was increased slightly from 94.5 to 96% with an increase in OT dosage from 30 to 50 mg L^{-1}, but with an increase in its initial concentration from 50 to 90 mg L^{-1}, the removal efficiency of the OT was reduced significantly. When the initial dosage of the pollutant was high (90 mg L^{-1}), the number of available active sites was decreased and the generation of hydroxyl radicals was reduced by OT molecules, because of their competitive adsorption on the TiO$_2$ surface. Only 55.5% of the pollutant was degraded after 90 min of reaction.

With an increase in the initial concentration of OT, active agents such as hydroxyl radicals that originated from the process were decreased because they reacted with a large number of pollutant molecules [26]. When the

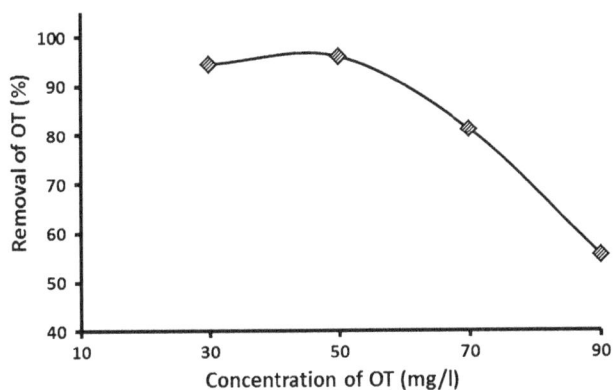

Fig. 5 Effect of initial dosage of OT on the degradation efficiency in the TiO$_2$/O$_3$ process (pH = 7, [TiO$_2$] = 1.2 g L^{-1}, time = 60 min)

concentration of OT was increased, the surface of TiO$_2$ was covered by pollutant molecules instead of ozone and subsequent production of the active agents for destroying the pollutant decreased. However, when the number of pollutant molecules was very low, their collisions with active sites were decreased and the degradation efficiency was reduced [27].

Degradation of OT by the TiO$_2$/O$_3$ and O$_3$ processes

Ernst et al. [28] proposed that the dissolved ozone is adsorbed first on the catalyst surface during the catalytic ozonation with Al$_2$O$_3$, and then degraded quickly; according to the existence of hydroxyl surface groups, O$_2^-$ was produced and then resulted in the production of hydroxyl radicals from a series of reactions. Also, Zhang et al. [29] recommended that O$_2^-$ and hydroxyl radicals were created by the ozone molecule with the hydroxyl group as catalyst. In this study, the dissolved molecular ozone was adsorbed on the TiO$_2$ surface at first and then decayed into O$_2^-$ and OH$^\bullet$ rapidly due to the presence of hydroxyl surface groups on the catalyst (Eqs. 6, 7). Furthermore, the produced O$_2^-$ could promote molecular ozone to decompose into hydroxyl radicals. Then the OT adsorbed on the surface of the TiO$_2$ would be degraded by OH$^\cdot$ and molecular ozone (Eqs. 4, 5). As ozonation along with TiO$_2$ can happen through either direct reaction with molecular ozone or indirect reaction with the produced hydroxyl radicals, the removal of the OT can be symbolized by the following simple reactions (Eqs. 4–9) [30]:

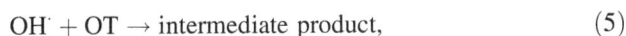

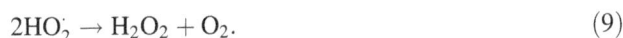

$$O_3 + OT \rightarrow \text{intermediate product}, \tag{4}$$

$$OH^\cdot + OT \rightarrow \text{intermediate product}, \tag{5}$$

$$O_3 + OH^- \rightarrow HO_2^\cdot + O_2^-, \tag{6}$$

$$O_3 + HO_2^\cdot \rightarrow OH^\cdot + 2O_2, \tag{7}$$

$$O_3 + OH^\cdot \rightarrow HO_2^\cdot + O_2, \tag{8}$$

$$2HO_2^\cdot \rightarrow H_2O_2 + O_2. \tag{9}$$

It has been suggested that ozone can be adsorbed on a catalyst surface to yield different oxidizing species [31]. According to the proposed mechanism, ozone and organic molecules were adsorbed on the catalyst surface concurrently, ozone degraded on the metallic sites and produced the surface bond radical (O$_2^-$), which are more reactive than ozone and lead to the production of hydroxyl radicals. Oxidation continues through some oxidized intermediates gradually, while O$_2^-$ radicals are continuously produced by dissolving ozone that is transferred to the catalyst surface. The attraction of the oxidation products to the surface of the catalyst reduces the repulsion of the final degradation products from the catalyst surface [32].

As it can be seen from Fig. 6, experiments were performed in 90 min for the comparative study of different processes at their optimum pH. Only 5% of OT disappeared when in contact with TiO$_2$ alone at pH = 7, as a result of the pollutant adsorption on the surface of the nanocatalyst at a catalyst concentration of 1.2 g L^{-1}. Therefore, the OT removal rate of the ozone along with TiO$_2$ was higher than the sum of the separate influences of single adsorption and single ozonation.

In runs with TiO$_2$/O$_3$ and O$_3$ processes, 96 and 89% of OT was removed, respectively. The corresponding results indicated that the presence of TiO$_2$ can accelerate the degradation of OT rather than ozone oxidation alone. In addition, TiO$_2$ has the ability to enhance the decomposition of ozone and promote the formation of hydroxyl radicals. According to preliminary experiments, about 1.2 g L^{-1} of TiO$_2$ was used because it can initiate the degradation of OT, and at high dosage of catalysts the aggregation of TiO$_2$ particles occurred.

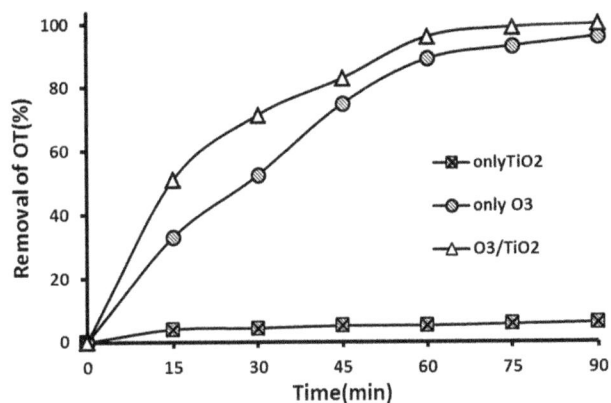

Fig. 6 Degradation of OT during ozonation and catalytic ozonation ([OT]$_o$ = 50 mg L^{-1}, [O$_3$]$_o$ = 16.6 mg L^{-1}, optimum pH for each process, [TiO$_2$] = 1.2 g L^{-1})

Reusability of the TiO$_2$ catalyst

The reusability of the TiO$_2$ catalyst was investigated for degradation of OT. The TiO$_2$ particles were separated from the solution by filtration, then washed with distilled water and regenerated by drying and heating at 150 °C for 2 h. At high temperatures, the adsorbed molecules on the surface were removed and some of the active sites were released.

An equal amount of regenerated catalyst was used for the degradation of OT at optimum experimental condition. After three runs of the experiment, the degradation efficiency decreased to 92, 84 and 66%. The following are the reasons for this decrease in efficiency: (I) adsorption of intermediates and side products of the process in the active sites and on the surface of the catalyst leads to decrease in the degradation efficiency [33]; (II) as Ti leaching occurred, the active sites on the catalyst surface were removed and the activity of the remaining catalyst was decreased; (III) the structure and morphology of the catalyst matrix was deformed gradually by continuous heating in the regeneration process. It was clear that using ozone along with TiO$_2$ prevented the catalyst deactivation to some extent [34].

Removal of COD

The removal efficiency of COD was studied by O$_3$ and O$_3$/TiO$_2$ processes at optimum pH for the degradation of each process. The removal of COD through the degradation of OT in 90 min is shown in Fig. 7.

In the O$_3$/TiO$_2$ process, the amount of COD was decreased sharply during the first 45 min and then decreased slowly. After 90 min of treatment, in the O$_3$ and O$_3$/TiO$_2$ process, the removal of COD was 47.5 and 73%, respectively.

It is clear that ozonation alone is a slow process, but the O$_3$/TiO$_2$ reaction is a rapid one. So, it can be inferred that

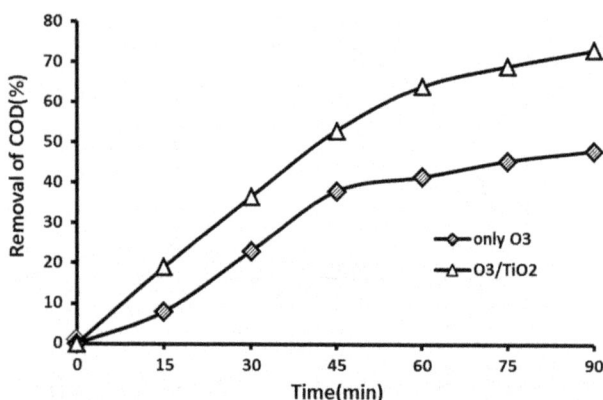

Fig. 7 Removal of COD in ozonation and catalytic ozonation of OT during 90 min (pH = 9 in O$_3$ and pH = 7 in O$_3$/TiO$_2$ process, [TiO$_2$] = 1.2 g L^{-1}, [O$_3$]$_o$ = 16.6 mg L^{-1}, [OT]$_o$ = 50 mg L^{-1})

OT was mineralized partially and some degradation intermediates were created during the process; OT cannot be totally mineralized even by the O$_3$/TiO$_2$ methods [35].

Kinetic study in the degradation of OT with the O$_3$/nano-TiO$_2$ process

The studies of other researchers established that the production of hydroxyl radicals according to the ozone decomposition on the surface of the catalyst was responsible for the enhancement of catalytic ozonation [36, 37]. But when the Mn–Ce–O catalyst was employed to improve the removal efficiency of phenolic acids, a dissimilar deduction was obtained from the study of Martins and Quinta-Ferreira [38]; in that case of the existence of radical scavengers had no influence on the catalytic ozonation efficiency. This finding showed that the reaction of their study was not a radical mechanism.

However, in our study, a kinetic study for degradation of OT in the O$_3$/TiO$_2$ process was performed at optimum pH (pH = 7). From the mentioned studies, it can be inferred that the probable mechanism for the catalytic ozonation comprised an indirect oxidation reaction with hydroxyl radicals and a direct oxidation reaction after the ozone and OT adsorbed on the surface of TiO$_2$. The kinetic relation for degradation of the OT by the pointed out process can be represented as:

$$\frac{-d[OT]}{dt} = ko_3[OT][TiO_2][O_3] + k_{OH\cdot}[OT][TiO_2][OH\cdot],$$
(10)

where [OT], [O$_3$], [OH·] and [TiO$_2$] are the concentrations of OT, ozone, hydroxyl radicals and TiO$_2$, respectively. Moreover $k_{OH\cdot}$ and ko$_3$ are the rate constants of OT with hydroxyl radicals and ozone. At neutral pH values, the nonselective reactions of hydroxyl radicals with OT were predominant [35], so the kinetic equation can be written as:

$$\frac{-d[OT]}{dt} = k_{OH\cdot}[OT][TiO_2][OH^\bullet].$$
(11)

In this process, the OT was degraded by reaction with TiO$_2$ and O$_3$, and the ratio of the concentration of OT to O$_3$, OH· or TiO$_2$ was low; so the concentration of hydroxyl radicals and catalyst can be considered constant. In these conditions, only the concentration of the OT was changed and the reaction was pseudo-first order [39]. So, the equation rate can be shown as:

$$\frac{-d[OT]}{dt} = k'_{OH}[OT],$$
(12)

where k'_{OH} is a pseudo-first-order rate reaction of OT with hydroxyl radicals originating from the O$_3$/nano-TiO$_2$ process. The integration of Eq. (12) results in:

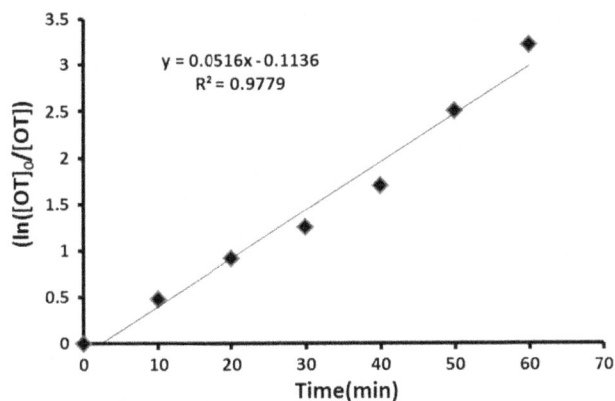

Fig. 8 The curve of $\ln \frac{[OT]_0}{[OT]_t}$ versus reaction time in the O_3/TiO_2 process (pH = 7, $[TiO_2]$ = 1.2 g L^{-1}, $[O_3]$ = 16.6 mg L^{-1} and $[OT]_0$ = 50 mg L^{-1})

$$-\ln \frac{[OT]}{[OT]_0} = k'_{OH}.t, \qquad (13)$$

where [OT] and $[OT]_0$ are the concentration of OT at time = t and time = 0, respectively. As seen from Fig. 8, the term $\ln \frac{[OT]_0}{[OT]_t}$ versus reaction time was plotted, and after linear regression analysis the apparent first-order rate constants (k'_{OH} = 51.6 × 10^{-3} min^{-1}) and half-life of degradation reaction ($t_{1/2}$ = 13.4 min were determined (Eq. 13). By comparing the rate constant of this study with the previous research (k = 54.8 × 10^{-3} min^{-1}) [40], it is obvious that the rate constant depends on the type and initial concentration of the pollutant, dosage of ozone, features of the reactor and catalyst characteristics.

Conclusion

The combination of ozone and TiO$_2$ catalyst has a significant effect on the removal of OT in aqueous solutions, and based on the experimental results the following conclusions are obtained:

The nanocatalyst of TiO$_2$ accelerates the decomposition of ozone under neutral condition (pH = 7). In the TiO$_2$/O$_3$ system, the pH has two direct effects on the process, one is ozone decomposition and the other is on the surface charge and characteristic of the TiO$_2$ nanocatalyst. The oxidation efficiency of OT is higher at neutral pH than at alkaline or acidic ones. The removal of OT is 96 and 89% at 60 min of reaction; also the removal of COD is 73 and 47.5% after 90 min of reaction in the TiO$_2$/O$_3$ and O$_3$ processes, respectively. The adsorption and the following reaction of OT on TiO$_2$ sites are responsible for the improvement of ozonation rate observed in catalytic runs. The OT removal rate in the ozone along with TiO$_2$ was higher than the sum of the separate influences of single adsorption of catalyst

(5%) at 1.2 g L^{-1} and single ozonation, especially in COD removal efficiencies (47.5 and 73% in O$_3$ and TiO$_2$/O$_3$ process). According to kinetic studies, it is clear that in O$_3$/TiO$_2$ process the rate equation for degradation of the OT is pseudo-first-order and after linear regression, R^2 is obtained at 0.9779 and a little deviation from 1, is because of assuming this concept that direct and selective reactions of ozone are negligible in neutral conditions. These experimental results established the hypothesis that the removal of OT by the O$_3$/nano-TiO$_2$ process followed a radical-type mechanism. The proposed mechanism mentioned that ozone and organic molecules are adsorbed on the catalyst surface simultaneously, ozone degrades and produce the surface bond radicals (O_2^-), that they are more reactive than ozone and lead to the production of OH· and hydroxyl radical was represented in kinetic equations.

Acknowledgements The authors wish to thank the Islamic Azad University of Arak, Iran, for the financial support.

References

1. Christie RM (2011) Color chemistry. Royal Society of Chemistry, London, p 2011
2. Ferlay J, Parkin DM, Steliarova-foucher E (2010) Estimates of cancer incidence and mortality in Europe in 2008. Eur J Cancer 46:765–781
3. Yahiat S, Fourcade F, Brosillon S, Amrane A (2011) Photo catalysis as a pre-treatment prior to a biological degradation of cyproconazole. Desalination 281:61–67
4. Gharbani P, Khosravi M, Tabatabaii SM, Zare K, Dastmalchi S, Mehrizad A (2010) Degradation of trace aqueous 4-chloro-2-nitrophenol occurring in pharmaceutical industry wastewater by ozone. Int J Environ Sci Technol 7:377–384
5. Ternes TA, Meisenheimer M, Mc Dowell D, Sacher F, Brauch HJ (2002) Removal of pharmaceuticals during drinking water treatment. Environ Sci Technol 36:3855–3863
6. Hua W, Bennett ER, Letcher RJ (2006) Ozone treatment and the depletion of detectable pharmaceuticals and atrazine herbicide in drinking water sourced from the upper Detroit River. Water Res 40:2259–2266
7. Kasprzyk B, Nawrocki J (2002) Preliminary results on ozonation enhancement by a perfluorinated bonded alumina phase. Ozone Sci Eng 24:63–68
8. Sui M, Xing S, Sheng L, Huang S, Guo H (2012) Heterogeneous catalytic ozonation of ciprofloxacin in water with carbon nanotube supported manganese oxides as catalyst. J Hazard Mater 15:227–236
9. Gracia R, Cortes S, Sarasa J, Ormad P, Ovelleiro JL (2000) TiO$_2$-catalysed ozonation of raw Ebro river water. Water Res 34:1525–1532
10. Farbod M, Khademalrasool M (2011) Synthesis of TiO$_2$ nano particles by a combined sol–gel ball milling method and investigation of nano particle size effect on their photo catalytic activities. Powder Technol 214:344–348
11. Saliby I, Okour Y, Shon HK, Kandasamy J, Lee WE, Kim J (2012) TiO$_2$ nano particles and nano fibres from TiCl$_4$ flocculated sludge: characterisation and photo catalytic activity. J Ind Eng Chem 18:1033–1038

12. Yang Y, Ma J, Qin Q, Zhai X (2007) Degradation of nitrobenzene by nano-TiO_2 catalyzed ozonation. J Mol Catal A: Chem 267:41–48

13. Rosal R, Gonzalo MS, Rodriguez A, Garcia-Calvo E (2009) Ozonation of clofibric acid catalyzed by titanium dioxide. J Hazard Mater 169:411–418

14. Rosal R, Gonzalo MS, Boltes K, Leton P, Vaquero JJ, Garcia-Calvo E (2009) Identification of intermediates and assessment of ecotoxicity in the oxidation products generated during the ozonation of clofibric acid. J Hazard Mater 172:1061–1068

15. Langlais B, Reckhow DA, Brink DR (1991) Ozone in water treatment: application and engineering. Lewis Publishers, Michigan- USA

16. Bader H, Hoigne AJ (1981) Determination of ozone in water by the indigo method. Water Res 15:449–456

17. APHA (1999) standard methods for examination of water and wastewater, twentieth edn. American Public Health Association, Washington

18. Beltran FJ (2004) Ozone reaction kinetics for water and wastewater systems. Lewis Publishers, Boca Raton

19. Muthukumar M, Sargunamani D, Selvakumar N, Rao VJ (2004) Optimization of ozone treatment for color and COD removal of acid dye effluent using central composite design experiment. Dyes Pigments 64:127–134

20. Hoigne J, Bader H (1983) Rate constants of reactions of ozone with organic and inorganic compounds in water—I: non-dissociating organic compounds. Water Res 17:173–183

21. Stumm W, Morgan JJ (1981) Aquatic chemistry:chemical equilibria and rates in natural waters. John Wiley & Sons Inc, New York

22. Rodea-Palomares I, Petre A, Boltes K et al (2010) Application of the combination index (CI)-isobologram equation to study the toxicological interactions of lipid regulators in two aquatic bioluminescent organisms. Water Res 44:427–438

23. Rosal R, Rodriguez A, Gonzalo MS, Garcia-Calvo E (2008) Catalytic ozonation of naproxen and Carbamazepine on titanium dioxide. Appl Catal B Environ 4:48–57

24. Halter WE (1999) Surface acidity constants of a-Al_2O_3 between 25 and 70 °C. Geochim Cosmochim Acta 63:3077–3085

25. Kasprzyk-Hordern B, Ziolek M, Nawrocki J (2003) Catalytic ozonation and methods of enhancing molecular ozone reactions in water treatment. Appl Catal B Environ 46:639–669

26. Nezamzadeh-Ejhieh A, Amiri M (2013) CuO supported clinoptilolite towards solar photo catalytic degradation of P-aminophenol. Powder Technol 235:279–288

27. Shokri A, Mahanpoor K, Soodbar D (2016) Degradation of 2-nitrophenol from petrochemical wastewater by UV/$NiFe_2O_4$/-clinoptilolite process. Fresen Environ Bull 25:500–508

28. Ernst M, Lurot F, Schrotter JC (2004) Catalytic ozonation of refractory organic model compounds in aqueous solution by aluminum oxide. Appl Catal B Environ 47:15–25

29. Zhang T, Ma J (2008) Catalytic ozonation of trace nitrobenzene in water with synthetic goethite. J Mol Catal A: Chem 279:82–89

30. Guzman-Perez CA, Soltan J, Robertson J (2011) Kinetics of catalytic ozonation of atrazine in the presence of activated carbon. Sep Purif Technol 79:8–14

31. Beltran FJ, Rivas FJ, Montero R (2002) Catalytic ozonation of oxalic acid in an aqueous TiO_2 slurry reactor. Appl Catal B Environ 39:221–231

32. Logeman FP, Annee JHJ (1997) Water treatment with a fixed bed catalytic ozonation process. Water Science Technol 35:353–360

33. Shokri A, Mahanpoor K, Soodbar D (2005) Evaluation of a modified TiO_2 (GO–B–TiO_2) photo catalyst for degradation of 4-nitrophenol in petrochemical wastewater by response surface methodology based on the central composite design. J Environ Chem Eng. 4:585–598

34. Crabtree RH (2011) Resolving heterogeneity problems and impurity artifacts in operationally homogeneous transition metal catalysts. Chem Rev 112:1536–1554

35. Goi A, Trapido M, Tuhkanen T (2004) A study of toxicity, biodegradability, and some by-products of ozonised nitrophenols. Ad Environ Res 8:303–311

36. Zhai X, Chen Z, Zhao S, Wang H, Yang L (2010) Enhanced ozonation of dichloroacetic acid in aqueous solution using nanometer ZnO powders. J Environ Sci 22:1527–1533

37. Tong S, Shi R, Zhang H, Ma C (2011) Kinetics of Fe_3O_4–CoO/Al_2O_3 catalytic ozonation of the herbicide 2-(2,4-dichloro phenoxy) propionic acid. J Hazard Mater 185:162–167

38. Martins RC, Quinta-Ferreira RM (2009) Catalytic ozonation of phenolic acids over a Mn–Ce–O catalyst. Appl Catal B Environ 90:268–277

39. Gottschalk C, Libra JA, Saupe A (2000) Ozonation of water and wastewater. Wiley-VCH Publisher, New York

40. Shokri A (2016) Degradation of 4-nitrophenol from industrial wastewater by nano catalytic ozonation. Int J Nano Dimens 7:160–167

8

Gasification of Iranian walnut shell as a bio-renewable resource for hydrogen-rich gas production using supercritical water technology

Farid Safari[1] · Ahmad Tavasoli[1,2] · Abtin Ataei[1]

Abstract Gasification in supercritical water (SCW) media is known as an efficient and promising technology for obtaining hydrogen-rich gas from dry and wet bio-renewable materials. Gasification of walnut shell as the main hard nutshell produced in Kurdistan Province of Iran was investigated using a stainless steel batch micro-reactor. Effects of reaction time in the range of 10–30 min, feed loading in the range of 0.06–0.18 g, and temperature in the range of 400–440 °C were investigated to determine the condition for maximum hydrogen yield. Furthermore, carbon gasification efficiency (CGE) and hydrogen gasification efficiency (HGE) were calculated according to the elemental analysis and the yields of gaseous products. Total gas yield and hydrogen yield were directly correlated with temperature. Steam reforming of walnut shell was favored at higher temperatures. Also, walnut shell loading was inversely correlated with total gas and hydrogen yields while production of methane was favored by higher loading of walnut shell. Furthermore, hydrogen yield increased first, when reaction time increased from 10 to 20 min, and then decreased. Maximum hydrogen yield of 4.63 mmol/g of walnut shell was obtained at 440 °C, walnut shell loading of 0.06 g and reaction time of 20 min.

Keywords Gasification · Supercritical water · Hydrogen · Walnut shell

✉ Farid Safari
f.safari@srbiau.ac.ir

[1] Department of Energy Engineering, Science and Research Branch, Islamic Azad University, Tehran, Iran

[2] School of Chemistry, College of Science, University of Tehran, Tehran, Iran

Introduction

Conventional fossil fuels are being depleted dramatically and the life of many living species has been threatened by the products of their combustion [1]. On the other hand, energy demand is increasing dramatically with population growth and industrialization of developing countries. Using renewable energy resources can be an alternative solution in response to climate change and energy demand related problems [2]. Production, storage and consumption of these renewable resources should be developed such that they positively affect people's preference. Therefore, it seems necessary to research and develop novel renewable energy technologies. Biomass as an organic matter is a rich source of carbon and hydrogen [3]. It is also a versatile resource for producing promising fuels and chemicals such as hydrogen and ethanol [4]. Hydrogen is one of the main products of gasification and regarded by many scientists as key energy carrier of the future with much higher energy density than other conventional fuels [5]. Hydrogen can be used for electricity generation in fuel cells and as an efficient fuel for transportation without any considerable emission [6]. Today most of the hydrogen around the world is produced from fossil fuels especially steam methane reforming. Therefore, biomass-based hydrogen can be a great leap forward in the utilization of renewable energy [7]. Million tons of agricultural wastes are discarded or burned annually around the world. According to the statistic report of Food and Agriculture Organization of the United Nations, Iran is the second largest producer of walnut in the world with a production of 452,000 tons between 2012 and 2013 [8]. These valuable resources have been mainly formed in lignocellulosic structure which consists of lignin, cellulose and hemicellulose [9]. Biomass processing technologies are being developed and some new methods have been proposed in recent years. The routes for hydrogen production from biomass have been presented by [10]. Two major routes are defined for biomass-based hydrogen production. Biochemical conversions include biological photosynthesis, biological water-gas shift (WGS) and biological fermentation. Thermochemical conversions include pyrolysis, gasification and supercritical water gasification (SCWG) [11]. Syngas is the main product of biomass gasification which mainly consists of CO, CO_2, CH_4 and H_2. Despite the low energy required for biological conversion, it takes too much time for completing the conversion and producing hydrogen. Moreover, experiments should be sustained in an equilibrium condition and any change in conditions disturbs the reactions [12]. SCWG is a hydrothermal treatment of biomass in water media with pressures above 220 bar and temperatures over 374 °C. SCW media prepares a unique condition for conversion of biomass in shorter reaction times with more efficiency [13, 14]. Water in this condition has much lower dielectric and ionic constant and becomes a unipolar solvent which can easily hydrolyze biomass [15, 16]. Lignocellulosic structure of biomass is composed of cellulose, hemicellulose and lignin. Cellulose is arranged in bundles while their molecules are interlinked with each other by another molecule named hemicellulose [17]. Lignin holds cellulose and hemicellulose in a network as a binder to form primary cell walls of the plants [18]. Cellulose is made of glucose subunits which have been linked with each other via β-1,4-glycosidic bonds [19]. Hemicellulose is a branched heteropolysaccharide consisting of C_5 sugars (usually xylose and arabinose) and C_6 sugars (galactose, glucose, and mannose) as subunits [20]. As mentioned in Eq. (1), hydrolysis of cellulose and hemicellulose takes place in the SCW media to produce fermentable sugars [9].

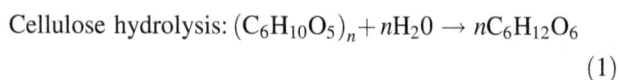

Cellulose hydrolysis: $(C_6H_{10}O_5)_n + nH_2O \rightarrow nC_6H_{12}O_6$

$$(1)$$

These fermentable sugars are also dehydrated into 5-HMF (5-hydroxymethylfurfural) [21]. Under a suitable condition, 5-HMFs are degraded into acids, alcohols, aldehydes and ketones [22, 23]. Eventually, these small molecules can be reformed to gaseous products.

Lignin is a natural polymer with no exact structure consisting of phenyl propane with ester bond links. Lignin content is one of the most influential factors limiting the hydrolysis [24]. Lignin's hydrolysis in SCW, followed by dealkylation, promotes the decomposition of lignin. As mentioned in Eq. (2), phenolic compounds such as syringols, guaiacols are formed to convert into polyphenols further [25].

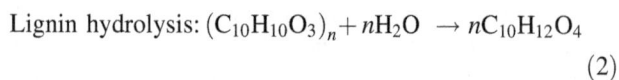

Lignin hydrolysis: $(C_{10}H_{10}O_3)_n + nH_2O \rightarrow nC_{10}H_{12}O_4$

$$(2)$$

Overall, the interaction between lignin, cellulose and hemicellulose in SCWG is not clearly specified. However, the amount of cellulose and hemicellulose and their availability in the structure can affect the conversion of lignocellulosic biomass into gaseous products.

It seems necessary to have a look at the main reactions occur during SCWG process to understand the effect of different parameters on gasification's procedure. It is reported by many researchers that biomass gasification in SCW media is a complex process, but the overall chemical conversion can be represented by the simplified net reaction (3) [15]:

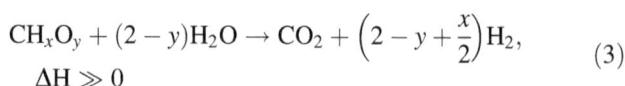

$$CH_xO_y + (2-y)H_2O \rightarrow CO_2 + \left(2 - y + \frac{x}{2}\right)H_2,$$
$$\Delta H \gg 0 \tag{3}$$

where x and y are the elemental molar ratios of H/C and O/C in biomass, respectively. In addition, a group of competing reactions take place to complete the successful gasification which are mentioned in Eqs. (4), (5) and (6) [15].

Steam reforming:

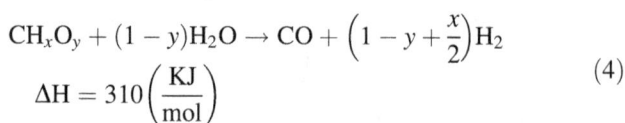

$$CH_xO_y + (1-y)H_2O \rightarrow CO + \left(1 - y + \frac{x}{2}\right)H_2$$
$$\Delta H = 310\left(\frac{KJ}{mol}\right) \tag{4}$$

Water-gas shift:

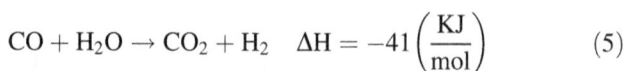

$$CO + H_2O \rightarrow CO_2 + H_2 \quad \Delta H = -41\left(\frac{KJ}{mol}\right) \tag{5}$$

Methanation:

$$CO + 3H_2 \rightarrow CH_4 + H_2O \quad \Delta H = -206\left(\frac{KJ}{mol}\right) \tag{6}$$

In the methanation reaction, 3 mol of produced hydrogen are consumed by Eq. (6) for producing 1 mol of methane. Also, water-gas shift reaction consumes CO and water and produces H_2 and CO_2. Therefore, for hydrogen-selective SCWG process, methanation reaction should be decelerated.

Many experiments have been conducted for SCWG of biomass model compounds [25–27]. However, some research has been done for gasification of agricultural wastes and real biomasses. Useful reviews for SCWG of biomass have been proposed by Kruse, Guo et al. and Tekin et al. [28–30]. Rashidi et al. performed SCWG of bagasse using the same reactor with and without Ni/CNTs catalysts. The effect of bagasse loading and reaction times on hydrogen yield was studied. Hydrogen yield of 3.84 mmol/g was observed for non-catalytic tests at the temperature of 400 °C, the reaction time of 20 min, and bagasse loading of 0.15 g [31]. Safari et al. reported the gasification performances of three different agricultural wastes including walnut shell and almond shell in a base case condition [32]. Madenoglu et al. investigated the subcritical and SCWG of some hard nutshells in the absence and presence of the catalyst. Effect of temperature and catalyst was investigated on gaseous, liquid and solid products. Hydrogen yield was enhanced by increasing the temperature from 400 to 600 °C [33]. Liu et al. reported the product identification and distribution from hydrothermal conversion of walnut shells into liquefied products using KOH and Na_2CO_3 catalysts [34]. However, none of the previous studies performed a holistic analysis of gaseous products of SCWG of walnut shell.

The objective of this study is to investigate the hydrothermal gasification of Iranian walnut shell in SCW media for hydrogen-rich gas production. All of the main important parameters including temperature, feed content and reaction time were studied to observe the variation of gaseous products and to determine the optimum condition for hydrogen yield. Also, the simultaneous effect of reaction time and feed concentration was studied. Gasification efficiencies are also calculated using elemental analysis of walnut shell and the yield of the gaseous products. There is no work in the literature with a holistic analysis of hydrothermal gasification of walnut shell and its gaseous products. Therefore, the present study aims to study the gaseous products of walnut shell holistically and determine the optimum condition for hydrogen production.

Materials and methods

Materials

Walnut shell was provided by agriculture gardens around Sanandaj, located in Kurdistan Province of Iran. It was first washed and then dried in atmospheric condition for 48 h. After that, they were ground and sieved in three steps to obtain the desired particle size (diameter <150 μm). The elemental analysis of the prepared walnut shell was conducted by a CHNS analyzer (Vario EL III by Elementar, Germany) for characterization. As presented in Table 1,

the C and H contents are 55.87 %. The mass fraction of oxygen in biomass is determined by Eq. (7):

$$O\% = 100 - C\% - H\% - N\% - S\% - Ash\%$$

$$(7)$$

Reaction setup and experimental outline

Schematic of the reactor system and experimental setup is indicated in Fig. 1. Stainless steel batch micro-reactor with total volume of 25 mL has been used in this work. 0.06, 0.12 and 0.18 g of feedstock were added to 6 g of deionized water to make three mixtures with different concentrations. Argon as an inert gas stream was used for vacating the reactor and removing air for several minutes. The mixtures were injected into the reactor by a syringe. The reactor was immersed in a molten salt bath containing a mixture of potassium nitrate, sodium nitrate, and sodium nitrite. The molten salt bath temperature was measured using a K-type thermocouple and was controlled by a PID temperature controller in determined temperature. Figure 2 indicates the temporal variation of pressure inside the reactor in the mentioned conditions. The SCW condition was gained after approximately 2 min after the immersion of reactor inside the molten salt bath. For each experiment, after a certain reaction time, the reactor was taken out of the molten salt bath and immersed in a water bath for cooling down to room temperature. The final pressure of reaction was measured using a low-pressure gage after opening the high-pressure valve. The volume of the gaseous product was measured using a gasometer. The gas

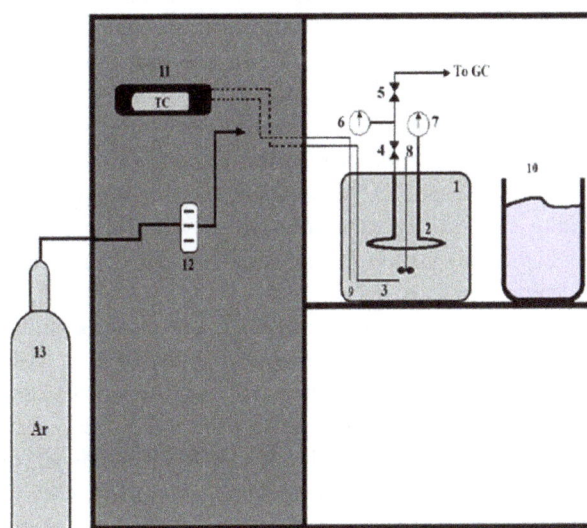

Fig. 1 Schematic of reactor system: *1* molten salt bath, *2* tubular batch reactor, *3* electrical heater, *4* high-pressure valves, *5* low-pressure valve, *6* low-pressure gage, *7* high-pressure gage, *8* mixer, *9* k-type thermocouple, *10* water bath, *11* temperature controller, *12* flow meter and *13* argon gas bottle

Table 1 Elemental analysis of Iranian walnut shell	Element	Mass fraction (%)
	C	50.2
	H	5.67
	N	0.84
	S	0
	O	42.64
	Ash	0.65

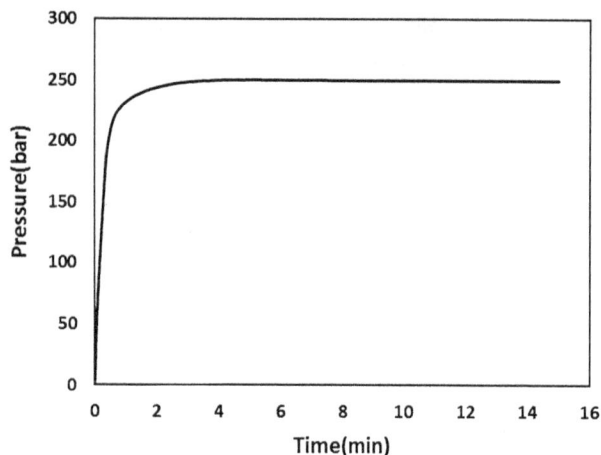

Fig. 2 Temporal variation of reactor's pressure (T: 440 °C, 0.06 g walnut shell, 6 g water)

volume was measured with ±5 % accuracy. Experiments were performed three times under the same experimental conditions and reported data are the averages of three replicates.

Product analysis method

Gas samples were taken using tight syringes and injected into the gas chromatograph's column. Gas chromatograph (Varian 3400 and Teyfgostar-Compact) was equipped with PORAPAK Q-S 80/100 (30 m long, 0.53 mm I.D) column, a methanizer and flame ionization detector (FID). Argon was used as carrier gas and oven temperature program was in the following: 40 °C isothermal for 5 min, increase in temperature from 40 to 75 °C in 17.5 min and isothermal in 75 °C for 5 min. The methanizer option enables the FID to detect levels of CO and CO_2. During analysis, methanizer is heated to 380 °C with the FID detector body. When the column effluent mixes with the FID hydrogen supply and passes through the methanizer, CO and CO_2 are converted to methane. GC was calibrated with standard gas mixture supplied by ROHAM Company in Tehran, Iran. The standard deviation for the results of gas composition was calculated to be ±2 %.

Reported data are the averages of several observations for each experiment for more reproducibility of data. Also, the accuracy of data collection and comparison between means of compositions and yields from the processing of two samples was studied through ANOVA (analysis of variance). The statistical level was 5 % with $p < 0.05$. Gaseous products have been presented in the form of gas yield which is the mmoles of the gaseous products in 1 g of walnut shell. Elemental analysis is used to determine the CGE which is the ratio of the amount of carbon in the produced gasses to the amount of carbon in the initial

walnut shell and HGE which is the ratio of the amount of hydrogen in the gaseous products to the amount of hydrogen in the initial walnut shell. Mathematically, CGE and HGE are defined by Eqs. 8 and 9 [25]:

$$CGE = \frac{(\text{Moles of carbon in gaseous products})}{(\text{Moles of carbon in walnut shell})} \qquad (8)$$

$$HGE = \frac{(\text{Moles of hydrogen in gaseous products})}{(\text{Moles of hydrogen in walnut shell})} \qquad (9)$$

Results and discussion

Results for SCWG of walnut shell will be presented in this section. The effect of operating conditions including temperature, reaction time and biomass loading on the gaseous product yields, CGE and HGE will be presented first, and then the combination of these effects will be analyzed. The total gaseous product presented in the results is the sum of main gaseous components including CO, CO_2, H_2 and CH_4. Table 2 presents the summary of the different conditions and the results of gasification. Temperature varied between 400 and 440 °C while biomass loading changed in the range of 0.06–0.18 g. Moreover, reaction times of 10, 20 and 30 min were considered for study the effect of the residence time of the biomass in the reactor.

Effect of temperature

As shown in Fig. 3, an increase in temperature from 400 °C to reactor's maximum temperature (440 °C), enhances endothermic reforming reaction of walnut shell in SCW and increases the total gas yield. Variations of individual gas yields against temperature are shown in Fig. 4. As seen, increasing temperature leads to increasing of H_2, CO_2, and CO yields while CH_4 decreases slightly. It can be concluded that increase in temperature has a significant effect on SCWG process and enhances total gas yield through promoting of reforming reaction. In addition, due to endothermic nature of overall reaction mentioned in Eq. (3), experiments should be performed at the allowable high temperatures. Thus, the best temperature for the conversion of walnut shell into gas products for our lab instruments is 440 °C.

Effect of reaction time

Reaction time increased from 10 to 30 min for investigating the effect of resident time of biomass in the reactor on gaseous products and gasification efficiencies. Experiments performed at least three times for each reaction time and the averages are reported in Figs. 5 and 6.

Table 2 Summary of experimental conditions and SCWG results

Test#	Temperature (°C)	Reaction time (min)	Feed/water (mass ratio)	Main gas yields (mmole gas/g of walnut shell)				
				CO	CH$_4$	CO$_2$	H$_2$	Total
1	400	20	0.01	1.63	2.46	7.98	2.44	14.51
2	410	20	0.01	2.02	2.21	9.41	3.17	16.81
3	420	20	0.01	2.53	2.14	11.59	3.34	19.6
4	430	20	0.01	2.71	1.83	12.81	4.12	21.47
5	440	20	0.01	2.62	1.62	13.48	4.63	22.35
6	440	10	0.01	3.12	0.76	12.28	3.86	20.02
7	440	30	0.01	2.31	2.77	14.22	3.92	23.22
8	440	10	0.02	2.77	1.48	8.63	2.68	15.56
9	440	20	0.02	1.45	3.13	9.47	2.89	16.94
10	440	30	0.02	1.82	2.28	11.8	3.47	19.37
11	440	10	0.03	1.99	2.85	5.92	1.94	12.7
12	440	20	0.03	1.38	3.32	6.32	2.14	13.16
13	440	30	0.03	1.18	3.99	8.64	1.66	15.47

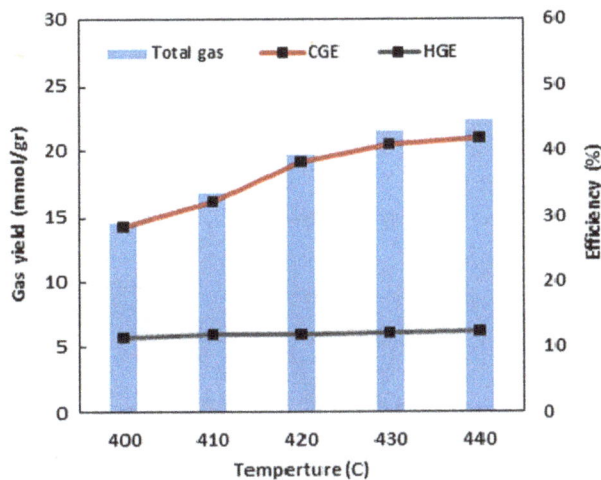

Fig. 3 Effect of temperature on total gas yield and efficiencies of walnut shell gasification (reaction time 20 min, walnut shell loading 0.06 g, water loading 6 g)

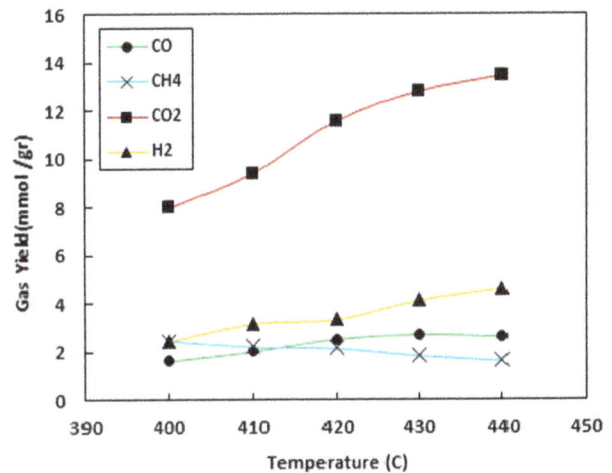

Fig. 4 Effect of temperature on the yield of main gaseous products of walnut shell gasification (reaction time 20 min, walnut shell loading 0.06 g, water loading 6 g)

Figure 5 depicts that hydrogen yield increases by increasing the reaction time, reaches a maximum at the reaction time of 20 min and then starts to decrease. The aim of this study is to optimize hydrogen production in SCWG of walnut shell. Hence, 20 min is the optimum reaction time for maximum yield of hydrogen when reaction time deviates from 10 to 30 min.

Methane yield also increased by a factor of 3.64. Extending the reaction time increases the methane yield from 0.39 to 1.22 (mmol gas/g of walnut shell). As reaction time increased from 10 to 20 min, the methane yield increased, while the total gasification yield increased by a factor of 1.28. Beyond 20 min of reaction time, the total

yield of the product gas was not changed significantly while the composition continued to change. The decrease in hydrogen yield and increase in the methane yield can be associated to the methanation process (Eq. 7). According to Eq. (7), consuming 3 mol H$_2$ generates 1 mol CH$_4$ and consumes 1 mol CO. Meanwhile, downward trend of H$_2$ yield should be sharper than the rising trend of CH$_4$ and CO yields. When the objective of biomass gasification in SCW is hydrogen production, reaction (7) must be restrained and CO reacting with water to form CO$_2$ and H$_2$ (Eq. 6) must be enhanced. Figure 5 also shows that the yield of CO$_2$ increases as reaction time increases from 10 to 30 min. This figure also shows that the CO yield

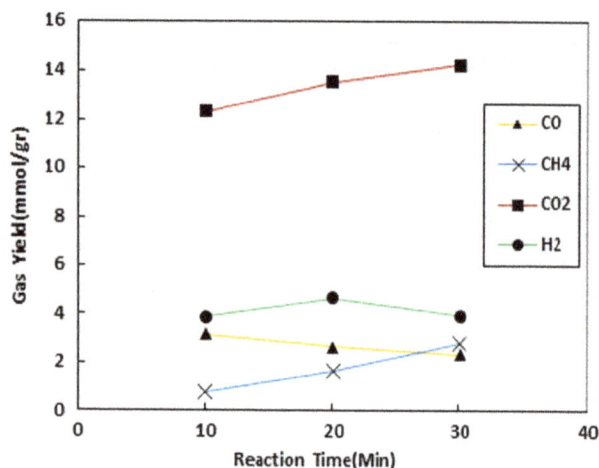

Fig. 5 Temporal variation of main gaseous products of walnut shell (*T*: 440 °C, walnut shell loading 0.06 g, water loading 6 g)

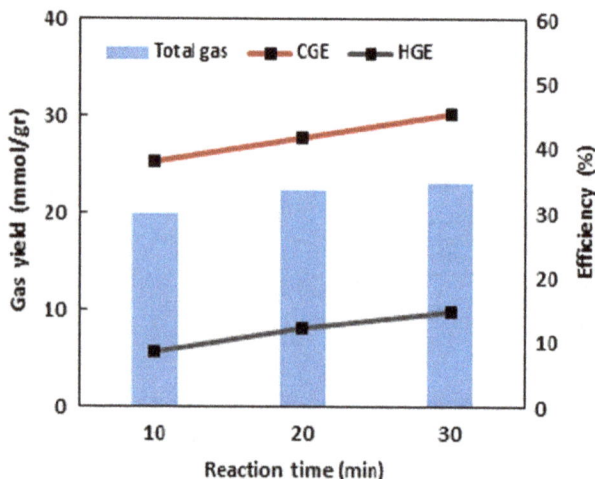

Fig. 7 Effect of biomass loading on total gas yield, CGE, and HGE of walnut shell gasification (*T*: 440 °C, reaction time 20 min, water loading 6 g)

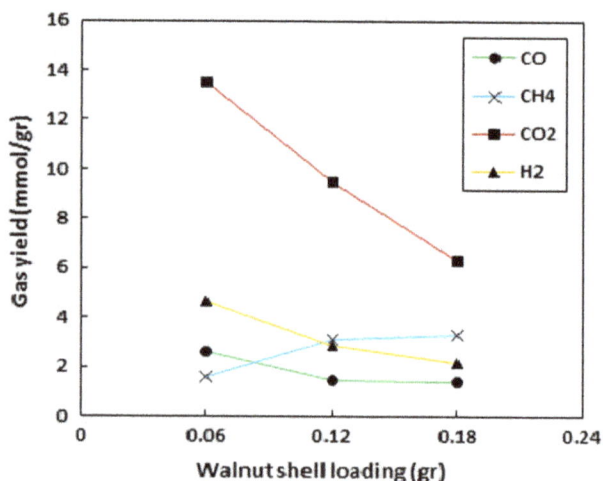

Fig. 6 Effect of reaction time on total gas yield and efficiencies of walnut shell gasification (*T*: 440 °C, walnut shell loading 0.06 g, water loading 6 g)

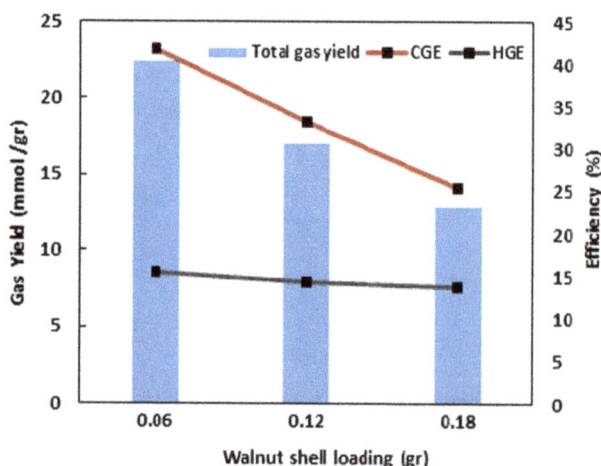

Fig. 8 Effect of feed loading on total gas yield, CGE and HGE of SCWG of walnut shell (*T*: 440 °C, reaction time 20 min, water loading 6 g)

decreases with time slightly which is due to the reaction of CO with water to form CO_2 and H_2 by increasing time. As shown in Fig. 6, the amount of total generated gas increased from 20.02 to 23.22 (mmol/g of walnut shell) when reaction time increased from 10 to 30 min.

Effect of walnut shell loading

Figure 7 shows the total gas yield as a function of walnut shell loading. The walnut shell loading changed from 0.06 to 0.18 g while the amount of water loading was fixed at 6 g. As walnut shell loading increased, H_2 and CO_2 yields decreased by factors of 2.17 and 2.13, respectively. In addition, CO yield decreased from 1.622 to 1.084 (mmol/g of walnut shell). The increment in the ratio of walnut shell to water decelerates the steam reforming reaction on

Eq. (5), which is followed by a decrease in walnut shell conversion and total gas yield. However, there is an increase in the methane yield which is due to the consumption of CO and H_2 to produce methane (Eq. 7) at lower water/walnut shell ratios.

As shown in Fig. 8, the total gas yield decreased dramatically by increasing the concentration of walnut shell. Increasing walnut shell concentration from 0.06 to 0.18 g decreased the total gas yield from 22.36 to 12.86 (mmol gas/g of walnut shell). Moreover, CGE and HGE decreased when walnut shell loading increased. CGE decreased sharply from 41.7 to 25.2 %, while HGE decreased slightly from 15.38 to 13.67 %.

Figure 9 indicates the simultaneous effect of reaction time and feed loading on hydrogen yield. The highest hydrogen yield was observed at the lowest walnut shell

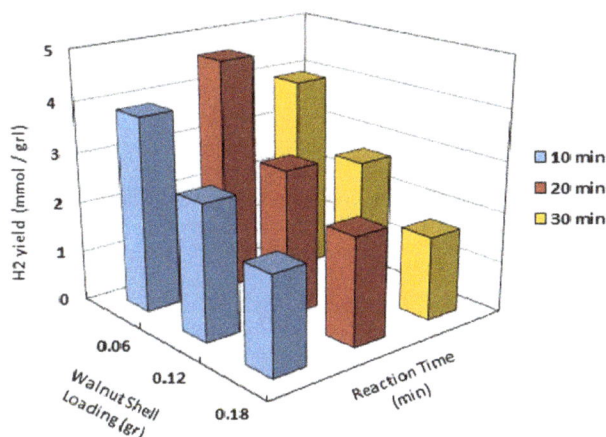

Fig. 9 Comparison of hydrogen yields at various reaction times and walnut shell loadings (T 440 °C, water loading 6 g)

loading of 0.06 g and at the optimum reaction time of 20 min. It can also be perceived that the effect of feed loading is much more significant than the effect of reaction time on hydrogen yield.

Compared to the previous study of SCWG of walnut shell, this study resulted in the H_2 yield of 2.44 mmol/g in 400 °C, which was slightly higher than that of Madenoglu et al. Also, the methane yield was nearly the same. There is no previous study on SCWG of Iranian walnut. Hence, this research presents a sustainable way for utilization of this bio-renewable agricultural waste.

Conclusion

The results of this study promise a sustainable process for making an alternative fuel from walnut shell which is an agricultural waste. Gasification of walnut shell in SCW media was performed under 13 different conditions to observe the effect of parameters including temperature, feed content and reaction time on the yields of main gaseous products considering the main reactions of the process. Promising results were obtained using holistic and comprehensive analysis of gaseous products and affecting parameters, for waste management sector and energy industry of Iran which has not been studied before. Ranges of variation of parameters were determined, considering the capacity of the reactor system and the obtained results from the literature. The temperature had a significant effect on the total gas and hydrogen yield so that as temperature increased, hydrogen yield increased while increasing the temperature may cause intensive energy consumption. So, the appropriate range of temperature was considered between 400 and 440 °C and the maximum hydrogen yield occurred in 440 °C. The effects of reaction time and walnut shell loading were investigated at this temperature. The

maximum hydrogen yield of 4.63 (mmol gas/g of walnut shell) was observed at the reaction time of 20 min, walnut shell loading of 0.06 g and temperature of 440 °C.

Acknowledgments The authors would like to thank the Iran Renewable Energy Organization (SUNA) for their kind support of this research.

References

1. Ahmadi P, Dincer I, Rosen MA (2012) Exergo-environmental analysis of an integrated organic Rankine cycle for trigeneration. Energy Convers Manag 64:447–453
2. Chiari L, Zecca A (2011) Constraints of fossil fuels depletion on global warming projections. Energy Policy 39:5026–5034
3. Ballarin A, Vecchiato D, Tempesta T, Marangon F, Troiano S (2011) Biomass energy production in agriculture: a weighted goal programming analysis. Energy Policy 39:1123–1131
4. Tavasoli A, Barati M, Karimi A (2015) Conversion of sugarcane bagasse to gaseous and liquid fuels in near-critical water media using K_2O promoted Cu/γ–Al_2O_3–MgO nanocatalysts. Biomass Bioenerg 80:63–72
5. Dincer I, Zamfirescu C (2012) Sustainable hydrogen production options and the role of IAHE. Int J Hydrog Energy 37:16266–16286
6. Dincer I (2012) Green methods for hydrogen production. Int J Hydrog Energy 37:1954–1971
7. Midilli A, Dincer I (2008) Hydrogen as a renewable and sustainable solution in reducing global fossil fuel consumption. Int J Hydrog Energy 33:4209–4222
8. Faostat annual report 2013. http://faostat3.fao.org/. Accessed Oct 2015
9. Pavlovi I, Knez Z, Skerget M (2013) Hydrothermal reactions of agricultural and food processing wastes in sub- and supercritical water: a review of fundamentals, mechanisms, and state of research. J Agric Food Chem 61:8003–8025
10. Hepbasli A, Kalinci Y, Dincer I (2009) Biomass-based hydrogen production: a review and analysis. Int J Hydrog Energy 34:8799–8817
11. Saxena RC, Seal D, Kumar S, Goyal HB (2008) Thermo-chemical routes for hydrogen rich gas from biomass: a review. Renew Sust Energ Rev 12:1909–1927
12. Das D, Veziroglu TN (2001) Hydrogen production by biological processes: a survey of literature. Int J Hydrog Energy 26:13–28
13. Basu P, Mettanant V (2009) Biomass gasification in supercritical water—a review. Int J Chem React Eng 7:1542–6580
14. JaranaMB Garcia, Sanchez-Oneto J, Portela JR, Nebot Sanz E, de la OssaEJ Martinez (2008) Supercritical water gasification of industrial organic wastes. J Supercrit Fluid 46:329–334
15. Barati M, Tavasoli A, Babatabar M, Dalai AK, Das U (2014) Hydrogen production via supercritical water gasification of bagasse using unpromoted and zinc promoted Ru/γ-Al2O3 nanocatalysts. Fuel Proc Tech 123:140–148
16. Calzavara Y, Dubien CJ, Boissonnet G, Sarrade S (2005) Evaluation of biomass gasification in supercritical water process for hydrogen production. Energ Convers Manag 46:615–631
17. Huber GW, Dumesic JA (2006) An overview of aqueous-phase catalytic processes for production of hydrogen and alkanes in a biorefinery. Catal Today 111:119–132
18. Waldner MH, Vogel F (2005) Renewable production of methane from woody biomass by catalytic hydrothermal gasification. Ind Eng Chem Res 44:4543–4551
19. Kobayashi H, Fukuoka A (2013) Synthesis and utilisation of sugar compounds derived from lignocellulosic biomass. Green Chem 15:1740–1763

20. Murphy JD, McCarthy K (2005) Ethanol production from energy crops and wastes for use as a transport fuel in Ireland. Appl Energy 82:148–166

21. Mohan D, Pittman CU, Steele PH (2006) Pyrolysis of wood/biomass for bio-oil: a critical review. Energy Fuel 20:848–889

22. Minowa T, Fang Z (1998) Hydrogen production from cellulose in hot compressed water using reduced nickel catalyst: product distribution at different reaction temperatures. J Chem Eng Jpn 31:488–491

23. Reddy SN, Nanda S, Dalai AK, Kozinski JA (2014) Supercritical water gasification of biomass for hydrogen production. Int J Hydrog Energy 39:6912–6926

24. Salimi M, Safari F, Tavasoli A, Shakeri A (2016) Hydrothermal gasification of different agricultural wastes in supercritical water media for hydrogen production: a comparative study. Int J Ind Chem 7:277–285

25. Resende FLP, Fraley SA, Berger MJ, Savage PE (2008) Non-catalytic gasification of lignin in supercritical water. Energy Fuel 22:1328–1334

26. Azadi P, Khan S, Stroble F, Azadi F, Farnood R (2012) Hydrogen production from cellulose, lignin, bark and model carbohydrates in supercritical water using nickel and ruthenium catalysts. Appl Catal B Environ 117–118:330–338

27. Susanti RF, Dianningrum LW, Yum T, Kim Y, Gwon B, Kim J (2012) High-yield hydrogen production from glucose by super-critical water gasification without added catalyst. Int J Hydrog Energy 37:11677–11690

28. Kruse A (2009) Hydrothermal biomass gasification. J Supercritic Fluid 47:391–399

29. Guo Y, Wang SZ, Xu DH, Gong YM, Ma HH, Tang XY (2010) Review of catalytic supercritical water gasification for hydrogen production from biomass. Renew Sust Energ Rev 14:334–343

30. Tekin et al (2014) A review of hydrothermal biomass processing. Renew Sust Energ Rev 40:673–687

31. Rashidi M, Tavasoli A (2015) Hydrogen rich gas production via supercritical water gasification of sugarcane bagasse using unpromoted and copper promoted Ni/CNT nanocatalysts. J Supercrit Fluid 98:111–118

32. Safari F, Tavasoli A, Ataei A, Choi JK (2015) Hydrogen and syngas production from gasification of lignocellulosic biomass in supercritical water media. Int J Recycl Org Waste Agric 4:121–125

33. Madenoglu TG, Yildirir E, Saglam M, Yuksel M, Ballice L (2014) Improvement in hydrogen production from hard-shell nut residues by catalytic hydrothermal gasification. J Supercrit Fluid 67:22–28

34. Liu WJ, Jiang H, Yu HQ (2015) Thermochemical conversion of lignin for functional materials: A review and future direction. Green Chem. doi:10.1039/C5GC01054C (accepted manuscript)

9

Multi-site phase transfer catalyzed radical polymerization of methyl methacrylate in mixed aqueous–organic medium: a kinetic study

Vajjiravel Murugesan[1] · Elumalai Marimuthu[1] · K. S. Yoganand[2] ·
M. J. Umapathy[2]

Abstract This work establishes the kinetics of radical polymerization of methyl methacrylate in an aqueous–organic two-phase system using 1,4-bis (triethylmethylammonium) benzene dichloride (TEMABDC) as multi-site phase transfer catalyst and potassium peroxydisulphate ($K_2S_2O_8$) as water-soluble initiator at 60 ± 1 °C under nitrogen atmosphere. The role of concentrations of monomer, initiator, catalyst, acid and ionic strength, temperature and volume fraction of aqueous phase on the rate of polymerization (R_p) was investigated. The rate of polymerization (R_p); $R_p \propto$ [MMA]$^{0.64}$, [TEMABDC]$^{1.24}$ and [$K_2S_2O_8$]$^{1.50}$. The rate of polymerization increases with an increase in the concentration of monomer, initiator, catalyst and temperature. A generalized reaction model was developed to explain the phase transfer catalyzed polymerization reaction. Based on the kinetic results, radical mechanism has been derived. The activation energy and other thermodynamic parameters were calculated. The FT-IR spectroscopy validates a band of 1732 cm^{-1} of ester group of the obtained polymer. The viscosity average molecular weight of the PMMA was found 1.6955×10^4 g/mol.

Keywords Kinetics · Multi-site phase transfer catalyst · Radical polymerization · Rate of polymerization · Aqueous–organic media

Introduction

Phase transfer catalysis (PTC) is presently a well mature and established technique to accelerate the reactions between mutually insoluble two or more reactants located in different phases. In this technique, the two mutually insoluble reactants, one being an organic liquid or substrate dissolved in an organic solvent and other being an organic or inorganic salt from a solid or aqueous phase, react with the help of a phase transfer catalyst. It has been applied over 600 processes in variety of industries such as intermediates, dyestuffs, agrochemicals, perfumes, flavors, pharmaceuticals and polymers and value exceeds twelve billion (US$) per year [1–4]. In polymer chemistry, they have been employed in synthesis of polymers [5–7], condensation polymerization [8], anionic polymerization [9, 10] and free radical polymerization [11–18].

In order to get the maximum desired product in a short duration of reaction period, the catalyst should be more efficient; with the aim of these requirements, novel "multi-site phase transfer catalysts" (multi-site PTC) have been developed which contain more than one catalytic active site per molecule. The concept of multi-sited phase transfer catalyst was introduced by Idoux et al. in which they have synthesized phosphonium and quaternary onium ions containing more than one active site per molecule [19]. The benefits of multi-site PTC are: enhance the rate of reaction with less time consumption and it transfers more number of active species from aqueous phase to organic phase during the reactions in contrast with single site—PTC. The reports on multi-site phase transfer catalyst aided radical polymerization of different alkyl methacrylates were gradually blooming in recent years [20–26]. The acrylic esters especially methyl methacrylate (MMA) are commercially fascinating and significant functional monomer for the

✉ Vajjiravel Murugesan
chemvel@rediffmail.com

[1] Department of Chemistry, B S Abdur Rahman Crescent University, Vandalur, Chennai 600 048, India

[2] Department of Chemistry, College of Engineering, Anna University, Chennai 600 025, India

synthesis of acrylic resins and various polymers based on poly(methyl methacrylate) (PMMA) with tunable properties. PMMA has good mechanical strength, acceptable chemical resistance and extremely good weather resistance. Further, it has favorable processing properties, good thermoforming and can be modified with pigments, flame retardant and UV absorbent additives [27, 28]. PMMA has vast profound and diverse applications that influence our lives every day. Radical polymerization is one of the best processes for the synthesis of polymers and the few important merits of radical polymerization are: it can be applied to all vinyl monomers under mild reaction condition with a wide range of temperature, it is water tolerant and its cost is relatively low. A curiosity on free radical polymerization has been stimulated to a great extent by the impressive progress made in several methods such as atom transfer radical polymerization (ATRP), nitroxyl radical-mediated polymerization (NMP), and reversible addition fragmentation transfer polymerization (RAFT). These methods and approaches were successfully introduced into polymerization process by different research groups [29–33]. Polymerization of MMA was effectively performed in ATRP [34, 35], NMP [36] and RAFT [37, 38]. The growth of a new kinetic model for the polymerization of methyl methacrylate (MMA) using novel catalyst and different methods at moderate temperature will be one of the major progresses in an industrial perspective.

The design, synthesis of novel catalysts and its applications in polymerization and organic reactions are a vital focus in the current research. Inspired by inherent characteristics of PTC technique and considering merits of water-soluble initiator, the present work endeavours to conduct a systematic investigation and explore the kinetics of free radical polymerization of methyl methacrylate (MMA) using potassium peroxydisulphate (PDS) as water-soluble initiator in the presence of synthesized multi-site phase transfer catalyst in cyclohexane/water two-phase system at 60 ± 1 °C. The role of various reaction variables on the rate of polymerization was studied, including the concentration of monomer, initiator, catalyst and temperature, aqueous phase variation. An extraction reaction model was proposed to explain the polymerization pathways and its significance was discussed.

Experimental

Chemicals and solvents

Methyl methacrylate (MMA, Sigma Aldrich, India) was first washed with 5% of aqueous sodium hydroxide to remove the inhibitor and washed with water to remove the

alkali and then dried over anhydrous calcium chloride at last distilled under reduced pressure. The middle fraction of the distillate was collected and stored in dark brown bottle at 5 °C in the refrigerator. The initiator, potassium peroxydisulphate ($K_2S_2O_8$, Merck, India), was purified twice by recrystallization in cold water. The solvents, cylclohexanone, cyclohexane, ethyl acetate, benzene and methanol (Avra, Merck, SRL, India) were used as received. The double distilled water was used to make an aqueous phase. The 1,4-bis (triethylmethylammonium) benzene dichloride (TEMABDC) was synthesized by adopting the reported procedure [39].

Synthesis of multi-sited phase transfer catalyst (TEMABDC)

Measured quantity of α-α'-dichloro-p-xylene (0.01 mol) was introduced into a 150 mL flask. Triethylamine (0.01 mol) in excess amount dissolving in ethanol (30 mL) was then introduced in the flask for the reaction with α-α'-dichloro-p-xylene under agitation speed 800 rpm at 60 °C for 24 h. Organic solvent ethanol and triethylamine were stripped in a vacuum evaporator. White precipitates of 1,4-bis (triethylmethylammonium) benzene dichloride (TEMABDC) were obtained. A white solid crystal of the product is obtained by recrystallizing the product in an ethanol solvent [39] (Scheme 1).

Characterization of multi-site PTC (TEMABDC): ^{1}HNMR analysis

^{1}HNMR spectra of 1,4-bis(triethylmethylammonium) benzene dichloride (TEMABDC) were recorded with BRUKER 400 MHz spectrometer using d-DMSO as a solvent and tetramethylsilance (TMS) as an internal reference. ^{1}HNMR: δ(400 MHz, d-DMSO): Benzyl, methylene and methyl protons are giving signals at 4.50, 3.0–3.50 and 1.30 ppm, respectively. The aromatic protons are well positioned at 7.6 ppm. An integrated total number of protons was good consistent with theoretical total protons of the catalyst (Fig. 1).

Polymerization of MMA

A polymerization experiment was carried out in annular glass ampoules with dimensions of 30 and 26 mm for outer and inner diameter, respectively, and 120 mm height. These ampoules have a surface area/volume ratio large enough for the heat transfer necessary to maintain the isothermal conditions during the polymerization. The total concentrations of 2.0 mol dm^{-3} of monomer (MMA) in the range of 4.5–9.5, 0.02 mol dm^{-3} of multi-site PTC (TEMABDC) from 1.5 to 2.5 mol dm^{-3} and the potassium

Scheme 1 Synthesis of multi-site phase transfer catalyst (TEMABDC)

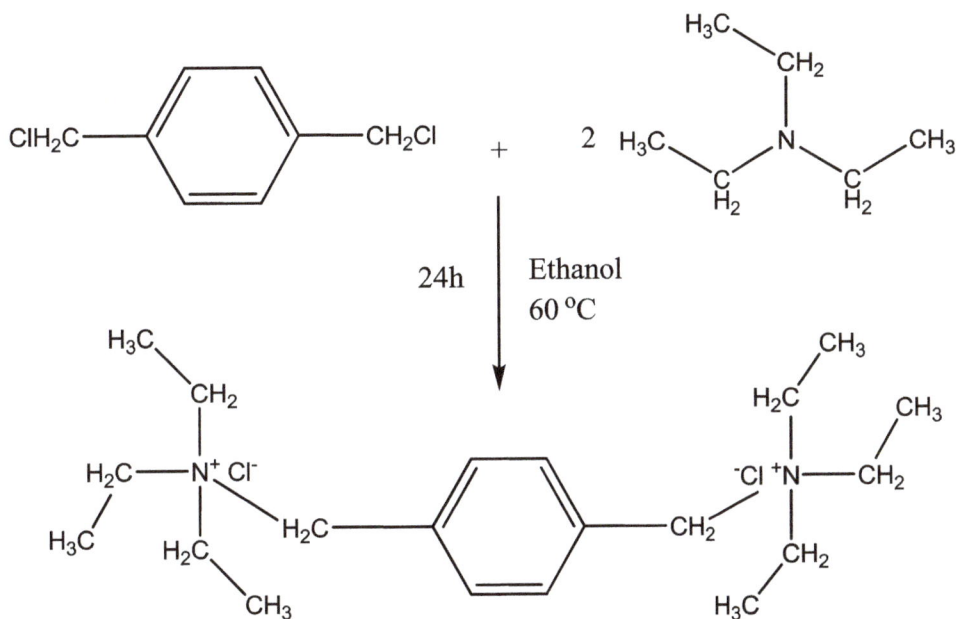

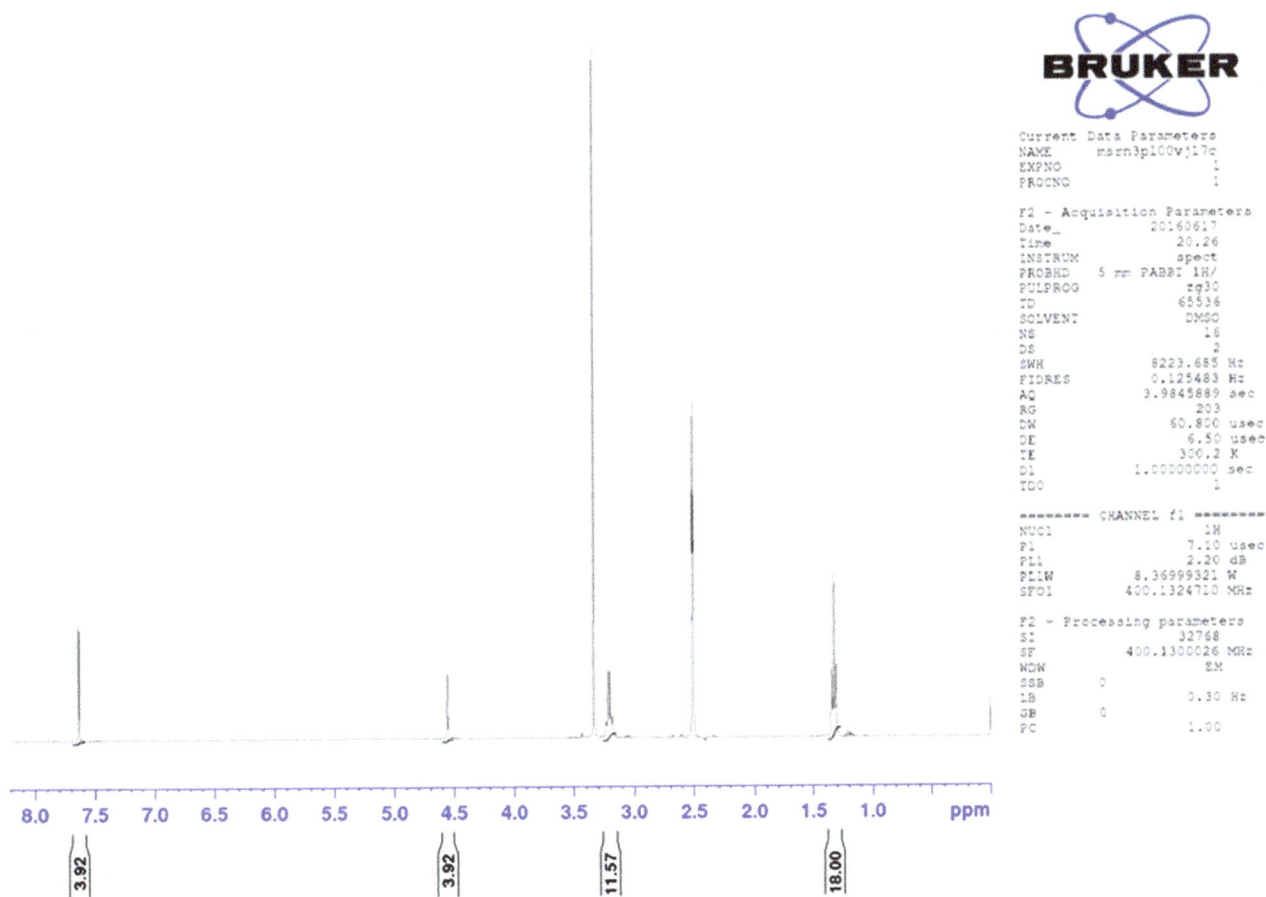

Fig. 1 ^{1}HNMR analysis of 1,4-bis (triethylmethylammonium) benzene dichloride (TEMABDC)

peroxydisulphate (PDS) initiator varied from 1.5 to 2.5 mol dm^{-3} was used in the polymerization reaction. The ratio of monomer and catalyst was 1:0.01. The polymerization ampoule consists of equal volumes of aqueous and organic phase (10 mL each). The monomer (MMA) in cyclohexane was the organic phase and the catalyst, sodium bisulfate (for adjusting the ionic strength [µ]) and sulfuric acid (maintaining the [H^{+}]) were in the aqueous phase. The ampoule was degassed using nitrogen gas continuously about 15 min after which it was sealed. Polymerizations were performed by placing the ampoules in a constant water bath at 60 ± 1 °C and the ampoules were removed from the water bath after a recorded time interval. The polymer was precipitated into large volume of ice cold methanol, filtered and dried at high vacuum until a constant weight was reached. The rate of polymerization (R_p) was calculated from the gravimetric determination of the polymer formed in a given time of polymerization. The R_p was calculated from the weight of polymer obtained using the formula:

$$R_p = 1000W/V \times t \times M,$$

where W is the weight of the polymer in gram; V is the volume of the reaction mixture in mL; t is the reaction time in seconds; M is the molecular weight of the monomer in g/mol. The kinetic experiment was carried out by changing the concentration of monomer, initiator, catalyst, temperature, etc., by adopting above stated polymerization procedure (Scheme 2). The average yield of polymer 65–70% was obtained in polymerization reaction on tuning of different reaction parameters.

Instruments

The FT-IR spectrum of poly (methyl methacrylate) was recorded on an FT-IR spectrometer (Perkin Elmer RX I) in the spectral region from 3500 to 500 cm^{-1}. A pellet of polymer sample was made with KBr on recording of spectrum. The viscosity average molecular weight (M_v) of the polymer was determined in benzene at 30 ± 1 °C with an Ubbelohde viscometer using Mark–Houwink equation [40].

Reaction model

The rate of reaction involving two immiscible reactants is low due to less molecule collisions between them. To enhance the reaction rate, the common way to solve this difficulty is to carry out the reaction at extreme conditions or in a co-solvent. However, these efforts are generally

limited because a few side reactions occur at the extreme conditions or a desired co-solvent is not existent at all. Generally, this kind of situation in two-phase system, the rate of reaction dramatically enhanced with the help of phase transfer catalyst (PTC). PTC is capable of transferring the reactants of aqueous phase into organic phase, where the reaction will take place. In this aqueous–organic two-phase system, the reaction of QX (phase transfer catalyst) and KY (initiator) in the aqueous phase produces QY at the interface between the two phases where it was decomposed and produced the radical ions which initiate the polymerization reactions at 60 ± 1 °C. The simple representation of this process is shown in Scheme 3.

Results and discussion

The radical polymerization of methyl methacrylate initiated by multi-site PTC-$K_2S_2O_8$ in cyclohexane-water two-phase systems was examined at 60 ± 1 °C with changing various reaction parameters, which influence the rate of polymerization.

Steady state rate of polymerization

The steady-state rate of polymerization for the methyl methacrylate was studied first by carrying out the experiments at regular intervals of time with fixed concentrations of all other parameters. The rate of polymerization (R_p) increases nicely to some extent, slightly decreases thereafter and reaches constant value [21]. The plot of R_p versus time shows that the steady-state rate of polymerization of MMA was obtained at 40 min. Hence, the polymerization reaction time was fixed at 40 min to carry out the experiments with changing various reaction parameters (Fig. 2).

Role of [MMA] on the rate of polymerization (R_p)

The role of [MMA] on the rate of polymerization (R_p) was studied by changing the concentrations in the range of 4.5–9.5 mol dm^{-3} by keeping the concentrations of

Scheme 2 Polymerization of methyl methacrylate (MMA) in two-phase system

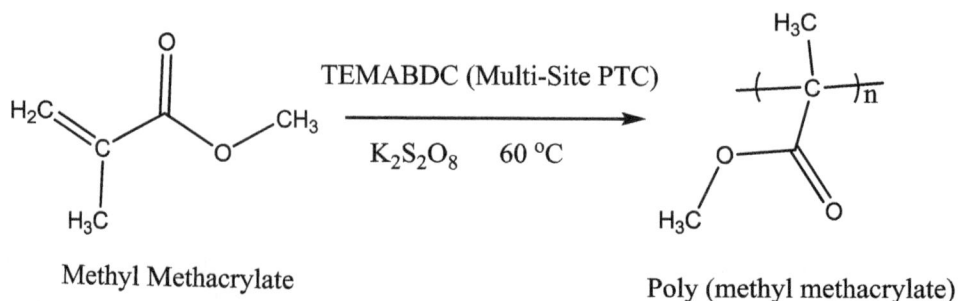

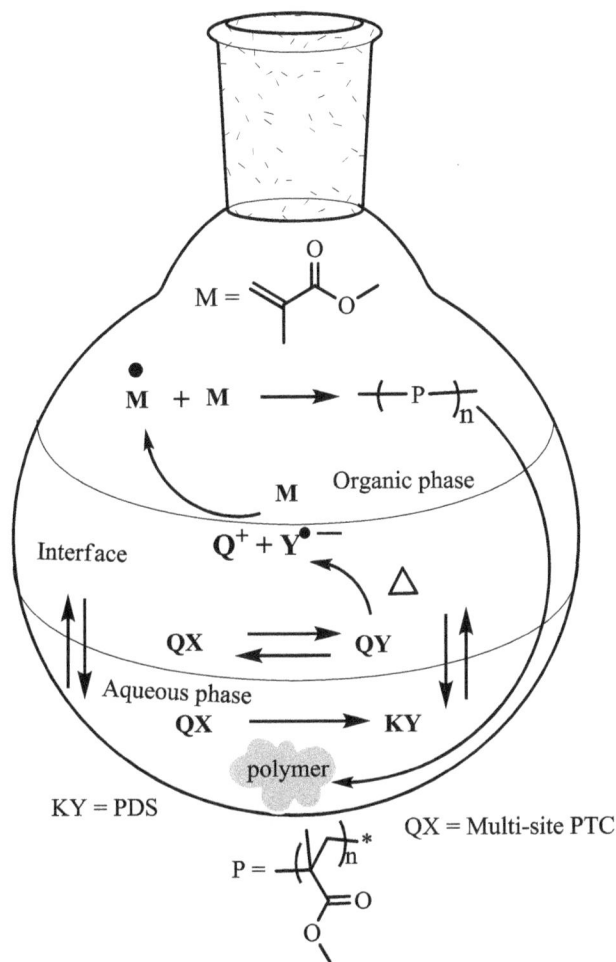

Scheme 3 Reaction model for polymerization of MMA in an aqueous–organic media

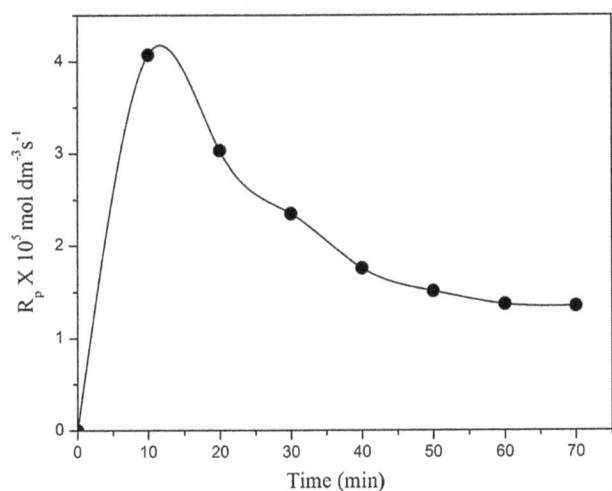

Fig. 2 Steady-state rate of polymerization. Reaction condition: [MMA]: 2.0 mol dm^{-3}; [K$_2$S$_2$O$_8$]: 2.0 × 10^{-2} mol dm^{-3}; [TEMABDC]: 2.0×10^{-2} mol dm^{-3}; [H$^+$]: 0.50 mol dm^{-3}; [μ]: 0.20 mol dm^{-3}; temperature: 60 ± 1 °C

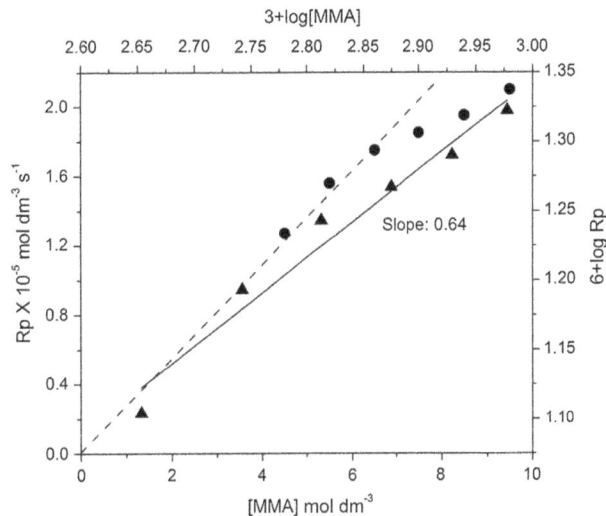

Fig. 3 Role of [MMA] on the R_p. Reaction condition: [K$_2$S$_2$O$_8$]: 2.0 × 10^{-2} mol dm^{-3}; [TEMABDC]: 2.0 × 10^{-2} mol dm^{-3}; [H$^+$]: 0.50 mol dm^{-3}; [μ]: 0.20 mol dm^{-3}; Temperature: 60 ± 1°C; Time: 40 min

potassium peroxydisulphate (initiator), multi-site phase transfer catalyst, ionic strength and pH constant. The R_p increases with increase in the concentration of the monomer. The reaction orders with respect to monomer concentration were determined from the slope of $6 + \log R_p$ versus $3 + \log$ [MMA] and the reaction order with respect to the monomer concentration was found to be 0.65. The plot of R_p versus [MMA] passing through the origin confirms the above observations with respect to [MMA] (Fig. 3). The half-order with respect to concentration of monomer has been reported for the polymerization of n-butyl methacrylate and ethyl methacrylate with other multi-site PTC using potassium peroxydisulphate as initiator [23–25].

Role of [K$_2$S$_2$O$_8$] on the rate of polymerization (R_p)

The role of [K$_2$S$_2$O$_8$] on the rate of polymerization was studied by varying its concentration in the range of 1.5–2.5 mol dm^{-3} at fixed concentrations of other parameters. The R_p increases with increasing concentration of initiator for MMA system. The initiator order value was calculated from the plot of $6 + \log R_p$ versus $3 + \log$ [K$_2$S$_2$O$_8$] the slope was found to be 1.50. As expected a plot of R_p versus [K$_2$S$_2$O$_8$] is linear passing through the origin supporting the above deduction (Fig. 4). Generally, the rate of polymerization is proportional to the square root of initiator concentration at a condition that the termination is bimolecular. In case, if the termination takes place by combination of primary radicals, the initiator order is expected to deviate from half order. However, in this study,

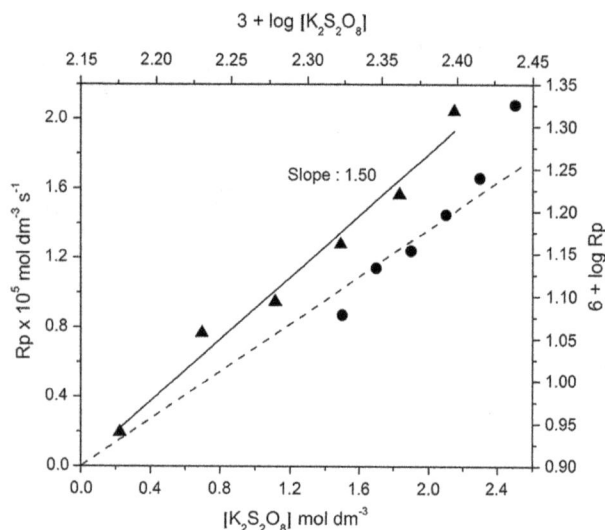

Fig. 4 Role of [K$_2$S$_2$O$_8$] on the R_p. Reaction condition: [MMA]: 2.0 mol dm^{-3}; [TEMABDC]: 2.0×10^{-2} mol dm^{-3}; [H$^+$]: 0.50 mol dm^{-3} [μ]: 0.20 mol dm^{-3}; temperature: 60 ± 1 °C; time: 40 min

Fig. 5 Role of [TEMABDC] on the R_p. Reaction condition: [MMA]: 2.0 mol dm^{-3}; [K$_2$S$_2$O$_8$]: 2.0×10^{-2} mol dm^{-3}; [H$^+$]: 0.50 mol dm^{-3}; [μ]: 0.20 mol dm^{-3}; Temperature: 60 ± 1 °C; Time: 40 min

the initiator order was found to be greater than half and it may be attributed to gel effect or diffusion-controlled termination constant [41].

Role of [TEMABDC] on the rate of polymerization (R_p)

The role of concentration of 1,4-bis (triethylmethylammonium) benzene dichloride (TEMABDC) on the rate of polymerization was studied by varying its concentration in the range from 1.5 to 2.5 mol dm^{-3} at fixed concentrations of other parameters. From the slope of linear plot obtained by plotting of 6 + log R_p versus 3 + log [TEMABDC], the order with respect to [TEMABDC] was found to be 1.24. The observed order was confirmed from the straight line passing through the origin in a plot of R_p versus [TEMABDC] (Fig. 5). An increase in the rate of polymerization with an increase in the concentration of catalyst may attribute to the number of active site (multi-site) of catalyst, thus giving an opportunity to collision between initiator and catalyst. Thereby, more reactive intermediates enhance the rate of polymerization [42].

Blank experiment: polymerization of methyl methacrylate was carried out by adopting mentioned polymerization procedure without adding catalyst at 60 ± 1°C for 40 min. The changes in the appearance of two-phase media were observed (slight turbid), but while pouring the reaction mixture into methanol it was disappeared. This observation was the proof for role of catalyst on the polymerization reaction. The polymerization did not occur in the absence of catalyst even after several mints.

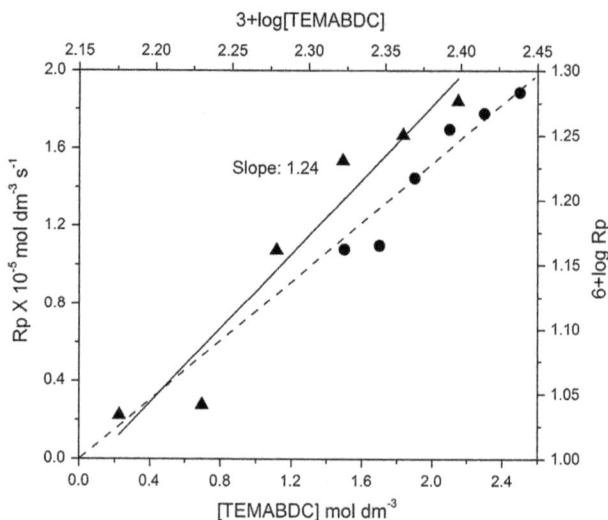

Role of temperature on the R_p

The role of variation of temperature 50–65 °C on the rate of polymerization was investigated by keeping other parameters constant. The rate of polymerization increases with an increase in temperature. This may be due to the fact that when the temperature is gradually raising the rate of initiator decomposition was also increased drastically and thus yields more radicals which are responsible to accelerate the polymerization process promptly thereby the rate of polymerization was increased significantly. The overall activation energy of polymerization (E_a) was found to be 18.51 k J/mol (Table 1; Fig. 6). The higher Ea value of 66.36 kJ/mol was reported for the phase transfer catalyzed radical polymerization of n-butyl acrylate in two-phase systems. From the E_a value, we believe that multi-site PTC polymerization occurs promptly than reported [43]. The thermodynamic parameters such as entropy of activation ($\Delta S^{\#}$), enthalpy of activation ($\Delta H^{\#}$) and free energy of activation ($\Delta G^{\#}$) have been calculated and presented in Table 2.

Role of acid [H$^+$] on R_p

The role of different concentrations of acid in the range 0.16–0.24 mol dm^{-3} on the rate of polymerization reaction was examined at fixed concentrations MMA, PDS, TEMABDC and at constant ionic strength. Rp was found to be almost independent of variation of acid strength in the range employed in this experiment (Table 3). A similar kind of observation was reported on phase transfer catalyzed polymerization reactions [20–23].

Table 1 Role of temperature on the rate of polymerization

Temperature, K	$R_p \times 10^5$ mol dm^{-3} s^{-1}	$1/T \times 10^{-3}$ K^{-1}	$6 + \log [R_p]$
323	1.2970	3.0959	1.1129
328	1.3801	3.0487	1.1399
333	1.4470	3.0030	1.1936
338	1.7460	2.9585	1.2420

Reaction condition: [MMA]: 2.0 mol dm^{-3}; [K$_2$S$_2$O$_8$]: 2.0 × 10^{-2} mol dm^{-3}; [TEMABDC]: 2.0 × 10^{-2} mol dm^{-3}; [H$^+$]: 0.50 mol dm^{-3}; [μ]: 0.20 mol dm^{-3}

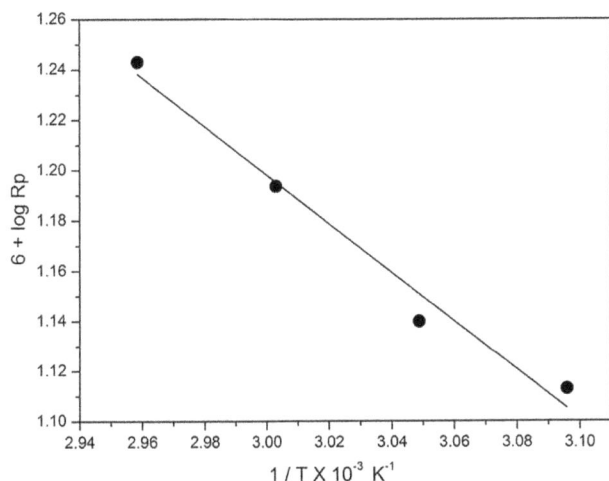

Fig. 6 Role of temperature on the R_p. Reaction condition: [MMA]: 2.0 mol dm^{-3}; [K$_2$S$_2$O$_8$]: 2.0×10^{-2} mol dm^{-3}; [TEMABDC]: 2.0 × 10^{-2} mol dm^{-3} [H$^+$]: 0.50 mol dm^{-3}; [μ]: 0.20 mol dm^{-3}; time: 40 min

Table 2 Thermodynamic parameters

E_a k J/mol	$\Delta G^{\#}$ k J/mol	$\Delta H^{\#}$ k J/mol	$\Delta S^{\#}$ eu
18.51	53.29	14.33	−117.00

Table 3 Role of [H$^+$] on the rate of polymerization

[H$^+$] mol dm^{-3}	$R_p \times 10^5$ mol dm^{-3} s^{-1}
0.16	1.3590
0.18	1.3731
0.20	1.3940
0.22	1.4238
0.24	1.4312

Reaction condition: [MMA]: 2.0 mol dm^{-3}; [K$_2$S$_2$O$_8$]: 2.0×10^{-2} mol dm^{-3}; [TEMABDC]: 2.0×10^{-2} mol dm^{-3}; [μ]: 0.60 mol dm^{-3}; Temperature: 60 ± 1 °C

Role of ionic strength (μ) on the R_p

To find out the role of ionic strength on the R_p, the ionic strength of the reaction medium was varied from 0.50 to 0.70 mol dm^{-3} by keeping other reaction parameters are

Table 4 Role of [μ] on the rate of polymerization

[μ] mol dm^{-3}	$R_p \times 10^5$ mol dm^{-3} s^{-1}
0.50	1.6023
0.55	1.5826
0.60	1.5982
0.65	1.6145
0.70	1.6210

Reaction condition: [MMA]: 2.0 mol dm^{-3}; [K$_2$S$_2$O$_8$]: 2.0 × 10^{-2} mol dm^{-3}; [TEMABDC]: 2.0×10^{-2} mol dm^{-3}; [H$^+$]: 0.20 mol dm^{-3}; Temperature: 60 ± 1 °C

constant. The results show that the R_p was not much influenced on the variation of concentrations of ionic strength of the medium in this investigation (Table 4). An independent nature of ionic strength on the rate of polymerization in the phase transfer catalyzed polymerization was reported [20–23].

Role of organic solvents polarity on R_p

The role of organic solvents polarity on the R_p was examined by carrying out the polymerization in three solvents cyclohexane, ethylacetate and cyclohexanone having the dielectric constants 2.02, 6.02 and 18.03, respectively. It was found that the R_p decreased in the following order: cyclohexanone > ethyl acetate > cyclohexane. An increase in the rate of polymerization was attributed to the increase in the polarity of the solvents, which facilitates greater transfer of peroxydisulfate ion from aqueous phase to organic phase [20–26, 42] (Table 5).

Role of variation of aqueous phase on R_p

Polymerization reactions were conducted with a constant volume of organic phase and changing the volumes of aqueous phase ($V_w/V_o = 0.29$–0.90) at fixed concentrations of all other parameters. The variation of aqueous phase was found to exert no significant change in the rate of polymerization. It has been reported that the variation of aqueous phase on R_p was not affected significantly on phase transfer catalyzed polymerization studies [24–26].

Table 5 Role of organic solvent polarity on the rate of polymerization	Experimental conditions	$R_p \times 10^5$ mol dm^{-3} s^{-1}		
		Cyclohexanone (18.3)	Ethylacetate (3.91)	Cyclohexane (1.13)
	[MMA]: 2.0 mol dm^{-3} [K$_2$S$_2$O$_8$]: 2.0 $\times 10^{-2}$ mol dm^{-3} [TEMABDC]]:2.0 $\times 10^{-2}$ mol dm^{-3} [H$^+$]: 0.50 mol dm^{-3} [μ]: 0.20 mol dm^{-3} Temperature: 60 $\pm$ 1 °C	2.15	1.52	1.06

Mechanism and rate law

Scheme 4 represents the reactions characterizing the polymerization of methyl methacrylate (M) initiated by K$_2$S$_2$O$_8$/MPTC in cyclohexane/water two-phase systems. It is assumed that dissociation of QX and K$_2$S$_2$O$_8$, formation of QS$_2$O$_8$ in aqueous phase, and initiation of monomer in organic phase occur along the reactions such as Eqs. (1)–(5).

The equilibrium constants (K_1 and K_2) in the reactions in Eqs. (1)–(3) and distribution constants (α_1 and α_2) of QX and QS$_2$O$_8$ are defined as follows, respectively.

$$K_1 = \frac{[Q^{2+}]_w[X^-]_w^2}{[QX]_w} \tag{6}$$

$$K_2 = \frac{[K^+]_w^2[S_2O_8^{2-}]_w}{[K_2S_2O_8]_w} \tag{7}$$

$$K_3 = \frac{[QS_2O_8]_w}{[Q^+]_w[S_2O_8^{2-}]_w} \tag{8}$$

$$\alpha_1 = \frac{[Q^{2+}X_2^-]_w}{[QX]_o} \tag{9}$$

$$\alpha_2 = \frac{[Q^{2+}S_2O_8^{2-}]_w}{[QS_2O_8]_o} \tag{10}$$

The initiation rate (R_i) of radical, SO$_4^-$ in Eq. (4), may be represented as follows; f is the initiator efficiency:

$$R_i = \frac{d[SO_4^{0-}]}{dt} = 2K_d f K_3[Q^+]_w[S_2O_8^{2-}]_w \tag{11}$$

Scheme 4 Polymerization pathways of MMA-TEMABDC-PDS in an aqueous–organic medium

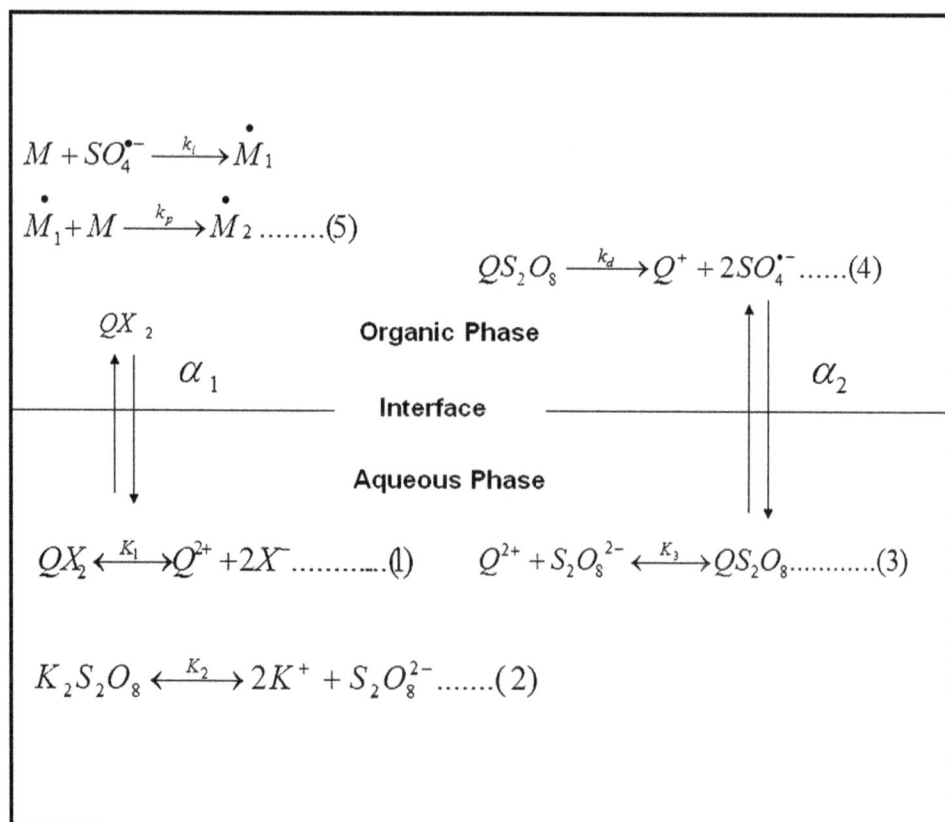

The growth of polymer chain occurs according to the reaction in Eq. (5) and the propagation step represented as follows

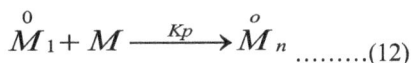

$$\overset{0}{M}_1 + M \xrightarrow{Kp} \overset{o}{M}_n \quad \ldots\ldots(12)$$

$$
\begin{array}{ccc}
\ldots.. & \ldots\ldots.. & \ldots\ldots \\
\ldots\ldots & \ldots\ldots & \ldots\ldots
\end{array}
$$

$$\overset{o}{M}_{n-1} + M \xrightarrow{Kp} \overset{o}{M}_n \quad \ldots\ldots\ldots(13)$$

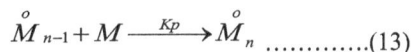

The rate of propagation (R_p) step in the reaction in Eq. (12) is

$$R_p = k_p \overset{0}{[M]}[M] \tag{14}$$

$$\overset{0}{[M]} = \frac{R_p}{k_p[M]}. \tag{15}$$

The termination occurs by the combination of two growing polymer chain radicals; it can be represented as

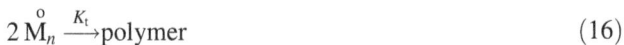

$$2\overset{o}{M}_n \xrightarrow{K_t} \text{polymer} \tag{16}$$

The rate equation of termination (R_t) process according to the Eq. (16) is

$$R_t = 2k_t\overset{0}{[M]}^2 \tag{17}$$

The steady state prevails; the rate of initiation equals to the rate of termination, i.e.

$$R_i = R_t \tag{18}$$

$$2K_d f K_3 [Q^+]_w [S_2O_8^{2-}]_w = 2k_t \overset{0}{[M]}^2 \tag{19}$$

$$\overset{0}{[M]}^2 = \frac{K_d f K_3 [Q^+]_w [S_2O_8^{2-}]_w}{k_t} \tag{20}$$

$$\overset{0}{[M]} = \left[\frac{K_d f K_3 [Q^+]_w [S_2O_8^{2-}]_w}{k_t}\right]^{1/2} \tag{21}$$

Using Eqs. (15) and (21), the rate of polymerization is represented as follows:

$$R_p = k_p \left[\frac{k_d K_3 f}{k_t}\right]^{1/2} [Q^{2+}]_w^{1.2} [S_2O_8^{2-}]_w^{1.50} [M]^{0.64} \tag{22}$$

The above equation satisfactorily explains all the experimental observations for multi-site phase transfer catalyzed radical polymerization of methyl methacrylate in an aqueous–organic two-phase system.

Determination of viscosity average molecular weight of polymer

The viscosity average molecular weight (M_v) of the PMMA was determined using the intrinsic and reduced viscosity of the polymer solution obtained from the viscosity measurement. Viscosity measurements were performed in an Ubbelohde viscometer. The principle behind capillary viscometry is the Poiseuille's law, which states that the time of flow of a polymer solution through a thin capillary is proportional to the viscosity of the solution. The flow time of the pure solvent and different concentration of the polymer solution was measured (Table 6). The few important and well-established terms related to viscosity of polymer solutions are defined as: Relative viscosity, $\eta_r = \eta/\eta_o$ (flow time of polymer solution/flow time of pure solvent); Specific viscosity, $\eta_{sp} = \eta_r - 1$; Reduced viscosity, $\eta_{red} = \eta_{sp}/C$; The intrinsic viscosity of the polymer is related to its molecular weight related by Mark–Houwink equation. $[\eta] = KM_v^a$ where 'K' and 'a' are constants for a polymer, solvent and at a temperature ($K = 5.2 \times 10^{-5}$; $a = 0.760$ for benzene at 30 ± 1 °C); M_v represents the viscosity average molecular weight of the polymer. The viscosity average molecular weight of the PMMA was found to be 1.6955×10^4 g/mol (Fig. 7).

Characterization of polymer: FT-IR analysis

The FT-IR spectra of PMMA with sharp band of ester stretching frequency at 1720 cm^{-1} confirm the atactic nature of PMMA [44]. The conformational assignment of the ester band was in the region of 1100–1300 cm^{-1} for isotactic and syndiotactic PMMA [45, 46]. The following bands were observed in the spectra 1124 cm^{-1} (C–O–C stretching band), 1458 cm^{-1} (C–H deformation), and 2924 cm^{-1} (C–H stretching band) (Fig. 8).

Table 6 Determination of viscosity average molecular weight (M_v) of PMMA	Concentration (C, g/mL)	Flow time (s) t_1	t_2	Average	$\eta r = t/t_o$	ln η_r	ln η_r/C	η_{sp}/C
	Benzene (solvent)	45.02	45.26	45.14 (t_0)				
	0.02	47.00	47.01	47.00	1.0142	0.0403	2.0189	0.0412
	0.04	48.78	48.72	48.75	1.0799	0.0769	1.9234	0.0799
	0.06	49.34	49.09	49.21	1.0901	0.0863	1.4388	0.0901
	0.08	50.15	50.18	50.16	1.1112	0.1054	1.3181	0.1112
	0.10	51.12	51.16	51.14	1.1329	0.1247	1.2479	0.1329

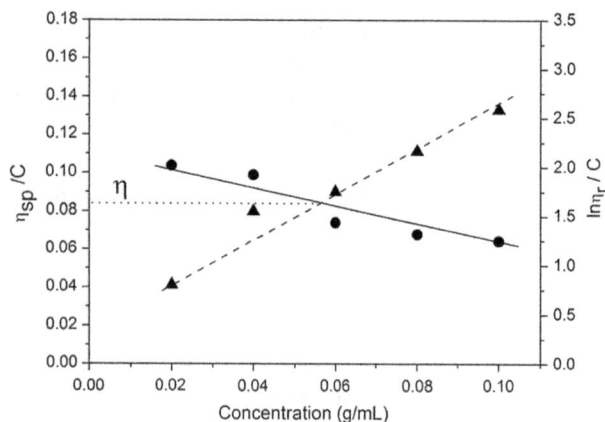

Fig. 7 Determination of viscosity average molecular weight of PMMA

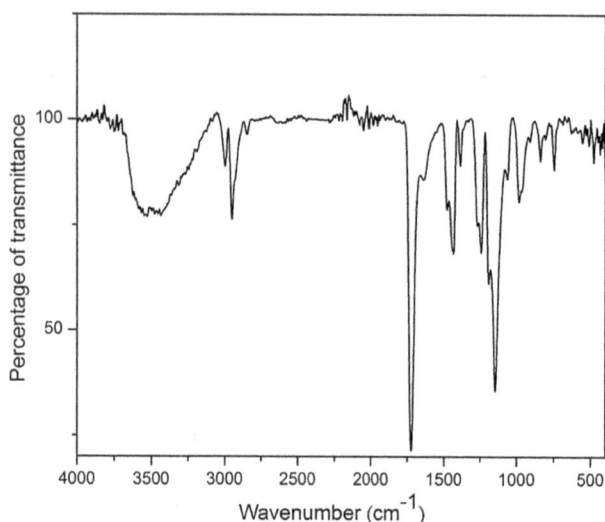

Fig. 8 FT-IR spectral analysis of poly (methyl methacrylate)

Conclusions

Free radical polymerization of methyl methacrylate was successfully performed with the help of 1,4-bis (triethyl methyl ammonium) benzene dichloride (TEMABDC) as multi-site phase transfer catalyst and potassium peroxydisulphate as initiator in two-phase system. The kinetic features, such as the rate of polymerization (R_p), increase with increasing concentration of [MMA], [$K_2S_2O_8$], [TEMABDC] and temperature. An absence of multi-site PTC polymerization did not occur; it reveals that the catalyst was responsible for polymerization reaction. The aqueous phase, acid and ionic strength of the medium do not show any appreciable effect on the R_p. The phase transfer catalyzed polymerization of methyl methacrylate follows a half order with respect to monomer, one and half-order with respect to initiator and order of unity for

catalyst. Based on the results obtained, a suitable mechanism has been proposed. The polymer obtained through radical polymerization of methyl methacrylate was confirmed by FT-IR analysis. The viscosity average molecular weight of the PMMA was found to be 1.6955×10^4 g/mol. Multi-site PTC accelerates the polymerization reaction promptly in two-phase system using water-soluble initiator. The development of feasible pathways for the synthesis of polymers is an important goal in processing industries. Hence, this methodology could be a great interest in synthesis of polymers or any industrially important compounds where the immiscible reactants located in different phase.

Acknowledgements VM is grateful to the Science and Engineering Research Board (DST-SERB), New Delhi, for the young scientist start-up research grant (SB/FT/CS-008/2014) and also (VM and EM) thanks the Management of B S Abdur Rahman Crescent University for the support.

Compliance with ethical standards

Conflict of interest The authors declare that there is no conflict of interest regarding the publication of this article.

References

1. Starks CM, Liotta C, Halpern M (1994) Phase transfer catalysis: fundamentals, applications and perspectives. Chapman and Hall, New York
2. Dehmlow EV, Dehmlow SS (1993) Phase transfer catalysis, 3rd edn. VCH, Weinheim
3. Sasson Y, Neuman R (1997) Handbook of phase transfer catalysis. Blackie Academic and Professional Edition, London
4. Dariusz B, Mateusz G, Grzegorz B, Jacek W, Stanislaw W (2010) Preparation of polymers under phase transfer catalytic conditions. Org Proc Res Develop 14:669–683
5. Tamani B, Mahdavi H (2002) Synthesis of thiocyanohydrins from epoxides using quaternized amino functionalized cross-linked polyacrylamide as a new solid–liquid phase-transfer catalyst. Tetradron Lett 43:6225–6228
6. Mahdavi H, Amirsadeghi M (2013) Synthesis and applications of quaternized highly branched polyacrylamide as a novel multi-site polymeric phase transfer catalyst. J Iran Chem Soc 10:791–797
7. Mahdavi H, Mahmoudian M (2011) Synthesis and applications of cross-linked poly(diallyldimethyl ammonium chloride) and its derivative copolymers as efficient phase transfer catalyst for nucleophilic substitution reactions. Chin J Polym Sci 29:165–172
8. Yamazaki N, Imai Y (1983) Phase-transfer catalyzed polycondensation of α,α'-dichloro-p-xylene with 2,2-bis(4-hydroxyphenyl)propane. Polym J 15:603–608
9. Yamada B, YasudaY Matsushita T, Otsu T (1976) Preparation of polyester from acrylic acid in the presence of crown ether. J Polym Sci Polym Lett Edn 14:277–281
10. Reetz MT, Ostarek R (1988) Polymeriztion of acrylic esters initiated by tetrabutylammonium alky-and aryl-thiolates. J Chem Soc Chem Commun 3:213–215
11. Savitha S, Vajjiravel M, Umapathy MJ (2006) Polymerization of butyl acrylate using pottassium peroxydisulfate as initiator in the presence of phase transfer catalyst—a kinetic study. Int J Polym Mater 55(8):537–548

12. Balakrishnan T, Damodarkumar S (2000) Phase transfer catalysis: free radical polymerization of acrylonitrile using peroxymonosulphate/tetrabutylphosphonium chloride catalyst system:a kinetic study. J Appl Polym Sci 76:1564–1571

13. Umapathy MJ, Mohan D (1999) Studies on phase transfer catalysed polymerization of acrylonitrile. Hung J Ind Chem 27(4):245–250

14. Dharmendirakumar M, Konguvelthehazhnan P, Umapathy MJ, Rajendran M (2004) Free radical polymerization of methyl methacrylate in the presence of phase transfer catalyst—a kinetic study. Int J Polym Mater 53:95–103

15. Umapathy MJ, Balakrishnan T (1998) Kinetics and mechanism of polymerization of methyl methacrylate initiated by phase transfer catalyst-ammonium perdisulfate system. J Polym Mater 15(3):275–278

16. Umapathy MJ, Mohan D (1999) Studies on phase transfer catalyzed polymerization of glycidyl methacrylate. J Polym Mater 16(2):167–171

17. Umapathy MJ, Mohan D (2001) Phase transfer polymerization of butyl methacrylate using potassium peroxydisulfate as initiator—a kinetic study. Ind J Chem Tech 8(6):510–514

18. Umapathy MJ, Malaisamy R, Mohan D (2000) Kinetics and mechanism of phase transfer catalyzed free radical polymerization of methyl acrylate. J Macromol Sci Pure Appl Chem A 37(11):1437–1445

19. Idoux JP, Wysocki R, Yong S, Turcot J, Ohlman C, Leonard R (1983) Polymer supported multi-site phase transfer catalyst. Synth Commun 13:139–144

20. Vajjiravel M, Umapathy MJ, Bharathbabu M (2007) Polymerization of acrylonitrile using potassium peroxydisulfate as an initiator in the presence of a multisite phase-transfer catalyst: a kinetic study. J Appl Polym Sci 105:3634–3639

21. Vajjiravel M, Umapathy MJ (2008) Synthesis and characterization of multi-site phase transfer catalyst: application in radical polymerisation of acrylonitrile—a kinetic study. J Polym Res 15(1):27–36

22. Vajjiravel M, Umapathy MJ (2008) Free radical polymerisation of methyl methacrylate initiated by multi-site phase transfer catalyst—a kinetic study. Colloid Polym Sci 286:729–738

23. Vajjiravel M, Umapathy MJ (2009) Kinetics and mechanism of multi-site phase transfer catalyzed radical polymerization of ethyl methacrylate. Int J Polym Mater 58:61–76

24. Vajjiravel M, Umapathy MJ (2010) Multi-site phase transfer catalyzed radical polymerization of n-butyl methacrylate: a kinetic study. Chem Eng Commun 197:352–365

25. Vajjiravel M, Umapathy MJ (2010) Synthesis, characterization and application of a multi-site phase transfer catalyst in radical polymerization of n-butyl methacrylate—a kinetic study. Int J Polym Mater 59:647–662

26. Vajjiravel M, Umapathy MJ (2011) Kinetics of radical polymerization of glycidyl methacrylate initiated by multi-site phase transfer catalyst–potassium peroxydisulfate in two-phase system. J Appl Polym Sci 120:1794–1799

27. Albrecht K, Stickler M, Rhein T (2013) Polymethacrylates in Ullmann's encyclopedia of industrial chemistry. Wiley-VCH Verlag GmbH & Co, Weinheim

28. Kine BB, Novak RW (1985) Acrylic and methacrylic ester polymers in encyclopedia of polymer science and engineering. Wiley, New York

29. Wang JS, Matyjaszewski K (1995) Controlled/living radical polymerization. Atom transfer radical polymerization in the presence of transition-metal complexes. J Am Chem Soc 117:5614–5615

30. Kato M, Kamigaito M, Sawamoto M, Higashimura T (1995) Polymerization of methyl methacrylate with the carbon tetrachloride/dichlorotris-(triphenylphosphine) ruthenium (II)/methylaluminum bis(2,6-di-tert-butylphenoxide) initiating system: possibility of living radical polymerization. Macromolecules 28:1721–1723

31. Percec V, Barboiu B (1995) Living radical polymerization of styrene initiated by arenesulfonyl chlorides and CuI (bpy) nCl. Macromolecules 28:7970–7972

32. TPT L, Moad G, Rizzardo E, Thang SH (1998) WO patent. 9801478

33. Moad G, Rizzardo E, Solomon DH (1982) Selectivity of the reaction of free radicals with styrene. Macromolecules 15:909–914

34. Zhu G, Zhang L, Zhang Z, Zhu J, Tu Y, Cheng Z, Zhu X (2011) Iron-mediated ICAR ATRP of methyl methacrylate. Macromolecules 44:3233–3239

35. Ding M, Jiang X, Peng J, Zhang L, Cheng Z, Zhu X (2015) An atom transfer radical polymerization system: catalyzed by an iron catalyst in PEG-400. Green Chem 17:271–278

36. Guillaneuf Y, Gigmes D, Marque SRA, Astolfi P, Greci L, Tordo P, Bertin D (2007) First effective nitroxide-mediated polymerization of methyl methacrylate. Macromolecules 40:3108–3114

37. Zhu J, Zhu X, Cheng Z, Liu F, Lu J (2002) Study on controlled free-radical polymerization in the presence of 2-cyanoprop-2-yl 1-dithionaphthalate (CPDN). Polymer 43:7037–7042

38. Yang L, Luo Y, Liu X, Li B (2009) RAFT miniemulsion polymerization of methyl methacrylate. Polymer 50:4334–4342

39. Wang ML, Hsieh YM (2004) Kinetic study of dichlorocyclopropanation of 4-vinyl-1-cyclohexene by a novel multisite phase transfer catalyst. J Mol Cat A Chem 210:59–68

40. Brandrup J, Immerugut EH, Grulke EA (1999) Polymer handbook, 4th edn. Wiley, New York

41. Balakrishnan T, Damodarkumar S (2000) Phase transfer catalyzed free radical polymerization: kinetics of polymerization of butyl methacrylate using peroxymonosulphate/tetrabutyl phosphonium chloride catalyst system. Ind J Chem 39A:751–755

42. Vivekanand PA, Wang ML, Hsieh Y-M (2013) Sonolytic and silent polymerization of methacrlyic acid butyl ester catalyzed by a new onium salt with bis-active sites in a biphasic system—a comparative investigation. Molecules 18:2419–2437

43. Murugesan V, Umapathy MJ (2016) Phase transfer catalyst aided radical polymerization of n-butyl acrylate in two phase system—a kinetic study. Int J Ind Chem 7(4):441–448

44. Prajapati K, Varshney A (2006) Free radical polymerization of methylmethacrylate using p-nitrobenzyltriphenyl phosphonium ylide as novel initiator. J Polym Res 13:97–105

45. Tretinmikov ON (2003) Backbone and ester group conformations of stereoregular poly (methyl methacrylate)s in the stereocomplex. Macromolecules 36:2179–2182

46. Daniel N, Srivastava AK (2007) Kinetics and mechanism of radical polymerization of methyl methacrylate using p-acetylbenzylidene triphenylarsonium ylide (p-ABTAY) as an initiator. J Polym Res 7(3):161–165

10

Comparative study of Levofloxacin and its amide derivative as efficient water soluble inhibitors for mild steel corrosion in hydrochloric acid solution

Turuvekere K. Chaitra[1] · Kikkeri N. Mohana[1] · Harmesh C. Tandon[2]

Abstract The influence of 8-fluoro-3-methyl-9-(4-methyl-piperazin-1-yl)-6-oxo-2,3-dihydro-6H-1-oxa-3a-aza-phenalene-5-carboxylic acid or levofloxacin (P1) and newly synthesized 8-fluoro-3-methyl-9-(4-methyl-piperazin-1-yl)-6-oxo-2,3-dihydro-6H-1-oxa-3a-aza-phenalene-5-carboxylic acid-(5-methyl-pyridin-2-yl)-amide (P2) on corrosion inhibition of mild steel in 0.5 M hydrochloric acid solution was studied using weight loss and electrochemical techniques. Inhibition efficiency of P1 and P2 increased with concentration and decreased with temperature in the concentration range 0.14–0.35 mM in the temperature range 303–333 K. Thermodynamic parameters for dissolution and adsorption process were studied. Increase in energy of activation after the addition of inhibitors indicated formation of barrier film which prevents charge and mass transfer. Free energy of adsorption showed that the type of adsorption was neither physical nor chemical but comprehensive. The adsorption of the P1 and P2 on the mild steel surface was found to obey the Langmuir isotherm. Impedance measurement showed that there is increase in the polarization resistance and decrease in double layer capacitance after the addition of inhibitors. From polarization study as the shift in corrosion potential is more than 85 mV, both P1 and P2 are anodic type of inhibitors. Scanning electron microscope images confirm the formation of inhibitory film on mild steel surface. Quantum chemical calculation results well correlated with experimental results. Lower values of energy gap, ionization potential and hardness, higher value of softness make P2 better inhibitor compared to P1.

Keywords Corrosion · Mild steel · Levofloxacin · Electrochemical techniques · Quantum chemical parameters · SEM

Introduction

Corrosion is a natural destructive phenomenon where pure metals interact with the environment to form non-desirable metallic compounds. Protection of metals from corrosion is one of the major economic issues. Mild steel (MS) is an important metal which is widely applied in oil wells, constructional materials, automobiles and many other industries due to its excellent mechanical properties and low cost [1]. Hydrochloric acid solutions (approximately 0.5–1 M) are widely used in several industrial processes, some of the important fields of application being acid pickling of steel, chemical cleaning and processing, ore production and oil well acidification [2]. Corrosion of MS when exposed to aggressive acids, such as hydrochloric acid and sulfuric acid results in such damage that needs either repair or replacement of the part leading to huge loss of resources. Corrosion of MS is worth investigating because such corrosions cause damage to pipelines, bridges, marine structures and construction materials bringing heavy economic losses worldwide. Corrosion is also one of the major concerns in the durability of materials and structures; and studies are continuously carried out to develop effective methods for corrosion control [3]. There are a number of methods for corrosion control but the choice depends on economics, safety requirements and

✉ Kikkeri N. Mohana
drknmohana@gmail.com

[1] Department of Studies in Chemistry, Manasagangotri, University of Mysore, Mysuru 570006, Karnataka, India

[2] Department of Chemistry, Sri Venkateswara College, Dhaula kuan, New Delhi 110021, India

technical considerations. Mitigation of MS corrosion is achieved through such means as galvanisation, organic coating (enamel, polymer, oils etc.) and using corrosion inhibitors [4–6] which form film by adsorbing on the metal surface. The organic corrosion inhibitors although proved to be the best for the protection of MS, but they are restricted in some cases because of their toxicity. Therefore, the best means of protection is to adapt an inhibitor which is eco-friendly, easily soluble and effective at low concentration. The class of organic compounds which satisfy these conditions are drugs and their derivatives which are highly water soluble.

Corrosion inhibition studies of many drugs, such as β-lactam antibiotics {penicillin G [7], ampicillin [8], amoxicillin [9]}, quinolones {ofloxacin [10], ciprofloxacin [11], quinoline [12]}, tetracyclines {doxycycline [13]}, sulphonamides {sulfamethazine [14], dapsone [15] antifungal {ketoconazole [16]}, antiviral {rhodanine [17]}, have been reported.

Levofloxacin is a member of the fluoroquinolone class of antibacterial used in the treatment of chronic bronchitis, respiratory tract infection, pneumonia, skin infection and urinary tract infection [18]. Its structure has extended π-electron systems, good number of hetero atoms and two electron donating methyl groups which facilitate its adsorption on the MS surface. Fluoroquilones have been established as potential class of inhibitors, P1 and P2 which belong to the same class are expected to give good inhibition because similar molecules tend to behave alike [19]. Eddy et al. [20] studied derivatives of fluoroquinolone (ofloxacin, amifloxacin, enofloxacin, pefloxacin) on MS corrosion in sulfuric acid medium by gravimetric technique supported by quantum chemical calculations and obtained inhibition efficiency up to 94 %. Levofloxacin was previously studied by Pang et al. [21] as MS corrosion inhibitor in sulfuric acid medium by weight loss and electrochemical methods and maximum inhibition efficiency obtained was 90 %. In this study, Levofloxacin is being studied as MS corrosion inhibitor in HCl medium using gravimetric and electrochemical techniques at lower concentrations and the results are supported by theoretical studies. Comparison of inhibition efficiency of Levofloxacin with its synthesized derivative has been made to study the effect of an extra heterocyclic ring and an amine group transformed into amide bond present in P2.

In continuation of our previous work [22–26] the present paper reports the comparative study of the anti-corrosion potential of levofloxacin (P1) and its newly synthesized amide derivative (P2) in 0.5 M HCl media using weight loss method, electrochemical impedance spectrosocpy (EIS) and potentiodynamic polarization measurements. Morphological study has been done using scanning electron microscope (SEM). Quantum chemical calculations were done and different parameters, such as energies of highest occupied molecular orbital (E_{HOMO}) and the lowest unoccupied molecular orbital (E_{LUMO}), the energy gap (ΔE), hardness (η), softness (σ), electron affinity (A), electronegativity (χ), ionization potential (I) of P1 and P2 were determined and correlated with experimental results.

Experimental

Materials and sample preparation

The chemical composition by wt% of MS coupons used for experiment was as follows: C, 0.051; Mn, 0.179; Si, 0.006; P, 0.005; S, 0.023; Cr, 0.051; Ni, 0.05; Mo, 0.013; Ti, 0.004; Al, 0.103; Cu, 0.050; Sn, 0.004; B, 0.00105; Co, 0.017; Nb, 0.012; Pb, 0.001 and the remaining is iron. Before the commencement of experiment, samples were mechanically cut into 2 cm × 2 cm × 0.1 cm, abraded with different grades of silicon carbide emery paper, washed with double distilled water, degreased, dried and stored in desiccator until use. For polarization and impedance measurements, the MS specimens were embedded in epoxy resin to expose a geometrical surface area of 1 cm^2 to the electrolyte. Solutions of P1 and P2 in optimized concentration range of 0.14–0.35 mM were prepared from stock solution made of using 0.5 M HCl. Melting range was determined using Veego Melting Point VMP III apparatus.

Synthesis of 8-fluoro-3-methyl-9-(4-methyl-piperazin-1-yl)-6-oxo-2,3-dihydro-6H-1-oxa-3a-aza-phenalene-5-carboxylic acid-(5-methyl-pyridin-2-yl)-amide (P2)

The reported procedure [26] of acid–amine coupling was used for the synthesis of P2. Scheme for the synthesis of P2 is given in Fig. 1. To a mixture of 8-fluoro-3-methyl-9-(4-methyl-piperazin-1-yl)-6-oxo-2,3-dihydro-6H-1-oxa-3a-aza-phenalene-5-carboxylic acid (1 equivalent) and 5-methyl-pyridin-2-ylamine (1 equivalent), triethylamine (2.1equivalent) and O-(Benzotriazol-1-yl)-N,N,N',N'-tetramethyluronium tetrafluoroborate (TBTU) (1.2 equivalent) were added in dichloromethane (MDC). The reaction mixture was stirred at room temperature for 16 h. Reaction completion was monitored by TLC using solvent system, ethyl acetate: methanol (1:1). Reaction mixture was washed with 1 M HCl, 10 % bicarbonate solution followed by water and brine, dried over sodium sulfate, and concentrated to remove solvent under reduced pressure. The solid obtained was crystallized using ethanol to get the pure form. The yield of the product was 90 % and melting range

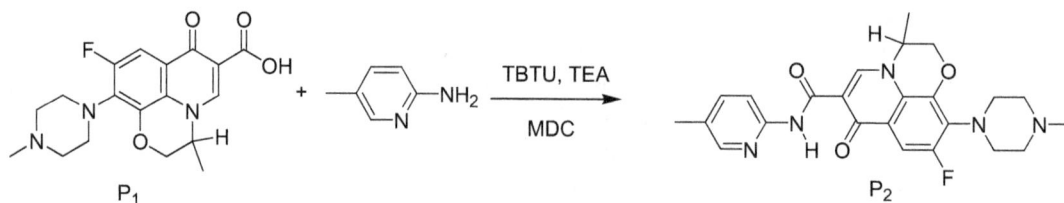

Fig. 1 Scheme for the synthesis of inhibitors

was 606–607 K. Spectral data: IR (cm^{-1}) 1673 (C=O, Amide), 3398 (N–H, Amide), 1673 (C=O, ketone), 1003–1095 (C–F), 1450–1599 (Ar C=C),^{1}H-NMR (400 MHz, DMSO-d$_6$) δ_H ppm:1.56 (s, 3H, CH$_3$), 1.57 (s, 3H, CH$_3$), 2.04–2.16 (m, 4H, piperazine), 2.17–2.35 (m, 4H, piperazine), 2.54 (s, 3H, N–CH$_3$), 4.42–4.38 (m, 3H, CH$_2$CH), 7.25–7.26 (d, 1H, phenyl), 7.48–7.51 (d, 1H, phenyl), 7.69 (s, 1H, phenyl), 8.188 (s, 1H, phenyl), 8.61 (s, 1H, phenyl), 12.55 (s, 1H, N–H). MS: 452.12 (M + 1), 453.12 (M + 2).

Weight loss measurements

MS specimens were immersed in 0.5 M HCl solution without and with varying amounts of inhibitor for 4 h in a thermostatically controlled water bath (with an accuracy of ±0.2 K) at constant temperature, under aerated condition (Weiber limited, Chennai, India). The coupons were taken out after 4 h of immersion, rinsed in water followed by drying in acetone. Weight loss of the specimens was recorded by analytical balance (Sartorius, precision ±0.1 mg). Experiment was carried out in triplicates and mean weight loss was calculated. The procedure was repeated for all other concentrations and temperatures.

Electrochemical measurements

Potentiodynamic polarization and EIS experiments were carried out using a CHI660D electrochemical workstation. A conventional three-electrode cell consisting of |Ag/AgCl| reference electrode, a platinum auxiliary electrode and the working MS electrode with 1 cm^2 exposed areas was used. Pre-treatment of the specimens was same as gravimetric measurements. The electrochemical tests were performed with P1 and P2 concentrations ranging from 0.14 to 0.35 mM at 303 K. Potentiodynamic polarization measurements were performed in the potential range from −850 to −150 mV with a scan rate of 0.4 mV s^{-1}. The exposure time before polarization measurements was 30 min. Prior to EIS measurement; 30 min were spent for making open circuit potential a stable value. The EIS data were taken in the frequency range of 10 kHz–1 Hz.

Quantum chemical calculations

The geometrical optimization of the investigated molecules was done by Ab initio method at 6-31G** basis set for all atoms. For energy minimization, the convergence limit at 1.0 and rms gradient 1.0 kcal/Amol was kept. The Polak–Ribiere conjugate gradient algorithm which is quite fast and precise was used for optimization of geometry. In DFR calculation B3LYP combined exchange-corelation potential functional has been used. The HYPERCHEM 7.52 (Hypercube Inc., Florida, USA, 2003) professional software was employed for all calculations.

Scanning electron microscopy (SEM)

The SEM experiments were performed using a Zeiss electron microscope with the working voltage of 15 kV and the working distance 10.5 mm. In SEM micrographs, the specimens were exposed to 0.5 M HCl in the absence and presence of inhibitors under optimum conditions after 4 h of immersion. The SEM images were taken for polished MS specimen and specimen immersed in acid solution with and without inhibitors.

Results and discussion

Weight loss measurement

Effect of inhibitor concentration

The effect of concentration of inhibitors P1 and P2 on MS corrosion in 0.5 M HCl was studied at concentrations from 0.14 to 0.35 mM and temperature range of 303 to 333 K by weight loss measurement and results are presented in Table 1. The corrosion rate and inhibition efficiency were calculated using the formulae (1) and (2).

$$C_R = \frac{\Delta W}{St} \quad (1)$$

$$IE\ (\%) = \frac{(C_R)_a - (C_R)_p}{(C_R)_a} \times 100 \quad (2)$$

where, ΔW is the weight loss, S is the surface area of the specimen (cm^2), t is the immersion time (h), and $(C_R)_a$,

Table 1 Weight loss data in the absence and presence of inhibitors in 0.5 M HCl at different concentrations and temperatures

Inhibitor	C (mM)		IE (%)		IE (%)		IE (%)		IE (%)
Blank	–	0.516	–	0.883	–	1.224	–	1.65	–
P1	0.14	0.087	83.0 ± 0.79	0.165	81.3 ± 0.44	0.244	80.1 ± 0.43	0.366	77.8 ± 0.41
	0.21	0.068	86.7 ± 0.58	0.136	84.5 ± 0.35	0.205	83.3 ± 0.32	0.338	79.5 ± 0.52
	0.28	0.047	90.8 ± 0.49	0.118	86.6 ± 0.50	0.183	85.0 ± 0.54	0.306	81.4 ± 0.86
	0.35	0.030	94.0 ± 0.68	0.097	89.0 ± 0.32	0.161	86.8 ± 0.55	0.297	82.0 ± 0.76
P2	0.14	0.072	86.0 ± 0.22	0.133	84.9 ± 0.65	0.220	82.0 ± 0.45	0.345	79.0 ± 0.82
	0.21	0.056	89.1 ± 0.52	0.115	87.0 ± 0.88	0.195	84.0 ± 0.38	0.313	81.0 ± 0.97
	0.28	0.035	93.2 ± 0.92	0.097	89.0 ± 0.84	0.173	85.8 ± 0.64	0.302	81.7 ± 1.06
	0.35	0.018	96.4 ± 0.88	0.078	91.1 ± 0.75	0.155	87.3 ± 0.88	0.287	82.6 ± 0.72

$(C_R)_p$ are corrosion rates in the absence and presence of the inhibitor, respectively.

As the concentration increases, IE (%) increases for both P1 and P2. As the concentration increases, availability of number of molecules to block reaction sites increases. The highest inhibition efficiency of 96 and 94 % were shown by P2 and P1 at 0.35 mM concentration, respectively. After that although concentration was raised and there was no much difference in inhibition efficiency.

Effect of temperature

Inhibition performance of P1 and P2 on MS in 0.5 M HCl was studied in the temperature range of 303–333 K (Table 1). The influence of temperature on corrosion reaction is very complex, because many changes take place on the metal surface, such as etching, desorption and inhibitor itself may undergo decomposition [27]. The corrosion rate of MS increases as the temperature of the surrounding solution increases both in the absence and presence of inhibitors. That is IE (%) decreases as the temperature increases. This is due to decrease in hydrogen evolution over potential which in turn increases the evolution of anodic hydrogen at higher temperatures. There is also possibility of desorption of adsorbed inhibitor film, as the inhibitor molecules gain sufficient energy to overcome interaction between metal empty orbital and inhibitor electrons at higher temperature.

The relationship between corrosion rate of MS and temperature of the environment is given by Arrhenius equation

$$C_R = k \exp\left(-\frac{E_a^*}{RT}\right) \quad (3)$$

Enthalpy and entropy of the activation are calculated based on transition state theory using the alternative form of Arrhenius equation, which takes the form as

$$C_R = \frac{RT}{Nh} \exp\left(\frac{\Delta S_a^*}{R}\right) \exp\left(\frac{-\Delta H_a^*}{RT}\right) \quad (4)$$

where, E_a^* is activation energy, ΔS_a^* is the entropy of activation, ΔH_a^* is the enthalpy of activation, k is Arrhenius pre-exponential factor, h is Planck's constant, N is Avogadro's number, T is the absolute temperature and R is the universal gas constant.

The plot of ln C_R versus $1/T$ is a straight line (Fig. 2), computing the values of slope and intercept, the values of E_a^* and k were calculated for both the inhibitors at various concentrations. Using the Eq. (4), another linear plot of ln C_R/T versus $1/T$ was drawn (Fig. 3) with slope $(-\Delta H_a^*/R)$ and intercept $[\ln(R/Nh) + \Delta S_a^*/R]$. This was used for the calculation of ΔH_a^* and ΔS_a^*. Different parameters involving Arrhenius equations were calculated and listed in Table 2.

Review of these data indicates that all the activation parameters of dissolution reaction of MS in 0.5 M HCl are higher in the presence of inhibitors than in their absence. E_a^* is higher for inhibited solution than for uninhibited solution and increases upon increasing concentration of inhibitors. Such a trend suggests that corrosion reaction will be further pushed to surface sites which are characterized by progressively higher value of E_a^* as concentration of the inhibitor becomes higher [28]. At higher concentration E_a^* increases further, because extent of surface coverage is close to saturation [29]. The higher values of E_a^* in inhibited solution might also be correlated with the increased thickness of double layer [30]. Further from the value of E_a^* which is greater than 20 kJ mol^{-1} for both inhibited and uninhibited solutions, it is confirmed that the whole process is surface controlled [31]. The positive sign of enthalpies shows endothermic nature of MS dissolution process i.e., dissolution of steel is difficult [32]. After the addition of P1 and P2 there is increase in the value of enthalpy of activation which suggests that the dissolution becomes more difficult. Entropy of activation is negative in the presence and absence of P1 and P2 which becomes

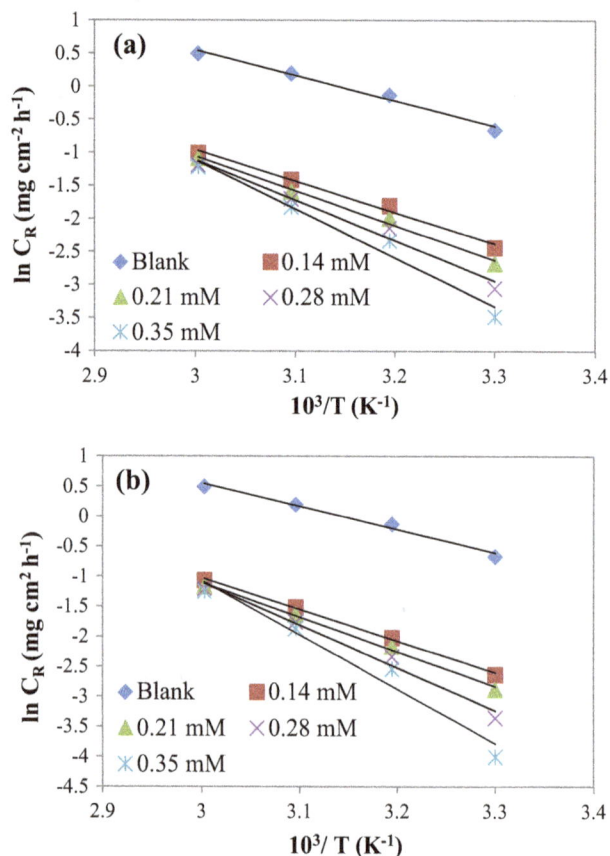

Fig. 2 Arrhenius plots in the absence and presence of different concentrations of **a** P1 and **b** P2

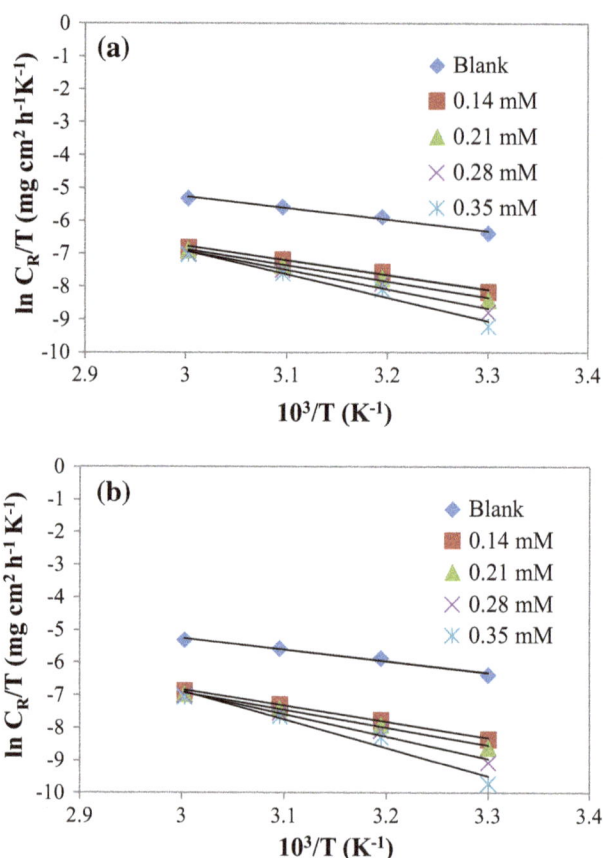

Fig. 3 Alternative Arrhenius plots in the absence and presence of different concentrations of **a** P1 and **b** P2

more positive on increasing concentration of inhibitor. This indicates that the dissolution of MS is characterized by an activated complex which is associative. This might be due to the adsorption of P1 and P2 molecules from HCl solution which could be regarded as quasi- substitution between the inhibitors in the aqueous phase and water molecules on MS electrode surface. In such a situation, the adsorption of P1 and P2 on MS surface which results in decrease in entropy was accompanied with desorption of water molecules from the electrode surface which increases the entropy. Here entropy is the sum of solvent entropy and solute entropy. Thus, increase in entropy of activation for P1 and P2 was attributed to predominance of solvent (H_2O) entropy [33].

Adsorption isotherm

The efficiency of a compound as successful corrosion inhibitor mainly depends on its adsorption ability on the metal surface. It is essential to know the mode of adsorption and the adsorption isotherm that can give valuable information on the interaction of inhibitor and metal surface [34]. The adsorption of an organic adsorbate at a

metal/solution interface can be presented as a substitution adsorption process between the organic molecules in aqueous solution and the water molecules on a metallic surface. All isotherms are having general formula

$$f(\theta, x)\exp(-2\alpha\theta) = K_{ads}C \qquad (5)$$

where, $f(\theta, x)$ is the configurational factor which depends upon the physical mode of adsorption. Here θ is the surface coverage, C is the concentration of the inhibitor, x is the size factor ratio, α is the molecular interaction parameter and K_{ads} is the equilibrium constant of the adsorption process [35]. Depending on the value of correlation coefficient the best fitter of the isotherm can be determined. After trying to fit the values to many isotherms like Temkin, Frumkin, Freundlich and Langmuir, the best fit was obtained with Langmuir adsorption isotherm for the inhibitors. A plot of C/θ versus C (Fig. 4) gave a straight line with an average correlation coefficient of 0.9995 and 0.9995, for P1 and P2, respectively and a slope of nearly unity (1.0706 and 0.9995 for P1 and P2, respectively) suggests that the adsorption of both the molecules obeys Langmuir isotherm, which can be expressed by the following equation:

Table 2 Activation parameters in the absence and presence of P1 and P2 in 0.5 M HCl

Inhibitor	C (mM)	E_a^* (kJ mol^{-1})	k (mg cm^{-2} h^{-1})	ΔH_a^* (kJ mol^{-1})	$\Delta H_a^* = E_a^* - RT$ (kJ mol^{-1})	ΔS_a^* (J mol^{-1} K^{-1})
Blank	–	32.1	186,465	29.5	29.5	−152.9
P1	0.14	39.3	564,671	36.7	36.8	−143.6
	0.21	43.5	2,317,501	40.9	40.9	−131.9
	0.28	50.9	31,077,666	48.2	48.3	−110.3
	0.35	61.5	1.41×10^9	58.8	58.9	−78.6
P2	0.14	43.8	2,610,363	41.1	41.2	−130.9
	0.21	47.8	10,201,038	45.1	45.2	−119.6
	0.28	59.2	6.35×10^8	56.6	56.6	−85.2
	0.35	75.2	2.11×10^{11}	72.6	72.6	−37.0

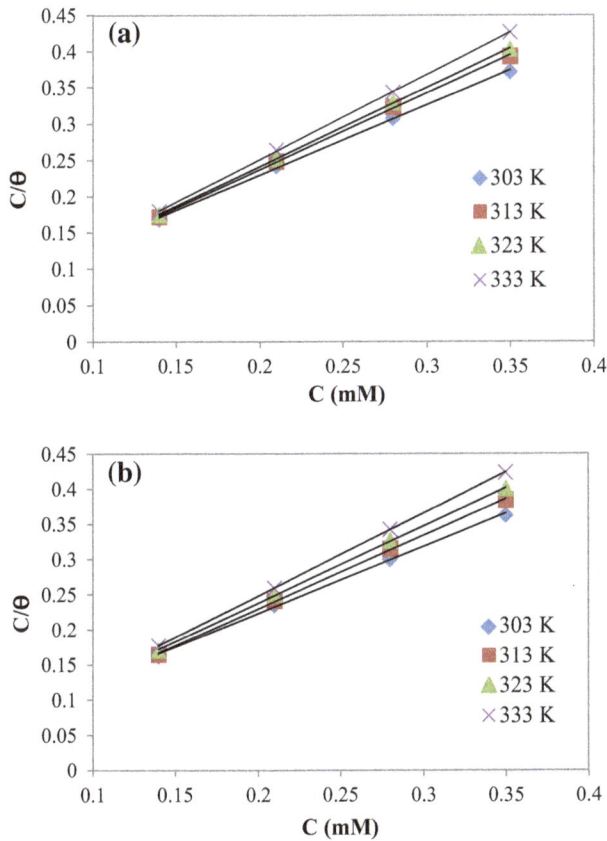

Fig. 4 Langmuir isotherm for the adsorption of **a** P1 and **b** P2 in 0.5 M HCl at different temperatures

$$\frac{C}{\theta} = \frac{1}{K_{ads}} + C \tag{6}$$

From Eq. (6), K_{ads} can be calculated from intercept of $\frac{C}{\theta}$ Vs C plot. Free energy of adsorption can be calculated from K_{ads} using the Eq. (7).

$$\Delta G_{ads}^o = -RT \ln(55.5 K_{ads}) \tag{7}$$

where, R is gas constant and T is the absolute temperature of the experiment and the constant value 55.5 is the concentration of water in solution in mol dm^{-3}.

Entropy of adsorption and enthalpy of adsorption process can be calculated using the following thermodynamic equation:

$$\Delta G_{ads}^o = \Delta H_{ads}^o - T\Delta S_{ads}^o \tag{8}$$

The values of all thermodynamic parameters are listed in Table 3. A plot of ΔG_{ads}^o versus T gives a straight line (Fig. 5) which can be used for the calculation of ΔH_{ads}^o and ΔS_{ads}^o. Calculated values of free energy of adsorption and adsorption equilibrium constant together represent spontaneity of the process and stability of the adsorbed layer on the metal surface. Large values of K_{ads} obtained for P1 and P2 indicate that they are efficient adsorbents which imply that they have better inhibition efficiency. In general, the values of ΔG_{ads}^o up to −20 kJ mol^{-1} are compatible with the electrostatic interaction between the charged inhibitor molecules and the charged metal surface (physisorption), and those which are more negative than −40 kJ mol^{-1} involve charge sharing or charge transfer from the inhibitor molecules to the metal surface (chemisorption) [36]. The calculated values of free energy of adsorption for P1 and P2 lies between −20 and −40 kJ mol^{-1} at lower temperatures and little higher than −40 kJ mol^{-1} at higher temperatures. Many authors reported that the adsorption of organic molecules on the solid surfaces cannot be considered as purely physical or as purely chemical [37–39]. Initially inhibitor may adsorb on the MS surface by electrostatic force of attraction but at later stage charge transfer also takes place. Therefore, it is concluded that the adsorption of P1 and P2 is neither totally physical nor chemical but complex comprehensive kind of adsorption involving both with a slight dominance of chemisorption. Similar type of observations have been reported in the literature [40–43].

Table 3 Adsorption thermodynamic parameters in the absence and presence of various concentrations of inhibitors

Inhibitor	T (K)	R^2	K_{ads} (L mol^{-1})	ΔG°_{ads} (kJ mol^{-1})	ΔS°_{ads} (J mol^{-1}K^{-1})	ΔH°_{ads} (kJ mol^{-1})	$\Delta G^{\circ}_{ads} = \Delta H^{\circ}_{ads} - T\Delta S^{\circ}_{ads}$ (kJ mol^{-1})
P1	303	0.9989	27,855	−36.2	−209	27.7	−35.8
	313	0.9996	38,461	−38.6			−37.8
	323	0.9999	43,290	−40.0			−40.0
	333	0.9999	60,606	−41.6			−42.1
P2	303	0.9987	30,769	−35.9	−201	24.7	−36.3
	313	0.9995	49,504	−37.9			−38.3
	323	0.9998	53,475	−39.5			−40.3
	333	1.0000	80,000	−42.4			−42.3

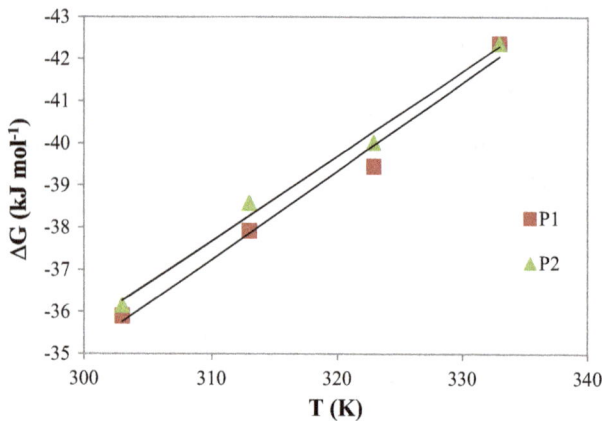

Fig. 5 Plot of ΔG_{ads} vs T for P1 and P2

Electrochemical impedance spectroscopy

The corrosion behavior of P1 and P2 was studied using EIS and results are tabulated in Table 4. Nyquist plot, bode modulus plot and phase angle plot of P1 and P2 are given in Figs. 6 and 7, respectively. Nyquist plots for all concentrations are characterized by one capacitive loop whose diameter increases on increasing concentration of the inhibitor. The capacitive loop was related to charge transfer in the corrosion process, whereas the depressed form of the higher frequency loop reflects the surface non-homogeneity of structural or interfacial origin, such as those found in adsorption processes [44]. Here contribution of all resistances correspond to the metal/solution interface, i.e., charge transfer resistance (R_{ct}), diffuse layer resistance (R_d), accumulation resistance (R_a), film resistance (R_f), etc. must be taken into account. So charge transfer resistance must be replaced by polarization resistance (R_p) [45]. Nyquist plots can be represented by equivalent circuit (Fig. 8) where solution resistance (R_s) is shorted by constant phase element (CPE) that is placed in parallel to polarization resistance (R_p).

The values of R_p were calculated from the difference in impedance at lower and higher frequencies [46]. The inhibition efficiency was calculated using polarization resistance according to the following equation:

$$\text{IE}\,(\%) = \frac{(R_p)_p - (R_p)_a}{(R_p)_p} \times 100 \qquad (9)$$

Table 4 Impedance parameters for the corrosion of MS in 0.5 M HCl in absence and presence of different concentrations of inhibitors at 303 K

Inhibitor	Concentration (mM)	R_p (Ω cm^2)	Y_o (μ Ω^{-1} s^n)	R_s (Ω cm^2)	n	C_{dl} (μF cm^{-2})	IE (%)
	Blank	205	275.6	2.47	0.7631	112.9	–
P1	0.14	739	73.88	2.58	0.8641	46.76	72.3
	0.21	919	55.66	4.54	0.8746	36.34	77.7
	0.28	1106	20.15	1.21	0.9207	14.52	81.5
	0.35	1519	45.46	2.14	0.8855	34.74	86.5
P2	0.14	939	55.9	1.303	0.8736	36.49	78.2
	0.21	1161	38.26	4.064	0.8558	22.63	82.3
	0.28	1372	19.32	1.194	0.9211	14.16	85.0
	0.35	1572	20.29	1.271	0.9064	35.89	87.1

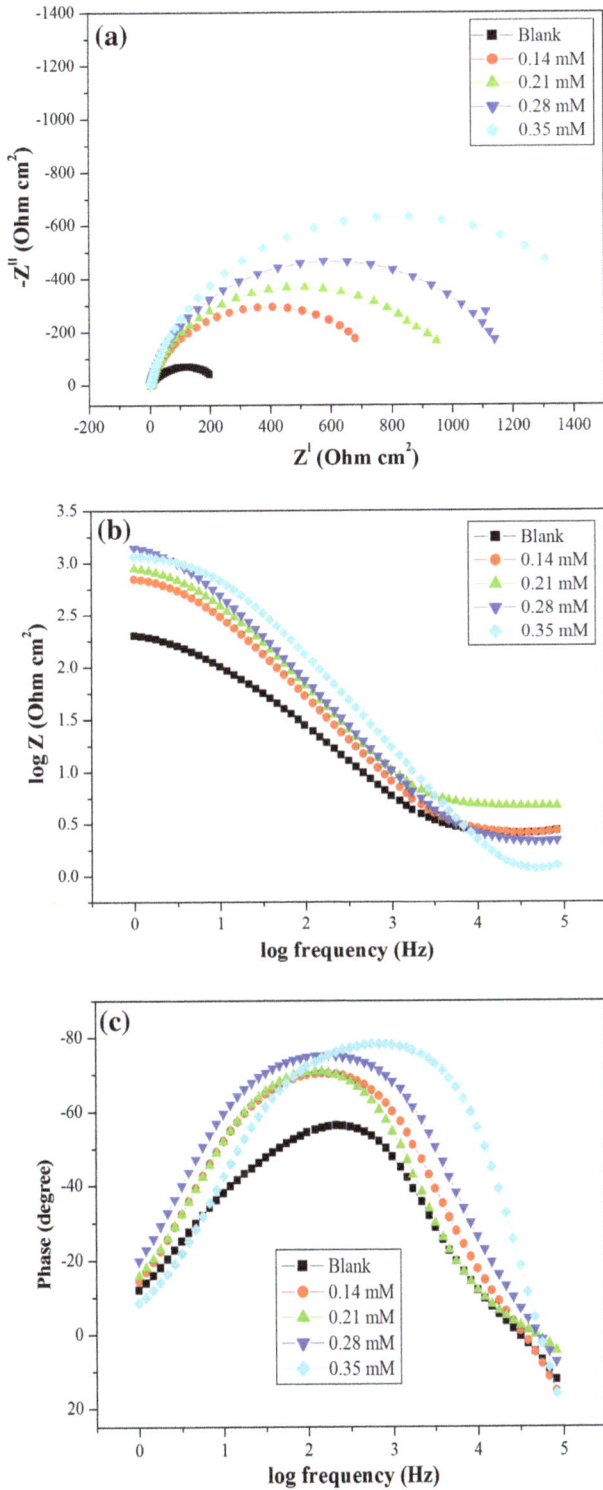

Fig. 6 **a** Nyquist plot. **b** Bode modulus plot. **c** Phase angle plot in the absence and presence of different concentrations of for P1

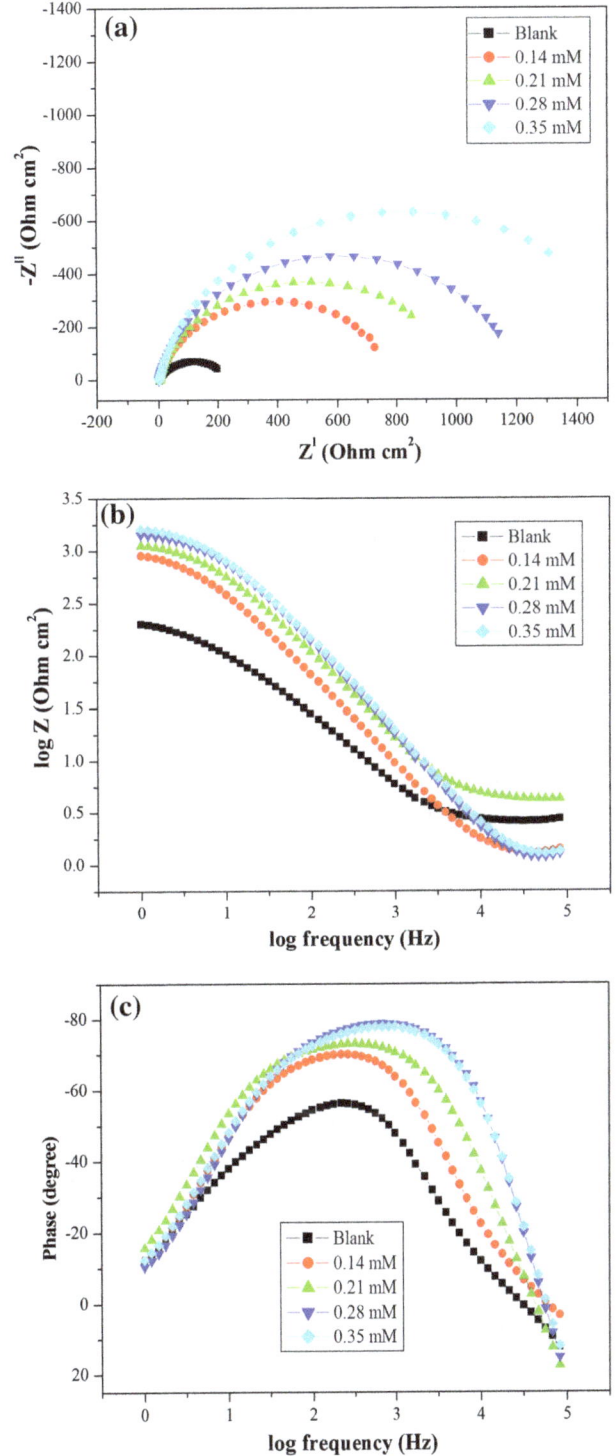

Fig. 7 **a** Nyquist plot. **b** Bode modulus plot. **c** Phase angle plot in the absence and presence of different concentrations of for P1

where $(R_p)_a$ and $(R_p)_p$ are the polarization resistances in the absence and presence of inhibitor, respectively. After the addition of inhibitors the polarization resistance (R_p) increases for both inhibitors. The increase in R_p value is due to the film formed on the steel surface which prevents the charge transfer. Such results were reported by many authors [47, 48]. Inhibition efficiency increases with the increase in the concentration of the inhibitors. As any other

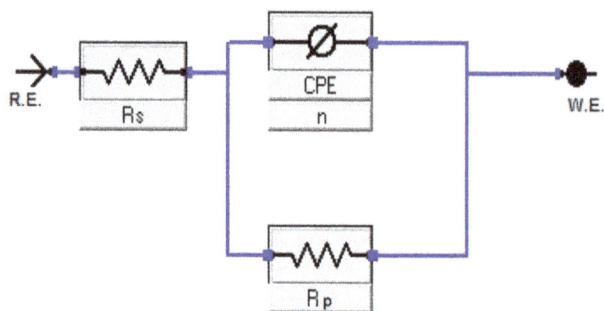

Fig. 8 Equivalent circuit model

electrochemical process corrosion also has two phases: (1) Oxidation of metal (charge transfer process). (2) The diffusion of metal ions from metal surface to the solution [49]. The film formed by the inhibitors P1 and P2 protects the metal by acting as a barrier to the diffusion of ions. Even though both P1 and P2 exhibit good inhibition efficiency, maximum efficiency is obtained for compound P2 because of the presence of plenty of electrons and an additional pyridine ring compared to P1.

CPE is a special element whose admittance is a function of angular frequency and whose phase is independent of frequency [50]. The impedance function of the CPE can be represented as follows:

$$Z_{CPE} = Y_o^{-1}(i\omega)^{-1} \tag{10}$$

where, Y_o is magnitude of CPE, ω is angular frequency (in rad s^{-1}), $i^2 = -1$ is the imaginary number, $n = \alpha/(\pi/2)$ in which α is the phase angle of CPE. The value of n signifies interphase parameter of working electrode. In this study n value for blank is 0.76, which increases by the addition of P1 (varying between 0.86 and 0.92) and P2 (0.87 and 0.92). Increase in the value of n by P1 and P2 addition represents capacitive behavior because for ideal capacitor n value is 1. The double layer capacitance (C_{dl}) can be calculated from CPE parameters using the equation,

$$C_{dl} = (Y_o R_{ct}^{1-n})^{1/n} \tag{11}$$

The decrease in C_{dl} is due to adsorption of inhibitors which displaces water molecules originally adsorbed on the MS surface which further decreases the active surface area [48]. Irregular trend in C_{dl} indicates the complexity of adsorption–desorption phenomenon. Such results were reported by many authors [51–53].

The single peak obtained in Bode plots for P1 and P2 indicates that the electrochemical impedance measurements fit well in one-time constant equivalent model. There is only one phase maximum in the Bode plot for both inhibitors, indicating only one relaxation process, which would be the charge transfer process, taking place at the metal–electrolyte interface [54]. The shift in phase angle is

due to the protective film formed on the MS surface. The shift increases with increase in concentration of the inhibitors. Phase angle value for P1 and P2 varies between 60° and 80°, whereas an ideal capacitor will be having by slope value of 90°.

Potentiodynamic polarization

Polarization measurements were performed to gain information regarding the kinetics of anodic and cathodic reactions. Tafel plots were drawn to study the mechanism of the inhibition process of P1 and P2 (Fig. 9). The electrochemical parameters, such as corrosion potential (E_{corr}), corrosion current density (i_{corr}), Tafel slopes (b_a, b_c) and linear polarization resistance are listed in Table 5.

Inhibition efficiency (IE %) will be known by corrosion current density (i_{corr}) calculated by the Tafel plot,

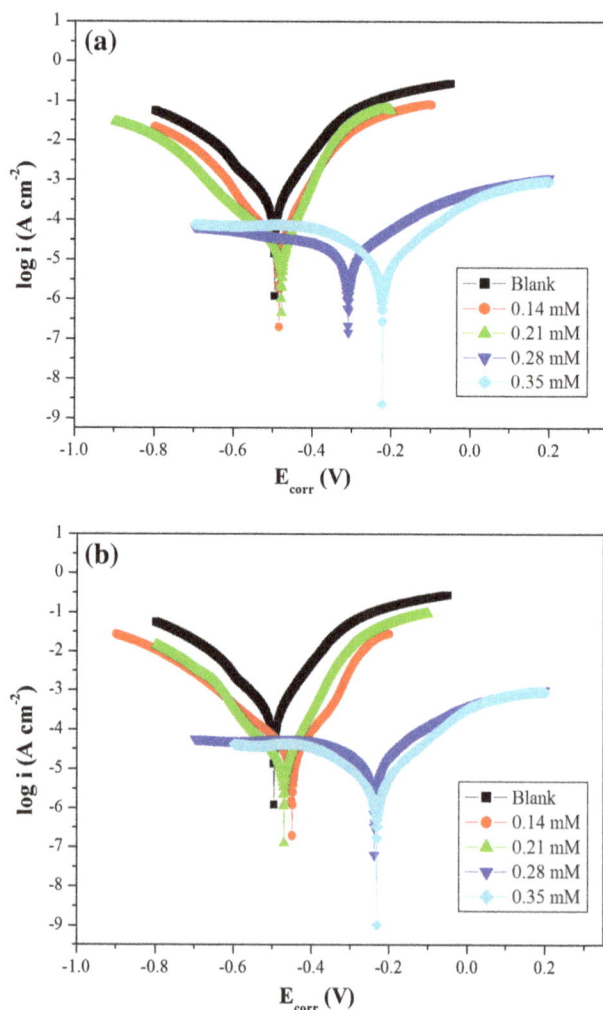

Fig. 9 Tafel plots for MS in 0.5 M HCl containing different concentration of **a** P1 and **b** P2

Table 5 Potentiodynamic polarization parameters for the corrosion of MS in 0.5 M HCl in absence and presence of different concentrations of P1 and P2 at 303 K

Inhibitor	Concentration (mM)	E_{corr} (mV)	i_{corr} (mA cm^{-2})	b_a (mV dec^{-1})	b_c (mV dec^{-1})	Linear polarization	IE (%)
	Blank	−496	0.2728	76	100	69	
P1	0.14	−484	0.0333	60	94	476	87.7
	0.21	−478	0.0232	47	120	636	91.5
	0.28	−308	0.0181	172	312	2668	93.4
	0.35	−222	0.0097	141	175	3525	96.4
P2	0.14	−449	0.0283	73	129	713	89.6
	0.21	−470	0.0173	51	85	807	93.6
	0.28	−237	0.0137	128	214	2546	94.9
	0.35	−231	0.0052	120	172	5927	98.0

$$IE (\%) = \frac{i^o_{corr} - i_{corr}}{i^o_{corr}} \times 100 \qquad (12)$$

where i^o_{corr} and i_{corr} are the uninhibited and the inhibited corrosion current densities, respectively.

Polarization is the shift in electrode potential from equilibrium value. By the addition of inhibitors there will be a shift in the corrosion potential value towards anodic region which is large in case of P1 and P2. This indicates that anodic reaction is predominantly hindered in the presence of inhibitors P1 and P2. Shift in anodic and cathodic Tafel slope values of P1 and P2 indicates that inhibitors are of mixed type promoting retardation of both anodic dissolution of C-steel and cathodic hydrogen discharge reaction. The irregular trends of b_a and b_c values indicate the involvement of more than one type of species adsorbed on the metal surface [55]. The corrosion current values decrease with increasing concentration of P1 and P2 as a result of decrease in corrosion rate after the formation of adsorbed film. Linear polarization resistance increases with increase in concentration of inhibitors. According to Ferreira et al. the displacement in E_{corr} is more than 85 mV relating to the corrosion potential of the blank, the inhibitor is considered as a cathodic or anodic type. If the change in E_{corr} is less than 85 mV, the corrosion inhibitor is regarded as a mixed type [56]. In the present case for both P1 and P2 shift is more than 85 mV and it is anodic hence, the inhibitors are anodic.

Quantum chemical calculations

Quantum chemical calculations were done with complete geometry optimizations to explore the theoretical–experimental consistency of P1 and P2 using Ab initio method. This method reveals the binding ability of an organic compound thereby it is possible to predict the ability to retard the dissolution of metal in aggressive media. Inhibition efficiency is correlated to the molecular and structural parameters that can be obtained through theoretical calculations, such as chemical selectivity, reactivity and charge distribution [57]. In this way, various quantum chemical parameters, such as energies of highest occupied molecular orbital (E_{HOMO}), and the energies of lowest unoccupied molecular orbital (E_{LUMO}),the energy gap(ΔE), hardness (η), softness (σ), electron affinity (A), electronegativity (χ) and ionization potential (I) of P1 and P2 were calculated and compared with results obtained by gravimetric and electrochemical methods. The computed parameters of P1 and P2 are listed in Table 6.

All the quantum chemical structures are given in Table 7. According to Frontier Molecular Orbital (FMO) theory of chemical reactivity, transition of electron is due to interaction between highest occupied molecular orbital (HOMO) and lowest unoccupied molecular orbital (LUMO) of reacting species [58]. High E_{HOMO} facilitates adsorption by influencing the transport process through the adsorbed layer, whereas low lying E_{LUMO} induces a back donation of charge from the metal to the molecule [59]. In

Table 6 List of quantum chemical parameters for P1 and P2

Quantum chemical parameters	P1	P2
Total energy (kJ mol^{-1})	−3,278,010	−3,927,246
Electronic kinetic energy (kJ mol^{-1})	3,268,423	3,917,446
Nuclear repulsion energy (kJ mol^{-1})	6,357,794	9,088,706
RMS gradient (kJ mol^{-1} Ang^{-1})	9.3766	5.02388
Dipole (debyes)	8.0509	2.3956
E_{HOMO} (eV)	−8.3873	−7.2715
E_{LUMO} (eV)	2.3311	2.2679
$\Delta E = E_{LUMO} - E_{HOMO}$ (eV)	10.7185	9.5394
Ionization potential, $I = -E_{HOMO}$	8.3873	7.2715
Electron affinity, $A = -E_{LUMO}$	−2.3311	−2.2679
Electronegativity (χ)	3.0281	2.5018
Hardness of the molecule (η)	5.3592	4.7697
Softness (σ)	0.1865	0.2096

Table 7 List of quantum chemical structures for P1 and P2

Quantum chemical structure	P1	P2
Optimized geometry		
Total Charge Density		
HOMO		
LUMO		

this study, P2 has higher E_{HOMO} value compared to P1, so it has better electron donating capacity. The better electron donation capacity of P2 can be correlated to the presence of nitrogen and oxygen atoms which contain lone pair of electrons and also aromatic electrons present on the ring. P2 has lower E_{LUMO} value compared to P1, which reflects better acceptance of electrons which helps in back donation and hence, emerges as a better inhibitor. The electron density of HOMO is mostly distributed around the delocalised electrons in both P1 and P2 which shows involvement of these in adsorption. Lower value of ΔE ensures better efficiency because the energy required to remove an electron from the last occupied orbital will be low [60]. The trend for ΔE follows the order P1 > P2 which suggests that P2 has higher interactions with the metal surface compared to P1. Molecular orbital (MO) theory can also be used to re-establish the results. As Fe atom is an electron pair acceptor it can be termed as Lewis acid whereas inhibitors P1 and P2 being electron donors act as base. According to MO concept, the overlap between LUMO of acid and HOMO of base acts as ruling factor for the formation of adsorption bond. The electron affinity of acid (Fe) is 0.2 eV, the negative of which is the E_{LUMO} of acid. The E_{HOMO} of P1 and P2 are −8.387 and −7.271 eV, respectively. Calculated values of energy gap between LUMO of acid and HOMO of two bases P1 and P2 works out to 8.187 and 7.071 eV, respectively. Lower the HOMO–LUMO energy gap, higher will be the HOMO–LUMO overlap and stronger is the acid–base bond formation and higher the inhibition efficiency [49]. As the energy gap of P2 is less compared to P1, it can be deduced that the interaction between P2 and Fe surface can be easily established compared to P1, so P2 is more efficient in inhibiting MS corrosion. Ionization potential is a fundamental descriptor of chemical reactivity of atoms and molecules [61]. If the ionization potential is high then molecule is more stable and it is difficult to remove an electron to form adsorption bond. Among the two inhibitors, P1 has higher value for ionization potential as compared to P2. Hence, P1 is chemically more inert compared to P2, therefore, P2 has better inhibition efficiency. The dipole moment value is inconsistent on the use of dipole moment as a predictor of the direction of a corrosion inhibition reaction. In literature also, significant relation has not been mentioned between dipole moment and inhibition efficiency [62, 63]. Electronegativity describes the tendency of an atom to attract electron density towards itself. According to Sanderson's electronegativity equalization principle [64], the molecule with high electronegativity quickly reaches equalization hence, low reactivity leads to lower inhibition efficiency. Among the studied molecules P1 has higher electronegativity hence, lower

reactivity. Chemical hardness measures the resistance of an atom to a charge transfer whereas softness is the measure of the capacity of an atom or group of atoms to receive electrons [65]. According to HSAB theory of chemical reactivity, hard acids prefer to co-ordinate with hard bases and soft acids with soft bases. It is a well-known fact that Fe being a transition metal acts as soft acid. P2 had the least hardness and maximum softness so it stands out as the better inhibitor.

Scanning electron microscopy

Morphologies of MS in the absence and presence of optimum concentrations of P1 and P2 at 300 K are presented in Fig. 10a–d. The polished and smooth surface of MS before immersion in HCl is shown in Fig. 10a. After exposure of MS to uninhibited solution, the corroded surface appears to be damaged with cracks and pits and it is shown in Fig. 10b. When the surface is treated with inhibitor solutions the adsorbed film formed on the steel surface retards corrosion. In Fig. 10c, d surface of MS covered with inhibitors P1 and P2, respectively can be observed.

Mechanism of inhibition

Inhibition mechanism can be proposed on the basis of adsorption of P1 and P2 on steel surface. As the studied inhibitors are basic (containing nitrogen atoms) there is a chance of protonation. So in acidic solution, both neutral and cationic forms of inhibitors exist. It is assumed that Cl^- ion first got adsorbed onto the positively charged metal surface by columbic attraction and then cationic form of inhibitor molecules can be adsorbed through electrostatic interactions between the positively charged molecules and the negatively charged metal surface [66]. The mechanism for the dissolution of Fe is reported by some authors [67].

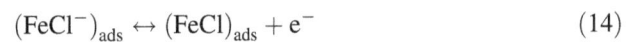

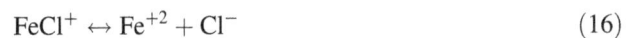

$$Fe + Cl^- \leftrightarrow (FeCl^-)_{ads} \tag{13}$$

$$(FeCl^-)_{ads} \leftrightarrow (FeCl)_{ads} + e^- \tag{14}$$

$$(FeCl)_{ads} \leftrightarrow FeCl^+ + e^- \tag{15}$$

$$FeCl^+ \leftrightarrow Fe^{+2} + Cl^- \tag{16}$$

Protonated form of P1 and P2 adsorbs on metal surface through $(FeCl^-)$ formed in step 13, forms $(FeCl^-InhH^+)$ and hence, steps 14, 15 and 16 can be prevented. So anodic reactions will be retarded.

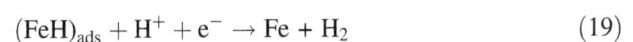

$$Fe + H^+ \leftrightarrow (FeH^+)_{ads} \tag{17}$$

$$(FeH^+) + e^- \leftrightarrow (FeH)_{ads} \tag{18}$$

$$(FeH)_{ads} + H^+ + e^- \rightarrow Fe + H_2 \tag{19}$$

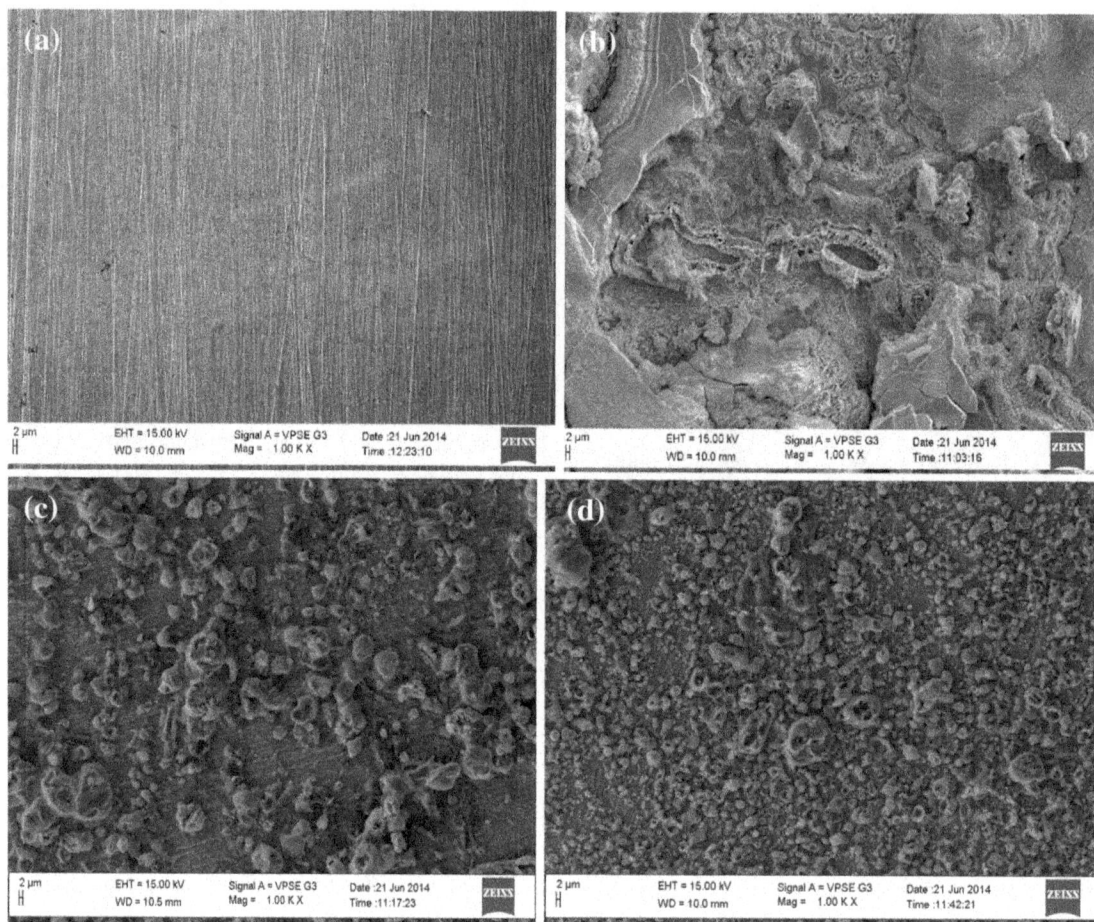

Fig. 10 SEM images of MS surface **a** polished, **b** immersed in 0.5 M HCl, **c** immersed in 0.5 M HCl in the presence 0.35 mM of P1, **d** immersed in 0.5 M HCl in the presence 0.35 mM of P2

Protonated inhibitors also adsorbs at the cathodic sites in competition with hydrogen ions so hydrogen evolution will be reduced.

Along with electrostatic force of attraction, inhibitors also adsorb on the MS surface through chemical adsorption. Both P1 and P2 contain hetero atoms (N and O), extended π electron system which helps in making effective co-ordinate bond with MS surface. Electron releasing methyl groups are also present which increases electron density on the ring. P2 contains an extended pyridine ring which also gives good contribution in decreasing corrosion rate. So inhibition takes place through both physisorption and chemisorption.

Conclusion

The following conclusions can be derived from the studies.

1. Levofloxacin derived amide (P2) and Levofloxacin (P1) showed very good activity in preventing corrosion of MS in hydrochloric acid media. P2 shows higher inhibition efficiency compared to P1.
2. Inhibition efficiency of MS varies directly with the concentration and inversely with the temperature.
3. Adsorption of both inhibitions follows Langmuir isotherm.
4. EIS measurements show that polarization resistance increases and double layer capacitance decreases on adding inhibitors.
5. Potentiondynamic polarization experiments reveal that P1 and P2 are anodic inhibitors.
6. Surface studies show the formation of adsorbed inhibitor layer on the metal surface.
7. From quantum chemical methods it is revealed that P2 has better electron donating ability compared to P1. This supports experimental results very well.

Acknowledgments One of the authors (T. K. C) received NON-NET fellowship from University grants Commission, New Delhi (ORDER NO. DV9/403/Misc/2014-15 dated 13/02/2015) and it is gratefully acknowledged.

References

1. Singh AK, Ebenso EE (2013) Cefotatan; a new corrosion inhibitor for corrosion of mild steel in hydrochloric acid solution. Int J Electrochem Sci 8:10903–10909

2. Zaafarany IA (2013) Corrosion inhibition of mild steel in hydrochloric acid solution using cationic surfactant olyel-amido derivatives. Int J Electrochem Sci 8:9531–9542

3. Loto RT, Loto CA, Popoola API (2012) Corrosion inhibition of thiourea and thiadiazole derivatives: a review. J Mater Environ Sci 3:885–894

4. Kumar RS, Danaee I, Avei RM, Vijayan M (2015) Quantum chemical and experimental investigations on equipotent effects of (+) R and (−) S enantiomers of racemic amisulpride as eco-friendly corrosion inhibitors for mild steel in acidic solution. J Mol Liq 212:168–186

5. Verma CB, Ebenso EE, Bahadur I, Obot IB, Quraishi MA (2015) 5-(phenylthio)-3H-pyrrole-4-carbonitriles as effective corrosion inhibitors for mild steel in 1 M HCl: Experimental and theoretical investigation. J Mol Liq 212:209–218

6. Gu T, Chen Z, Jiang X, Limei Z, Liao Y, Duan M, Wang H, Pu Q (2015) Synthesis and inhibition of N-alkyl-2-(4-hydroxybut-2-ynyl) pyridinium bromide for mild steel in acid solution: Box-Behnken design optimization and mechanism probe. Corros Sci 90:118–132

7. Eddy NO, Odoemelam SA, Ekwumemgbo P (2009) Inhibition of the corrosion of mild steel in H_2SO_4 by penicillin G. Sci Res Essays 4:33–38

8. Eddy NO, Ebenso EE, Ibok UJ (2010) Adsorption, synergistic inhibitive effect and quantum chemical studies of ampicillin (AMP) and halides for the corrosion of mild steel in H_2SO_4. J Appl Electrochem 40:445–456

9. Abdallah M (2004) Antibacterial drugs as corrosion inhibitors for corrosion of aluminium in hydrochloric acid solution. Corros Sci 46:1981–1996

10. Pang XH, Guo WJ, Li WH, Xie JD, Hou BR (2008) Electrochemical, quantum chemical and SEM investigation of the inhibiting effect and mechanism of ciprofloxacin, norfloxacin and ofloxacin on the corrosion for mild steel in hydrochloric acid. Sci China Ser B Chem 51:928–936

11. Pang X, Ran X, Kuang F, Xie J, Hou B (2010) Inhibiting effect of ciprofloxacin, norfloxacin and ofloxacin on corrosion of mild steel in hydrochloric acid. Chin J Chem Eng 18:337–345

12. Ebenso EE, Obot IB, Murulana LC (2010) Quinoline and its derivatives as effective corrosion inhibitors for mild steel in acidic medium. Int J Electrochem Sci 5:1574–1586

13. Shukla SK, Quraishi MA (2010) The effects of pharmaceutically active compound doxycycline on the corrosion of mild steel in hydrochloric acid solution. Corros Sci 52:314–321

14. El-Naggar MM (2007) Corrosion inhibition of mild steel in acidic medium by some sulfa drugs compounds. Corros Sci 49:2226–2236

15. Singh A, Singh AK, Quraishi MA (2010) Dapsone: a novel corrosion inhibitor for mild steel in acid media. Open Corros J 2:43–51

16. Obot IB (2009) Synergistic effect of nizoral and iodide ions on the corrosion inhibition of mild steel in sulphuric acid solution. Port Electrochim Acta 27:539–553

17. Solmaz R, Kardas G, Yazici B, Erbil M (2005) Inhibition effect of rhodanine for corrosion of mild steel in hydrochloric acid solution. Prot Met 41:628–632

18. Noel GJ (2009) A review of levofloxacin for the treatment of bacterial infections. Clin Med Ther 1:433–458

19. Gece G (2011) Drugs: a review of promising novel corrosion inhibitors. Corros Sci 53:3873–3898

20. Eddy NO, Stanislav R, Stoyanovand Ebenso EE (2010) Fluoroquinolones as corrosion inhibitors for mild steel in acidic medium; experimental and theoretical studies. Int J Electrochem Sci 5:1127–1150

21. Pang X, Zhang Y, Jie Z, Xie J, Baorong H (2011) Corrosion inhibition and mechanisms study on pipemidic acid, levofloxacin and ciprofloxacin for mild steel in 0.5 mol/L H_2SO_4. Acta Chim Sin 69:483–491

22. Gurudatt DM, Mohana KN (2014) Synthesis of new pyridine based 1, 3, 4-oxadiazole derivatives and their corrosion inhibition performance on mild steel in 0.5 M hydrochloric acid. Ind Eng Chem Res 53:2092–2105

23. Chaitra TK, Mohana KN, Tandon HC (2015) Thermodynamic, electrochemical and quantum chemical evaluation of some triazole Schiff bases as mild steel corrosion inhibitors in acid media. J Mol Liq 211:1026–1038

24. Mohana KN, Badiea AM (2008) Effect of sodium nitrite–borax blend on the corrosion rate of low carbon steel in industrial water medium. Corros Sci 50:2939–2947

25. Chaitra TK, Mohana KN, Tandon HC (2016) Study of new thiazole based pyridine derivatives as potential corrosion inhibitors for mild steel: theoretical and experimental approach. Int J Corros 2016:21. doi:10.1155/2016/9532809

26. Harish KP, Mohana KN, Mallesha L, Veeresh B (2014) Synthesis and in vivo anticonvulsant activity of 2-methyl-2-[3-(5-piperazin-1-yl-[1,3,4]oxadiazol-2-yl)-phenyl]-propionitrile derivatives. Arch Pharm Chem Life Sci 1:1–14

27. Bentiss F, Lebrini M, Legrenee M (2005) Thermodynamic characterization of metal dissolution and inhibitor adsorption processes in mild steel/2,5-bis(n-thienyl)-1,3,4-thiadiazoles/hydrochloric acid system. Corros Sci 47:2915–2931

28. Obi-Egbedi NO, Obot IB (2013) Xanthione: a new and effective corrosion inhibitor for mild steel in sulphuric acid solution. Arab J Chem 6:211–223

29. Herrag L, Hammouti B, Elkadri S, Aountini A, Jama C, Vezin H, Bentiss F (2010) Adsorption properties and inhibition of mild steel corrosion in hydrochloric solution by some newly synthesized diamine derivatives: experimental and theoretical investigations. Corros Sci 52:3042–3051

30. Singh AK, Quraishi MA (2011) Investigation of the effect of disulfiram on corrosion of mild steel in hydrochloric acid solution. Corros Sci 53:1288–1297

31. Fouda AS, Al-Sarawy AA, El-Katori EE (2006) Pyrazolone derivatives as corrosion inhibitors for C-steel in hydrochloric acid solution. Desalination 201:1–13

32. Guan NM, Xueming L, Fei L (2004) Synergistic inhibition between o-phenanthroline and chloride ion on cold rolled steel corrosion in phosphoric acid. Mater Chem Phys 86:59–68

33. Yadav DK, Quraishi MA, Maiti B (2012) Inhibition effect of some benzylidenes on mild steel in 1 M HCl: an experimental and theoretical correlation. Corros Sci 55(2012):254–266

34. Ghazoui A, Saddik R, Benchat N, Hammouti B, Guenbour M, Zarrouk A, Ramdani M (2012) The role of 3-amino-2-phenylimidazo[1,2-a]pyridine as corrosion inhibitor for C38 steel in 1 M HCl. Der Pharma Chem 4:352–364

35. Obi-Egbedi NO, Obot IB (2011) Inhibitive properties, thermodynamic and quantum chemical studies of alloxazine on mild steel corrosion in H_2SO_4. Corros Sci 53:263–275

36. Behpour M, Ghoreishi SM, Soltani N, Salavati-Niasari M, Hamadanian M, Gandomi A (2008) Electrochemical and theoretical investigation on the corrosion inhibition of mild steel by thiosalicylaldehyde derivatives in hydrochloric acid solution. Corros Sci 50:2172–2218

37. Solmaz R, Altunbas E, Kardas G (2011) Adsorption and corrosion inhibition effect of 2-((5 mercapto-1, 3, 4-thiadiazol-2

ylimino) methyl) phenol Schiff base on mild steel. Mat Chem Phys 125:796–801

38. Singh A, Ebenso EE, Quraishi MA (2012) Theoretical and electrochemical studies of metformin as corrosion inhibitor for mild steel in hydrochloric acid solution. Int J Electrochem Sci 7:4766–4779

39. Solmaz R (2014) Investigation of corrosion inhibition mechanism and stability of vitamin B1 on mild steel in 0.5 M HCl solution. Corros Sci 81:75–84

40. Solmaz R (2010) Investigation of the inhibition effect of 5-((E)-4-phenylbuta-1,3-dienylideneamino)-1,3,4-thiadiazole-2-thiol Schiff base on mild steel corrosion in hydrochloric acid. Corros Sci 52:3321–3330

41. Heakal FE, Fouda AS, Radwan MS (2011) Inhibitive effect of some thiadiazole derivatives on C-steel corrosion in neutral sodium chloride solution. Mat Chem Phys 125:26–36

42. Ayati NS, Khandandela S, Momeni M, Moayed MH, Davoodi A, Rahimizadeh M (2011) Inhibitive effect of synthesized 2-(3-pyridyl)-3, 4-dihydro-4-quinazolinone as a corrosion inhibitor for mild steel in hydrochloric acid. Mat Chem Phys 126:873–879

43. Benabdellah M, Tounsi A, Khaled KF, Hammouti B (2011) Thermodynamic, chemical and electrochemical investigations of 2-mercapto benzimidazole as corrosion inhibitor for mild steel in hydrochloric acid solutions. Arab J Chem 4:17–24

44. Tao Z, Zhang S, Li W, Hou B (2010) Adsorption and corrosion inhibition behavior of mild steel by one derivative of benzoic-triazole in acidic solution. Ind Eng Chem Res 49:2593–2599

45. Solmaz R, Kardas G, Culha M, Yazici B, Erbil M (2008) Investigation of adsorption and inhibitive effect of 2-mercaptothiazoline on corrosion of mild steel in hydrochloric acid media. Electrochim Acta 53:5941–5952

46. Tsuru T, Haruyama S, Gijutsu B (1978) Corrosion inhibition of iron by amphoteric surfactants in 2 M HCl. J Jpn Soc Corros Eng 27:573–581

47. Ozcan M, Dehri I, Erbil M (2004) Organic sulphur-containing compounds as corrosion inhibitors for mild steel in acidic media: correlation between inhibition efficiency and chemical structure. Appl Surf Sci 236:155–164

48. Khaled KF (2008) New synthesized guanidine derivative as a green corrosion inhibitor for mild steel in acidic solutions. Int J Electrochem Sci 3:462–475

49. Ahamed I, Prasad R, Quraishi MA (2010) Experimental and theoretical investigations of adsorption of fexofenadine at mild steel/hydrochloric acid interface as corrosion inhibitor. J Solid State Electrohem 14:2095–2105

50. Singh AK (2012) Inhibition of mild steel corrosion in hydrochloric acid solution by 3-(4-((z)-indolin-3-ylideneamino) phenylimino) indolin-2-one. Ind Eng Chem Res 51:3215–3223

51. Karthik R, Muthukrishnan P, Chen S, Jeyaprabha B, Prakash P (2015) Anti-corrosion inhibition of mild steel in 1 M hydrochloric acid solution by using tiliacoraaccuminata leaves extract. Int J Electrochem Sci 10:3707–3725

52. Ergun U, Yuzer D, Emregu KC (2008) The inhibitory effect of bis-2,6-(3,5-dimethylpyrazolyl) pyridine on the corrosion behaviour of mild steel in HCl solution. Mat Chem Phys 109:492–499

53. Quraishi MA, Sudheer Ansari KR, Ebenso EE (2012) 3-Aryl substituted triazole derivatives as new and effective corrosion inhibitors for mild steel in hydrochloric acid solution. Int J Electrochem Sci 7:7476–7492

54. Yadav M, Kumar S (2014) Experimental, thermodynamic and quantum chemical studies on adsorption and corrosion inhibition performance of synthesized pyridine derivatives on N80 steel in HCl solution. Surf Interface Anal 46:254–268

55. Fouda AS, Elewady YA, El-Aziz HKA (2012) Corrosion inhibition of carbon steel by cationic surfactants in 0.5 M HCl solution. J Chem Sci Technol 1:45–53

56. Ferreira ES, Giancomlli C, Giacomlli FC, Spinelli A (2004) Evaluation of the inhibitor effect of L-ascorbic acid on the corrosion of mild steel. Mater Chem Phys 83:129–134

57. Oguike RS, Kolo AM, Shibdawa AM, Gyenna HA (2013) Density functional theory of mild steel corrosion in acidic media using dyes as inhibitor: adsorption onto Fe(110) from gas phase. ISRN Phys Chem Article no. 175910. doi:10.1155/2013/175910

58. Musa AY, Kadhum AH, Mohamad AB, Rohoma AB, Mesmari H (2010) Electrochemical and quantum chemical calculations on 4, 4-dimethyloxazolidine-2-thione as inhibitor for mild steel corrosion in hydrochloric acid. J Mol Struc 969:233–237

59. Oguzie Onuoha GN, Onuchukwu AI (2005) Inhibitory mechanism of mild steel corrosion in 2 M sulphuric acid solution by methylene blue dye. Mat Chem Phys 89:305–311

60. Zarrok H, Zarrouk A, Salghi R, Oudda H, Hammouti B, Assouag M, Taleb M, Touhami ME, Bouachrine M, Boukhris S (2012) Gravimetric and quantum chemical studies of 1-[4-acetyl-2-(4-chlorophenyl)quinoxalin-1(4H)-yl]acetone as corrosion inhibitor for carbon steel in hydrochloric acid solution. J Chem Pharmac Res 4:5056–5066

61. Udayakala P, Rajendran TV, Gunashekaran S (2015) Theoretical approach to the corrosion inhibition efficiency of some pyrimidine derivatives using DFT method. J Comput Methods in Mol Design 2:1–12

62. Gao G, Liang C (2007) Electrochemical and DFT studies of β-amino-alcohols as corrosion inhibitors for brass. Electrochim Acta 52:4554–4559

63. Khalil N (2003) Quantum chemical approach of corrosion inhibition. Electrochim Acta 48:2635–2640

64. Geerlings P, Proft FD (2002) Chemical reactivity as described by quantum chemical methods. Int J Mol Sci 3:276–309

65. Senet P (1997) Chemical hardnesses of atoms and molecules from frontier orbitals. Chem Phys Lett 275:527–532

66. Zhou X, Yang H, Wang F (2011) [BMIM]BF$_4$ ionic liquids as effective inhibitor for carbon steel in alkaline chloride solution. Electrochim Acta 56:4268–4275

67. Yousefi A, Javadian S, Dalir N, Kakemam J, Akbari J (2015) Imidazolium-based ionic liquids as modulators of corrosion inhibition of SDS on mild steel in hydrochloric acid solutions: experimental and theoretical studies. RSC Adv 5:11697–11713

11

Biosorption of cationic dye from aqueous solutions onto lignocellulosic biomass (*Luffa cylindrica*): characterization, equilibrium, kinetic and thermodynamic studies

Noureddine Boudechiche[1] · Hassiba Mokaddem[1] · Zahra Sadaoui[1] · Mohamed Trari[2]

Abstract In the present study, biomass fiber (*Luffa cylindrica*) has been successfully used as biosorbent for the removal of a cationic dye namely, methylene blue, from aqueous solution using a batch process. The characterization of the biosorbent was carried out by the infrared spectroscopy (FTIR) and scanning electron microscopy (SEM). The chemical composition has been established by the energy dispersive X-Ray spectroscopy (EDS). The effects of various parameters such as the contact time (0–160 min), solution pH (2–10), biosorbent dose (0.5–8 g L^{-1}), particle size, initial MB concentration (20–300 mg L^{-1}) and temperature (20–60 °C) were optimized. The biosorption isotherms were investigated by the Langmuir, Freundlich, Dubinin–Radushkevich and Tempkin models. The data were well fitted with the Langmuir model, with a maximum biosorption capacity of 49.46 mg g^{-1} at 20 °C. The kinetics data were analyzed by the pseudo-first-order and pseudo-second-order models. The mass transfer model in terms of interlayer diffusion was applied to examine the mechanisms of the rate-controlling step ($R^2 = 0.9992$–0.9999). The thermodynamic parameters: free energy ($\Delta G° = -5.428$ to -3.364 kJ mol^{-1}), enthalpy ($\Delta H° = -20.547$ kJ mol^{-1}) and entropy ($\Delta S° = -0.052$ kJ mol^{-1} K^{-1}) were determined over the temperatures range (20–60 °C). The results indicate that

Luffa cylindrica could be an interesting biomass of alternative material with respect to more costly adsorbents used nowadays for dye removal.

Keywords Biosorption · *Luffa cylindrica* · Methylene blue · Characterization · Kinetic · Isotherm

Introduction

Dyes are widely used in various industries such as textile, leather, paper, printing, food, cosmetics, paint, pigments, petroleum, solvent, rubber, plastic, pesticide, wood preserving chemicals, and pharmaceutical industry. Over 10,000 of different commercial dyes and pigments exist currently and more than 7×10^5 tonnes are produced annually worldwide [1–3]. Discharge of dye-bearing wastewaters into the natural environment from textile, paper and leather industries causes a serious threat for the aquatic life [4]. On the other hand, limited aquatic resources and increasing demand for safe water require efficient water treatment methods [5]. Synthetic dyes are generally resistant to biodegradation and physicochemical techniques for their removal [6, 7], such as adsorption, chemical oxidation, electrocoagulation and advanced oxidation processes (AOPs) have been extensively used to comply with more and more stringent legislation regarding the maximum allowable dye concentration in wastewaters [7–10]. Methylene blue (MB) is a thiazine cationic dye with widespread applications, including coloring paper, dyeing cottons, wools and coating for paper stock. It is also used in microbiology, surgery and diagnostics and as a sensitizer in photo-oxidation of organic pollutants. Although it has low toxicity, it can cause some specific harmful effects for the human health such as heartbeat

✉ Mohamed Trari
solarchemistry@gmail.com

[1] Laboratory of Engineering Reaction, Faculty of Engineering Mechanic and Engineering Processes, USTHB, BP 32, Algiers, Algeria

[2] Laboratory of Storage and Valorization of Renewable Energies, Faculty of Chemistry, USTHB, BP 32, Algiers, Algeria

increase, vomiting, shocks, cyanosis, jaundice and tissue necrosis [11, 12]. Hence, its removal from wastewaters is an important issue for the environmental protection [13]. The conventional methods have been extensively used for treating waters contaminated with heavy metal and dyes [14–16]. However, these methods present some disadvantages such as high cost, low removal efficiency and production of excessive toxic sludge [17]. Recently, inexpensive, ecofriendly and not pathogenic organisms have been used for the dye removal [18]. In this respect, the biosorption process has attracted a great interest in this context, and seems a good alternative for the removal of dyes and other pollutants from wastewaters [19, 20], as a replacement for costly commercially biosorbents [21]. It can be defined as sequestering of organic or inorganic compounds by alive or dead biomasses or their derivatives; the biomass can consist of bacteria [22], fungal [19], yeasts [22], algae [23], seaweeds and even industrial or agricultural wastes [24, 25]. Different vegetal biomasses have been used such as Opuntia ficus indica [26], Sugar beet pulp [21], Stoechospermum marginatum [24], *Scolymus hispanicus* L. [27], Palm kernel [28], Pinus brutia Ten. [29], Waste orange peel [30], Posidonia oceanica L. [31], Cyperus rotundus [32], Date stones and Palm-trees waste [33].

The present study examines a new dye biosorbent namely the *Luffa cylindrica* fiber and its feasibility for the removal of methylene blue from aqueous solution. It is inexpensive and easily available in many regions of Algeria. *Luffa cylindrica* is composed of 60 % cellulose, 30 % hemicelluloses and 10 % lignin and is classified as lignocellulosic material [34]; the *Luffa* products are natural and biodegradable. The biosorption of methylene blue onto *Luffa cylindrica* fiber is carried out by batch biosorption experiments. The influence of the contact time, initial pH, biosorbent dose, initial MB concentration, particle size and temperature is investigated. Furthermore, the isotherm and kinetic models are evaluated and the thermodynamic data are determined.

Materials and methods

Preparation of the biosorbent

The *Luffa cylindrica* plant was naturally collected in July, from Algeria. The plant was repeatedly washed with distilled water to remove dirt particles, dried at 80 °C for 48 h, crushed in grinder and sieved to obtain particle sizes in the range (63–630 μm). The powdered biosorbent was stored in an airtight container until use.

Point of zero charge (pH_{pzc})

The point of zero charge (pH_{pzc}) of the *Luffa cylindrica* fiber was evaluated by the solid addition method using KNO_3 (0.01 M) solution [36]. The experiments were carried out in 100 mL erlenmeyer flasks with stopper cork containing 50 mL of KNO_3 solution (10^{-2} M). The initial pH (pH_i) in each flask was adjusted between 3 and 11 by adding NaOH or HCl solutions (0.1 M). Then, 0.5 g of the *Luffa cylindrica* was added to each flask which are kept for 48 h with intermittent manual shaking to reach the equilibrium. The difference of the initial and final pH (pH_i, pH_f) was plotted against the initial pH. The point of intersection of the resulting curve with the abscissa axis, for which $\Delta pH = 0$, gives pH_{pzc} (Fig. 1).

Methylene blue solution

The dye used in all experiments was methylene blue, a basic cationic dye supplied by (Biochem company, Algeria). MB was chosen because of its various applications. MB has a molecular weight of 319.85 g mol^{-1}, which corresponds to methylene blue hydrochloride with three water molecules, the structure is shown in Fig. 2.

The FT-IR spectra were recorded over the range (400–4000 cm^{-1}) using a Shimadzu FTIR-8400S spectrometer. The scanning electron microscopy (SEM) was performed with a JEOL-JSM 6360 Microscope.

Batch biosorption experiments

The biosorption was conducted in Pyrex 500 mL conical flasks at a constant agitation speed. The experiments were carried out by varying the biosorbent particle size over the range (63–630 μm), contact time (5–160 min), biosorbent dosage (0.5–8 g L^{-1}), pH (2–10), initial dye concentrations (20–300 mg L^{-1}) and temperatures (20–60 °C). The

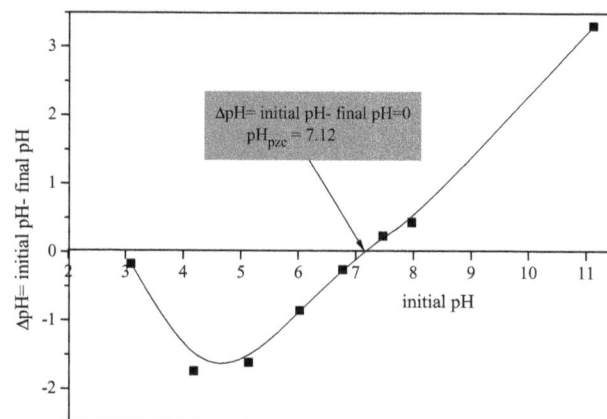

Fig. 1 The chemical structure of the methylene blue

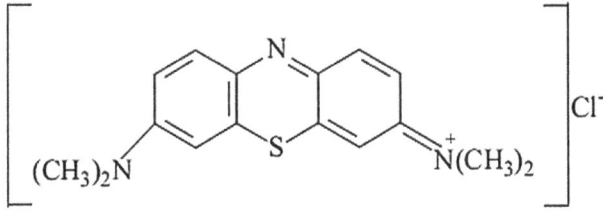

Fig. 2 The determination of the point of zero charge (pH$_{PZC}$)

temperature was controlled with an isothermal shaker. After each biosorption test, the sample was centrifugated (6000 rpm, 10 min) for solid–liquid separation; the residual MB concentration was analyzed by a UV–Vis spectrophotometer (2120 UV Optizen III, South Korea) at $\lambda_{max} = 663$ nm. The equilibrium, kinetic and thermodynamic study were performed by determining the optimum biosorption conditions. The amount of MB biosorbent q_t (mg g^{-1}) was calculated from the relation (1):

$$q_t = \frac{(C_0 - C_t)}{m} V \tag{1}$$

where C_0 is the initial dye concentration (mg L^{-1}), C_t the concentration of dye at time t (mg L^{-1}), V the volume of the solution (L) and m the mass of biosorbent (g). The dye removal percentage is calculated as:

$$R\,(\%) = \frac{(C_0 - C_t)}{C_0} 100 \tag{2}$$

Statistical evaluation of the kinetic and isotherm parameters

To determine the best-fit model for the biosorption, the linear curve fitting by the software OriginPro 8.5 was employed to simulate and to confirm the fitting of the biosorption kinetic and isotherm models to the experimental data. The statistical significance of variables was evaluated from the analysis of variance ANOVA (Fisher function, F value, and probability, P value), while the adjusted correlation coefficient (Adjusted R^2) was used to assess the adequacy of the fitting [35]. F value and Adjusted R^2 were calculated as:

$$F\text{ value} = \frac{\left(\sum_{i=1}^{n} \left(q_{i,cal} - \bar{q}_{i,exp}\right)^2\right)/p - 1}{\left(\sum_{i=1}^{n} \left(q_{i,exp} - q_{i,cal}\right)^2\right)/n - p} \tag{3}$$

$$\text{Adjusted } R^2 = 1 - \frac{\left(\sum_{i=1}^{n} \left(q_{i,exp} - q_{i,cal}\right)^2\right)/n - p}{\left(\sum_{i=1}^{n} \left(q_{i,exp} - \bar{q}_{i,exp}\right)^2\right)/n - 1} \tag{4}$$

where $q_{i,exp}$ is each value of q_i measured experimentally, $q_{i,cal}$ is each value of q_i predicted by the fitted model, $\bar{q}_{e,exp}$ is the average of q_i experimentally measured, n is the number of experiments performed and p is the number of parameter of the fitted model.

Desorption

MB solution (100 mg L^{-1}) was mixed with *Luffa cylindrica* at pH 6 for 4 h. The residual MB concentration was measured. The MB loaded *Luffa cylindrica* was dried at 80 °C. Four eluting solvents (100 mL): H$_2$O, HCl (0.1 M), NaOH (0.1 M), and NaCl (0.1 M) each one containing 0.2 g of MB loaded *Luffa cylindrica* at room temperature. The percentage of desorbed dye from the adsorbent was calculated (=100× desorbed mass/adsorbed mass).

Results and discussion

Characterization

FT-IR analysis of the biosorbent

The FT-IR spectrum of the *Luffa cylindrica* was plotted to obtain information about the nature of functional groups at the surface. The spectrum (Fig. 3) shows a dominant peak at 3450 cm^{-1} attributed to O–H stretching vibrations in hydroxyl groups, involved in hydrogen bonds. The bands observed at 2944 cm^{-1} are assigned to asymmetric C–H bonds, present in alkyl groups. The absorption peaks at 1737 cm^{-1} correspond to stretching of carboxyl groups. The strong absorption band at 1639 cm^{-1} is indicative of OH bending vibrations, while that at 1401 cm^{-1} is due to C–O stretching. The band at 1322 cm^{-1} is assigned to C–O groups on the biomass surface, whereas that at 1160 cm^{-1} corresponds to antisymmetric bridge C–OR–C stretching

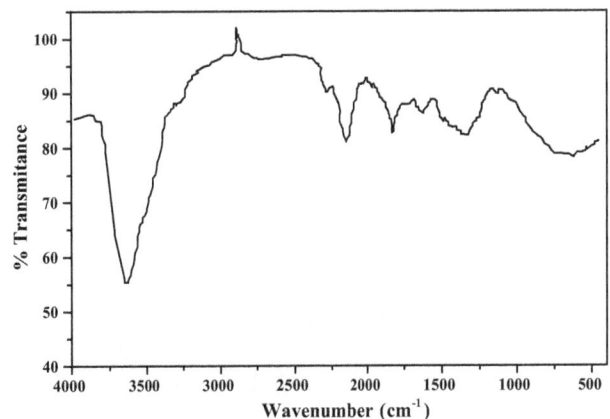

Fig. 3 FTIR spectrum of *Luffa cylindrica*

(cellulose) [37, 38]. The band at 1107 cm^{-1} is attributed to anhydroglucose ring (cellulose) [38]. The peaks at 1058 cm^{-1} are indicative of C-OR stretching (cellulose), while the band at 884 cm^{-1} could be attributed to antisymmetric, out of phase ring stretching [37].

SEM–EDS analysis

The morphology of the *Luffa cylindrica* was observed by SEM. The fibers, formed by fibrils glued, are disposed in a multi-directional array, forming a natural mat (Fig. 4a); the diameters of single fibers are in the range (63–125 μm). To observe the inner fibrils and further investigate the complicated physical structures in the natural *Luffa cylindrica*, a crude fiber was observed at high magnification (Fig. 4b). The SEM image shows that the fiber has a heterogeneous appearance with an outer rich lignin layer around the fibers. The internal fibrils cannot be seen due to the lignin layer. At higher magnification (Fig. 4c, d), the SEM image displays a rougher surface with lots of waxy and gummy substances on the untreated *Luffa cylindrica* fiber; the internal fibrils cannot be observed [38]. The EDS spectrum is shown in Fig. 5 and the contents of each element are listed in Table 1. The energy dispersive X-Ray microanalysis (SEM/EDS) of the *Luffa cylindrica* fibers indicates mainly the presence of carbon (65.68 %) and oxygen (30.13 %). However, as the EDS analysis is less sensitive for light elements ($Z \leq 10$) [39], the carbon and oxygen content were quantified by ultimate analysis. Their concentrations suggest the presence of high amount of different oxygenated groups on the carbon surfaces, such as Cl, Ca, Na, Cu, Mg, K, Ni, Si and P whose contents are between 0.09 and 1.21 %. Similar results (carbon: 64.0 %, oxygen: 34.9 %) were already obtained by Tanobe et al. [38].

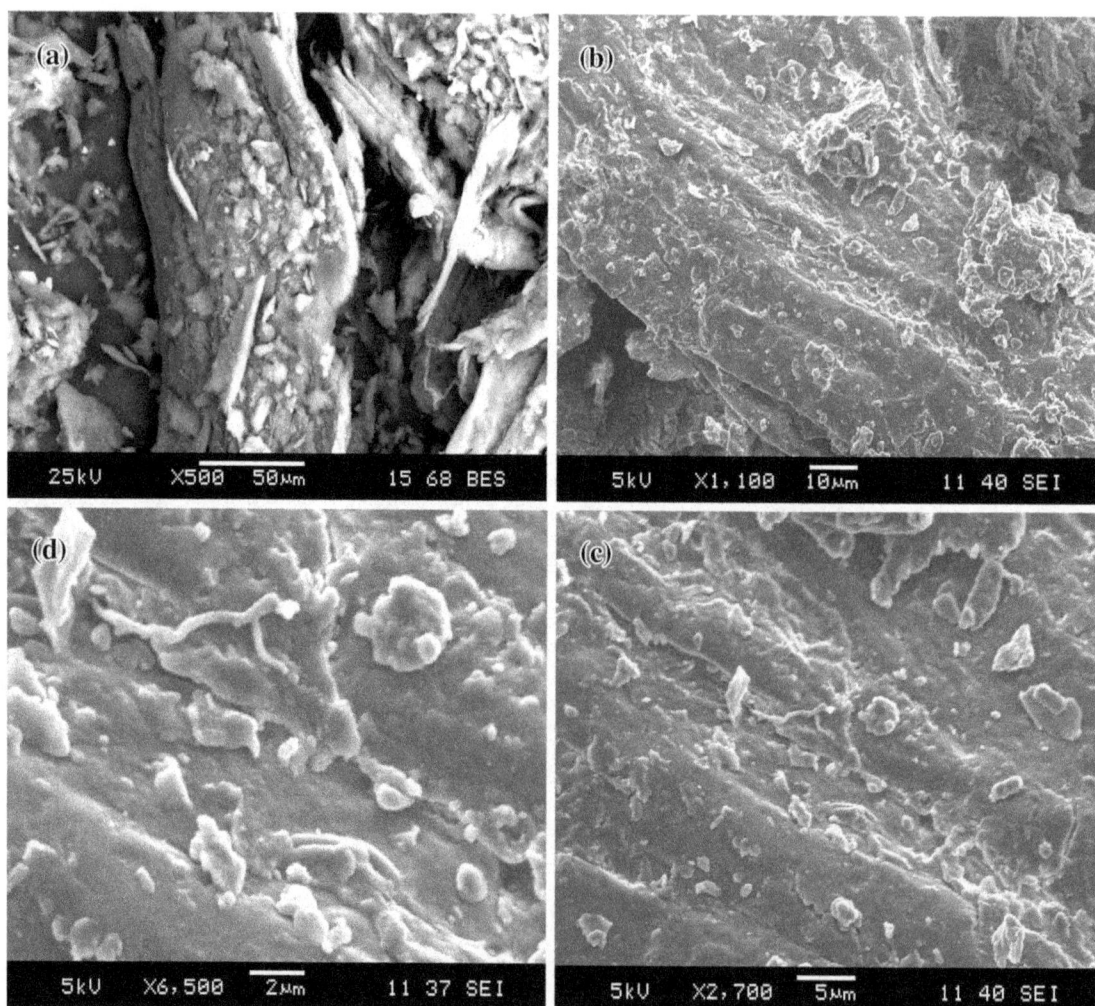

Fig. 4 SEM micrographs of *Luffa cylindrica*

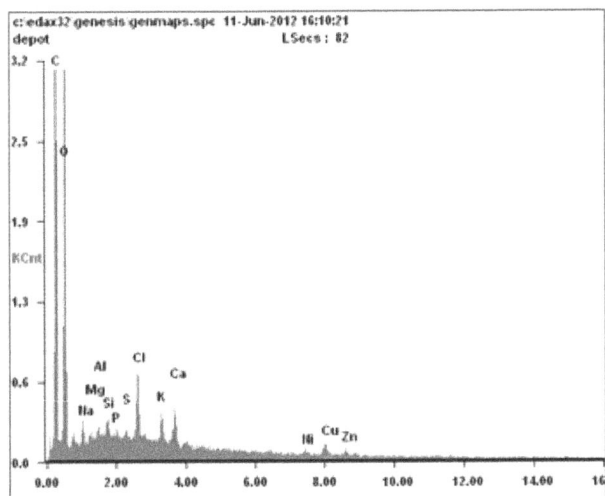

Fig. 5 EDS spectrum from the *Luffa cylindrica*

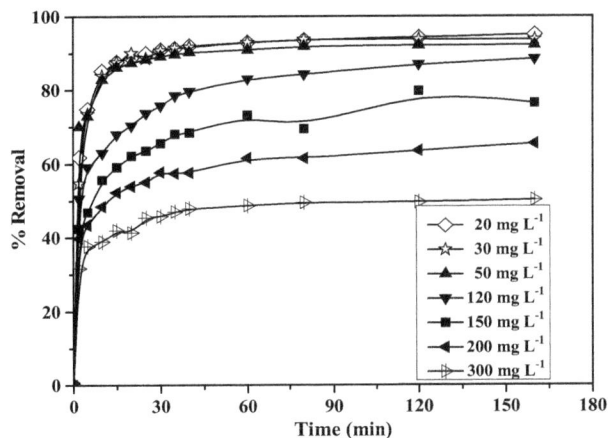

Fig. 6 Effect of contact time on the biosorption kinetics of MB by *Luffa cylindrica* (biosorbent dose $= 3$ g L^{-1}, initial pH $= 5.80$ and $T = 20$ °C)

Biosorption

Effect of contact time and initial dye concentration

Experiments were undertaken to study the effect of the initial concentration of MB over the range (20–300 mg L^{-1}) at 20 °C on the biosorption onto *Luffa cylindrica* at regular interval times. The rate of the MB removal by *Luffa cylindrica* was rapid, the maximum uptake was achieved in the first 20 min, accounting for 90–42 % biosorption, respectively, for MB initial concentrations of 20–300 mg L^{-1} (Fig. 6). The biosorption rate after this initial fast phase slows down significantly until it reaches a plateau after 60 min, indicating equilibrium of the system. The initial rapid phase may be due to an increase in the number of available vacant sites. The increase of the biosorption with raising the MB concentration is attributed to the fact that at higher concentrations, the ratio of the initial number of MB molecules to the available surface area is large; consequently, the fractional biosorption becomes dependent on the initial concentration. By contrast, at low concentrations, the available sites of biosorption are fewer and hence the MB removal depends upon their concentration [40].

Effect of solution pH

The pH of the solution is a crucial controlling parameter in the biosorption [41, 42]. This is possibly due to its impact on both the surface binding sites of the biosorbent and

ionization status of the MB molecule in water. Since the MB biosorption can dramatically change with changing pH, it has been stressed that not only it should be accurately reported but also the data for all comparative studies must be obtained at the same pH values. The effect of pH on MB biosorption was studied over the pH range (2–10) and the results are shown in Fig. 7. The equilibrium biosorption uptake presents a minimum at pH ~ 2 (6.16 %) and increases up to 5, then remains nearly constant (80.86 %) over the initial pH ranges (6–10). At low pHs, the surface charge is positively charged, and the H$^+$ ions compete effectively with dye cations causing a decrease in the amount of adsorbed dye. At higher pH, the *Luffa cylindrica* fibers, mainly lignin and cellulose chains, become negatively charged, thus enhancing the cationic dye by electrostatic attraction forces [43, 44].

Effect of biosorbent dose

The biosorbent dose is an important parameter because it determines the capacity of biosorbent for a given concentration of the adsorbate [45]. The effect of the biomass dosage (0.5–8 g L^{-1}) on the MB biosorption was studied in 1 L MB solution (50 mg L^{-1}) under optimized conditions of pH and contact time. The removal percentage of MB increases drastically from 12.77 to 96.16 % for biosorbent dosage of 0.5 and 8 g L^{-1}, respectively (Fig. 8). This is due to the availability of more binding sites as the dose of biosorbent increases. It is due to the high number of unsaturated biosorption sites during the biosorption process

Table 1 Principal elements identified on the biomass surface by SEM/EDS

Element	C	O	Na	Mg	Si	P	Cl	K	Ca	Ni	Cu
Content (%)	65.68	30.13	0.71	0.19	0.11	0.09	1.21	0.36	0.77	0.16	0.50

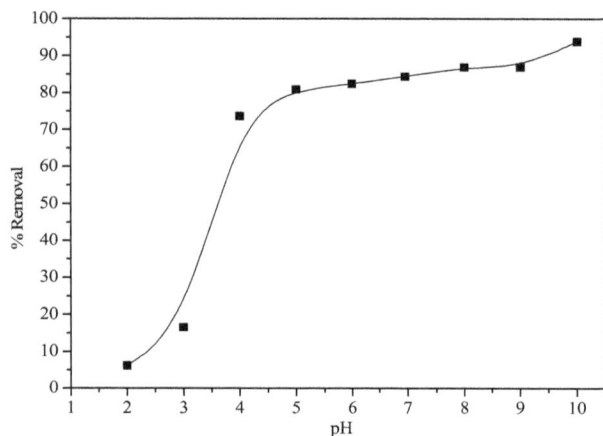

Fig. 7 Effect of the solution pH on the MB removal ($C_0 = 20$ mg L^{-1}, biosorbent dose = 1 g L^{-1} and $T = 20$ °C)

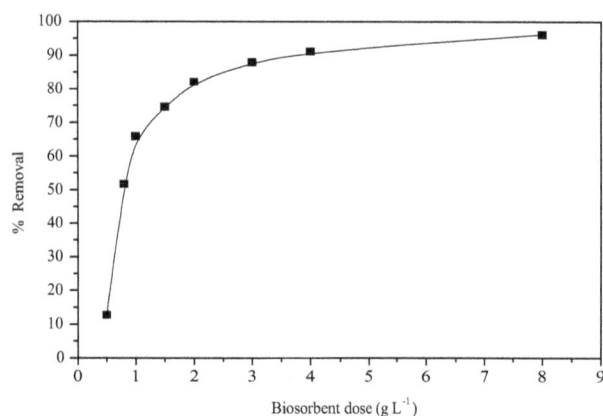

Fig. 8 Effect of the biosorbent dose on the MB biosorption by *Luffa cylindrica* ($C_0 = 20$ mg L^{-1}, initial pH 5.80 and $T = 20$ °C)

[46]. Similar results were previously reported by some researchers [45, 47].

Effect of biosorbent particle size

The particle size of the biosorbent can greatly influence the external surface of the biosorbent, thus impacting on its interaction with the solution through the effect of resistance to the film diffusion. As a consequence, a variation in the biosorbent particle size modifies the accessibility and the availability of reactive groups present on its surface [13]. The biosorption of MB was studied at four different domains (63–125, 125–250, 250–400 and 400–630 μm) of the biomass fibers. As expected, it was found that the MB biosorption decreases with increasing the size of the biosorbent (Fig. 9). This is due to larger surface area of smaller particles for the same amount of the biosorbent. For larger particles, the diffusion resistance to the mass transport is higher, and most of the internal surface of the

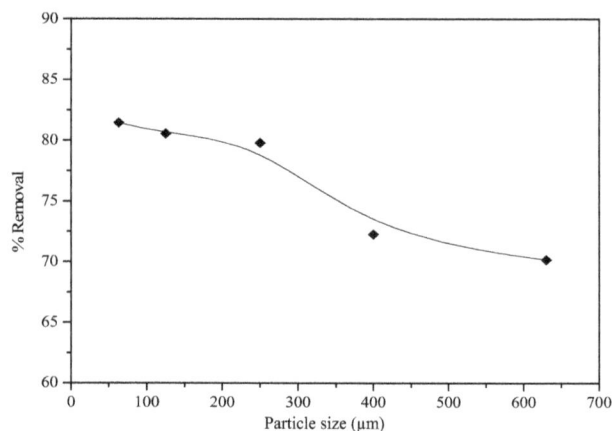

Fig. 9 Effect of the particule size on the MB removal ($C_0 = 10$ mg L^{-1}, initial pH 5.80, biosorbent dose = 0.5 g L^{-1} and $T = 20$ °C)

particle is not utilized for biosorption. Consequently, the amount of MB adsorbed is small. Similar results were reported by other researchers with coniferous brown macroalga Stoechospermum marginatum [24] and *Scolymus hispanicus* L. [27], Pinus brutia Ten. [29].

Effect of temperature

The temperature is well known to play an important role in the biosorption process [48]. The biosorption of MB on *Luffa cylindrica* fiber was investigated over the range (20–60 °C). A slight decrease in the dye biosorption with raising temperature was observed from Fig. 10, suggesting an exothermic process.

Biosorption isotherms

The isotherm describes the equilibrium between the concentration of the adsorbate on the solid phase and the concentration in the liquid phase. The equilibrium biosorption data have been analyzed using the Langmuir,

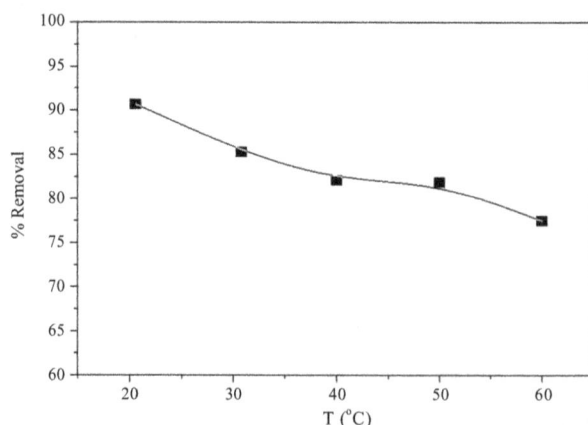

Fig. 10 Effect of the temperature on the MB biosorption ($C_0 = 50$ mg L^{-1}, initial pH 5.80, and biosorbent dose = 0.5 g L^{-1})

Table 2 Constants of isotherm models for the biosorption of MB onto *Luffa cylindrica* fiber at various initial MB concentrations

Langmuir				
K_L (L mg^{-1})	q_{max} (mg g^{-1})	R^2	F value	P value
0.120	49.456	0.9969	1899.657	1.200×10^{-7}
Tempkin				
A_T (L mg^{-1})	b_T (J mol^{-1})	R^2	F value	P value
2.049	276.986	0.9591	141.613	7.383×10^{-5}
Freundlich				
K_F (mg g^{-1}) (mg L^{-1})$^{-1/nF}$	n_F	R^2	F value	P value
7.572	2.342	0.8957	52.522	7.811×10^{-4}
Dubinin–Radushkevich (D–R)				
E (kJ mol^{-1})	q_{D-R} (mg g^{-1})	R^2	F value	P value
0.968	31.838	0.7201	16.438	0.978×10^{-2}

Freundlich, Dubinin–Radushkevich and Tempkin models. Such analysis is important to develop a relation that accurately represents the experimental results and could be used for design purposes [49].

The Langmuir model is based on an the assumption that the biosorption occurs on specific homogeneous sites of the biosorbent and the monolayer biosorption onto a surface containing a finite number of uniform sites with no trans-migration of adsorbate in the plane of the surface [50]; the isotherm is expressed by Eq. (5).

$$q_e = \frac{q_{max} K_L C_e}{(1 + K_L C_e)} \tag{5}$$

where C_e is the equilibrium dye concentration (mg L^{-1}), q_e the amount of biosorbed dye (mg g^{-1}), q_{max} the amount for a complete biosorption monolayer (mg g^{-1}), and K_L the constant related to the affinity of the binding sites and energy of biosorption (L mg^{-1}).

$$\frac{C_e}{q_e} = \frac{1}{q_{max} K_L} + \frac{C_e}{q_{max}} \tag{6}$$

A dimensionless constant separation factor (R_L) of the Langmuir isotherm was used to determine the favorability of the biosorption process. R_L is defined using Eq. (7); its value indicates the type of isotherm: irreversible ($R_L = 0$), favorable ($0 < R_L < 1$), linear ($R_L = 1$) or unfavorable ($R_L > 1$) [50].

$$R_L = \frac{1}{(1 + K_L C_0)} \tag{7}$$

The Freundlich expression is an empirical equation based on the biosorption onto a heterogeneous surface. The equation generates an exponential shaped theoretical equilibrium curve [51] and is represented as follows:

$$q_e = K_F C_e^{\frac{1}{n_F}} \tag{8}$$

$$\ln q_e = \ln K_F + \frac{1}{n_F} \ln C_e \tag{9}$$

where K_F (mg g^{-1} L$^{(1/n)}_F$ mg$^{-(1/n)}_F$) is the Freundlich constant and $(1/n_F)$ the heterogeneity factor, related to the capacity and the biosorption intensity.

The Dubinin–Radushkevich (D–R) model does not assume a homogeneous surface or a constant biosorption potential [52]. The biosorption characteristic is related to the porous structure of the biosorbent [53].

$$q_e = q_{D-R} \exp(-\beta \varepsilon^2) \tag{10}$$

The Polanyi potential (ε) is equal to:

$$\varepsilon = RT \ln\left(1 + \frac{1}{C_e}\right) \tag{11}$$

where ε is a constant related to the mean free energy of biosorption per mole of biosorbate (mol^2 J^{-2}), q_{D-R} (mg g^{-1}) the theoretical saturation capacity, R (J mol^{-1} K^{-1}) is the universal gas constant, and T (K) the absolute temperature.

The energy E is defined as the free energy change (kJ mol^{-1}), required to transfer 1 mol of ions from the solution to the solid:

$$E = (2\beta)^{-1/2} \tag{12}$$

$$\ln q_e = \ln q_{D-R} - \beta \varepsilon^2 \tag{13}$$

Tempkin and Pyzhev have considered the effects of indirect adsorbate/adsorbate interactions on the biosorption isotherms and suggested that the heat of biosorption of all molecules on the layer should decrease linearly with the coverage [26]. The Temkin isotherm is shown in Eq. (14) [54, 55]:

$$q_e = \left(\frac{RT}{b_T}\right) \ln(A_T C_e) \tag{14}$$

Equation (14) can be expressed in its linear form :

$$q_e = \frac{RT}{b_T} \ln(A_T) + \frac{RT}{b_T} \ln(C_e) \tag{15}$$

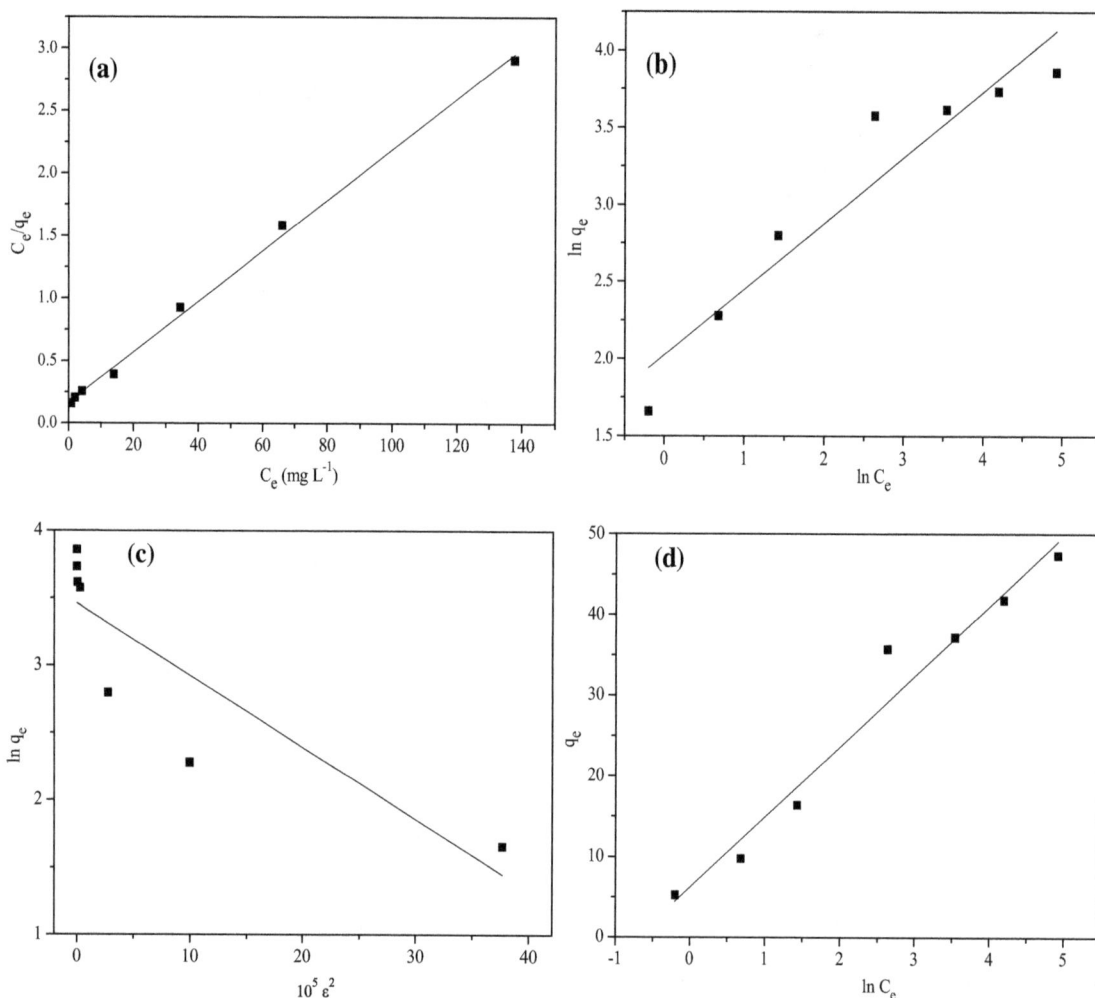

Fig. 11 The isotherm plots: Langmuir biosorption isotherm (**a**), Freundlich biosorption isotherm (**b**), Dubinin–Radushkevich (**c**) and Tempkin biosorption isotherm (**d**)

Table 3 Comparison of the maximum biosorption capacity of dyes for different absorbents

Absorbents	Adsorbates	q_{max} (mg g^{-1})	References
Raw *Luffa cylindrica*	MB	49.46	Present work
Activated *Luffa cylindrica* by NaOH (0.1 M)	MB	49	[57]
Lignite	MB	41.49	[58]
Activated *Luffa cylindrica* by H$_3$PO$_4$ (20 %) and ZnCl$_2$ (50 %)	Reactive Orange	38.31	[59]
Activated *Luffa cylindrica* by NaOH (2 %)	Malachite Green	29.4	[60]
Olive stone	MB	13.2	[61]
Defatted *Scenedesmus* sp. biomass	MB	7.73	[62]
Wood millet carbon	MB	4.94	[63]
Activated peanut stick	MB	2.54	[64]

where A_T is the equilibrium binding constant corresponding to the maximum binding energy (L mg^{-1}) and b_T (J mol^{-1}) the Tempkin isotherm constant related to the heat of biosorption.

The biosorption isotherms are useful to describe the interaction adsorbate/biosorbent of any system. The parameters obtained from different models provide information on the biosorption mechanisms, the surface properties and affinities of the biosorbent [56]. Table 2 and Fig. 11 illustrate the isotherms for 160 min of contact time, initial MB concentration in the range (20–300 mg L^{-1}), a pH of 5.80, a biosorbent dose of 3 g L^{-1} and a temperature of 20 °C. Based on the linear regression correlation coefficient (R^2), F and P values, the isotherm models fit well the experimental data in the following order:

- Langmuir R^2 > Tempkin R^2 > Freundlich R^2 > (D–R) R^2.
- Langmuir F value > Tempkin F value > Freundlich F value > (D–R) F value.
- Langmuir P value < Tempkin P value < Freundlich P value < (D–R) P value.

Table 3 presents the comparison of the maximum biosorption capacity (q_{max}) of MB onto *Luffa cylindrica* fiber with those obtained by other researchers. It is clear that the *Luffa cylindrica* used in this work without any treatment has a relatively suitable biosorption capacity compared to other biosorbents in the literature. Therefore, raw *Luffa cylindrica* fibers seem to be competitive to other methylene blue sorbents and some optimizing treatments on this biomass might be interesting for further studies.

Biosorption kinetics

The kinetic is important for understanding the treatment of aqueous solutions because it provides valuable information about the mechanism of biosorption processes and potential rate-controlling steps, such as the mass transport [56]. Experimental data of MB biosorption using *Luffa cylindrica* fibers were evaluated by the pseudo-first and pseudo-second-order kinetics and intra-particle diffusion models to understand the mechanisms of the biosorption process.

The pseudo-first-order rate expression of Lagergren [65] is generally described by the following equation [66]:

$$\log (q_e - q_t) = \log q_e - \frac{k_1 t}{2.303} \qquad (16)$$

where q_e and q_t are the amounts of dye adsorbed at equilibrium and at time t (mg g^{-1}), respectively, and k_1 the pseudo-first-order rate constant (min^{-1}), k_1 is obtained from the slope of the linear plot of $\log (q_e - q_t)$ against t.

The pseudo-second-order kinetic model is expressed as [67]:

$$\frac{t}{q_t} = \frac{1}{k_2 q_e^2} + \frac{1}{q_e} t \qquad (17)$$

where k_2 is the rate constant of second-order biosorption (g mg^{-1} min^{-1}). If the second-order kinetic is applicable, the plot of t/q_t against t of Eq. (17) should give a linear plot. The initial biosorption rate "h" (mg g^{-1} min^{-1}) is expressed as [68]:

Table 4 Kinetic parameters for the biosorption of MB onto *Luffa cylindrica* fiber at various initial MB concentrations

C_0 (mg L^{-1})	$q_{e,exp}$ (mg g^{-1})	Pseudo-first order					Pseudo-second-order						Intra-particle diffusion		
		k_1 (min^{-1})	q_e theo (mg g^{-1})	R^2	F value	P value	k_2 (g mg^{-1} min^{-1})	h (mg g^{-1} min^{-1})	q_e theo (mg g^{-1})	R^2	F value	P value	k_{int} (mg g^{-1} min$^{-1/2}$)	C (mg g^{-1})	R^2
20	5.243	2.754×10^{-2}	0.878	0.8118	48.439	3.901×10^{-5}	1.294×10^{-1}	3.600	5.274	0.9999	390658.062	0	0.736	2.407	0.9807
30	9.780	6.119×10^{-2}	1.431	0.8860	70.915	3.013×10^{-5}	7.918×10^{-2}	7.727	9.879	0.9999	449105.026	0	1.730	3.457	0.9228
50	16.387	4.535×10^{-2}	1.526	0.9634	290.858	1.012×10^{-8}	5.175×10^{-2}	14.114	16.515	0.9999	654225.822	0	0.305	14.162	0.9862
120	35.705	2.653×10^{-2}	2.984	0.9753	435.846	1.410×10^{-9}	6.017×10^{-3}	7.980	36.417	0.9992	14246.969	0	2.123	18.955	0.9913
150	37.165	2.517×10^{-2}	3.002	0.7352	28.760	4.546×10^{-4}	5.241×10^{-3}	7.753	38.462	0.9992	3300.665	0	1.990	21.034	0.9857
200	41.793	2.117×10^{-2}	2.950	0.9201	127.592	5.149×10^{-7}	5.736×10^{-3}	10.229	42.230	0.9964	11448.789	5.44×10^{-15}	2.058	25.293	0.9316
300	47.372	2.754×10^{-2}	3.251	0.8937	93.484	2.163×10^{-6}	5.148×10^{-3}	12.003	48.286	0.9999	32124.553	0	2.427	28.789	0.9668

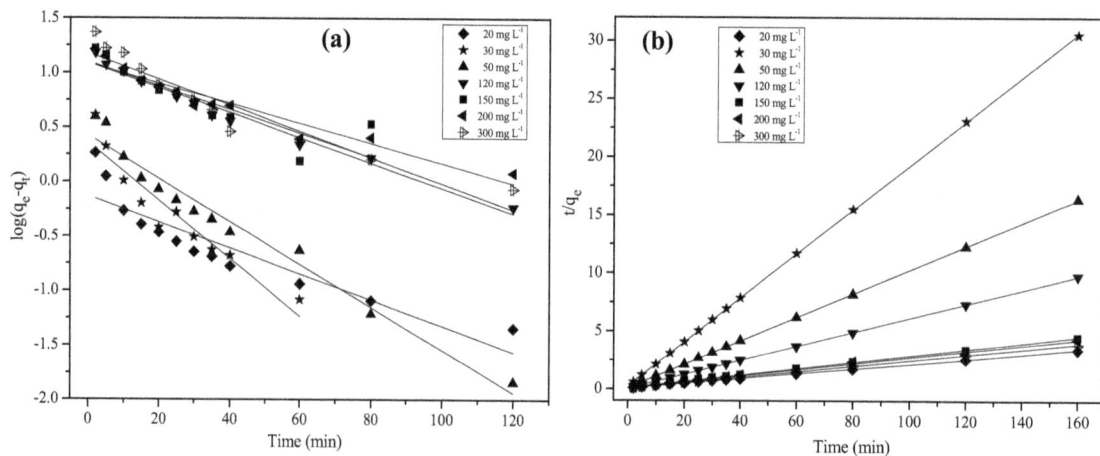

Fig. 12 The kinetic plots, pseudo-first-order (**a**) and pseudo-second-order (**b**) models

$$h = k_2 q_e^2 \qquad (18)$$

The intra-particle diffusion model is used by Weber and Morris [69] and the rate constant (k_{int}, mg g^{-1} min$^{-\frac{1}{2}}$) is given by [41, 67]:

$$q_t = k_{int} t^{1/2} + C \qquad (19)$$

C (mg g^{-1}) is the intercept. The relation gives information about the thickness of the boundary layer and the plot of q_t versus $t^{\frac{1}{2}}$ should yield a straight line passing by the origin if the biosorption process obeys to the intra-particle diffusion model [46, 70].

The kinetic parameters for the biosorption of MB onto *Luffa cylindrica* fiber are calculated and summarized in Table 4 and Fig. 12. We can observe that only the pseudo-second-order model gives the best fit, with low error probability (5.440×10^{-15} to zero), High F values of pseudo-first-order and high adjusted R^2 (0.9964 to 0.9999). Moreover, the calculated biosorption amount q_e (cal) fits well with experimental one q_e (exp).

An intra-particle diffusion model was used to identify the diffusion mechanism. The plots of q_t versus $t^{1/2}$ (Fig. 13), are multi-linear, indicating the existence of three different stages during the biosorption process. The first sharp stage represents the transfer of MB from the solution to the outer surface of the biosorbent; the second gradual stage can be attributed to the penetration of MB into the interlayer of the biosorbent where the intra-particle diffusion is rate limiting. The third stage corresponds to the equilibrium phase and the weak biosorption is ascribed to the residual low MB concentration [70]. The intra-particle diffusion rate constants (k_{int}) are gathered in Table 4. As the initial MB concentration increases, the amount of MB reaching the biosorbent surface increases and the intra-particle diffusion rate increases [40]. It can also be observed that the lines do pass by the origin ($C = 0.737$ to

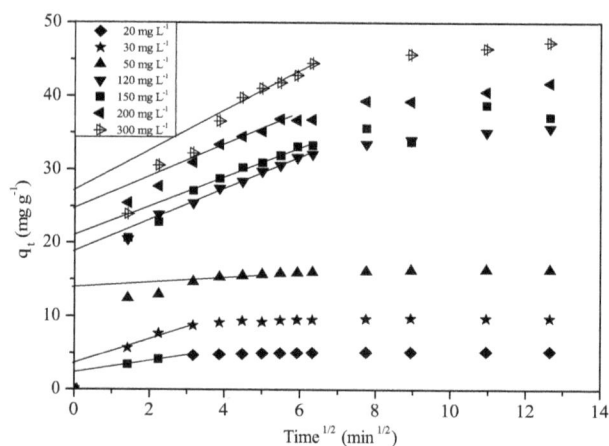

Fig. 13 The intra-particle diffusion model of MB removal by *Luffa cylindrica* fiber at various initial MB concentrations

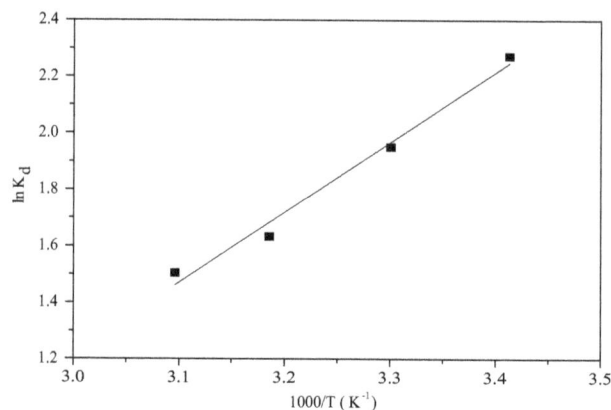

Fig. 14 The Vant Hoff's plot for the determination of thermodynamic parameters

28.789), and this indicates that the transfer mechanism is controlled not only by intra-particle diffusion but also by other mechanisms, such as boundary layer [57]. Similar

Table 5 Thermodynamic parameters for the biosorption of MB onto *Luffa cylindrica* fiber

$\Delta G°$ (kJ mol^{-1})					$\Delta H°$ (kJ mol^{-1})	$\Delta S°$ (kJ mol^{-1} K^{-1})
293 K	303 K	313 K	323 K	333 K		
−5.428	−4.912	−4.396	−3.88	−3.364	−20.547	−0.052

results have been reported for the biosorption of MB onto activated carbons prepared from NaOH-pretreated rice [71], *Luffa cylindrica* fiber-activated carbons [72], sugar beet pulp [21] and low cost biomass material lotus leaf [73].

Thermodynamic studies

The temperature presents a notable effect on the biosorption and the thermodynamic parameters such as change in the standard free energy ($\Delta G°$), standard enthalpy ($\Delta H°$), and standard entropy ($\Delta S°$) are determined [74]:

$$\Delta G° = -RT \ln K_d \tag{20}$$

$$\ln K_d = \frac{-\Delta G°}{RT} = \frac{\Delta S°}{R} - \frac{\Delta H°}{RT} \tag{21}$$

where R is the universal gas constant (8.314 J mol^{-1} K^{-1}), T (K) the absolute temperature and K_d (L g^{-1}) the distribution coefficient for the biosorption calculated from the following relation [27]:

$$K_d = \frac{q_e}{C_e} \tag{22}$$

The plot of $\ln K_d$ versus of $1/T$ yields a straight line form; $\Delta H°$ and $\Delta S°$ are calculated from the slope and intercept of the plot, respectively (Fig. 14, Table 5). The negative values of $\Delta G°$ and $\Delta H°$ indicate that the biosorption is spontaneous, exothermic and physical in nature, thus confirming the affinity of the biosorbent toward the MB molecule [75]. The negative entropy $\Delta S°$ reflects the decreased randomness at the solid/solution interface during the MB biosorption [75, 76]. Similar results were reported by Barka et al. [27] and Han et al. [77] where MB was adsorbed on *Scolymus hispanicus* L. and Fallen phoenix tree's leaf, respectively.

Desorption study

Desorption studies help in deciding the mechanism of the biosorption process and recovery of adsorbent for the reuse. The MB desorption on the *Luffa cylindrica* (Fig. 15) is low for the four solvents (<10 %) at 293 K. The undesorbed MB in the biosorbate is due to the complex formation (MB—active site) of the biomass, and hence the inability of the eluting solvent to completely desorb the dye [78].

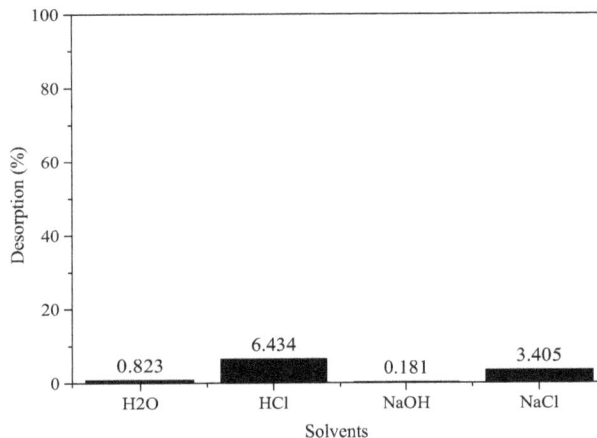

Fig. 15 Batch desorption of MB from biomass using different eluting solvents

Conclusion

The results obtained in the present work showed that the biomass derived from locally available material (*Luffa cylindrica*) can be readily used for the removal of methylene blue from aqueous solutions. In batch studies, the biosorption was strongly dependent on operating parameters such as the contact time, solution pH, particle size, biosorbent dose, initial MB concentration and temperature. The parameters were optimized and the experimental data were analyzed by various isotherm models; the results showed that the isotherm data were well correlated by the Langmuir model. The kinetic studies indicated that the pseudo-second-order model fits suitably the experimental data and suggest that the interlayer diffusion is not the rate-determining step in the MB biosorption mechanism. The maximum monolayer biosorption capacity was found to be 49.46 mg g^{-1} at 20 °C. Moreover, the thermodynamic parameters showed that the biosorption was spontaneous, exothermic and physical in nature. The biosorption experiments indicated that the *Luffa cylindrica* was an efficient biosorbent for the removal of MB and favorably compared with respect to most biomasses reported nowadays.

References

1. Javadian H, Angaji MT, Naushad M (2014) Synthesis and characterization of polyaniline/γ-alumina nanocomposite: a comparative study for the adsorption of three different anionic dyes. J Ind Eng Chem 20:3890–3900

2. Vinod K, Gupta VK, Rajeev Jain R, Arunima Nayak A, Agarwal S, Shrivastava M (2011) Removal of the hazardous dye—Tartrazine by photodegradation on titanium dioxide surface. Mat Sci Eng C 31:1062–1067

3. Mittal A, Mittal J, Malviya A, Gupta VK (2009) Adsorptive removal of hazardous anionic dye "Congo red" from wastewater using waste materials and recovery by desorption. J Colloid Interface Sci 340:16–26

4. Demirbaş Ö, Turhan Y, Alkan M (2014) Thermodynamics and kinetics of adsorption of a cationic dye onto sepiolite. Desalin Water Treat 1–8

5. Mittal A, Mittal J, Malviya A, Kaur D, Gupta VK (2010) Decoloration treatment of a hazardous triarylmethane dye, Light Green SF (Yellowish) by waste material adsorbents. J Colloid Interface Sci 342:518–527

6. Mittal A, Kaur D, Malviya A, Mittal J, Gupta VK (2009) Adsorption studies on the removal of coloring agent phenol red from wastewater using waste materials as adsorbents. J Colloid Interface Sci 337:345–354

7. Gupta VK, Ali I, Saleh TS, Nayak A, Agarwal S (2012) Chemical treatment technologies for waste-water recycling—an overview. RSC Adv 2:6380–6388

8. Benadjemia M, Millière L, Reinert L, Benderdouche N, Duclaux L (2011) Preparation, characterization and Methylene Blue adsorption of phosphoric acid activated carbons from globe artichoke leaves. Fuel Process Technol 92:203–1212

9. Gupta VK, Jain R, Mittal A, Saleh TA, Nayak A, Agarwal S, Sikarwar S (2012) Photo-catalytic degradation of toxic dye amaranth on TiO2/UV in aqueous suspensions. Mat Sci Eng C 32:12–17

10. Karthikeyan S, Gupta VK, Boopathy R, Titus A, Sekaran G (2012) A new approach for the degradation of high concentration of aromatic amine by heterocatalytic Fenton oxidation: kinetic and spectroscopic studies. J Mol Liq 173:153–163

11. Zhou Q, Gong W, Xie C, Yuan X, Li Y, Bai C, Chen S, Xu N (2011) Biosorption of Methylene Blue from aqueous solution on spent cottonseed hull substrate for Pleurotus ostreatus cultivation. Desalin Water Treat 29:317–325

12. Aksu Z, Ertuğrul S, Dönmez G (2010) Methylene Blue biosorption by Rhizopus arrhizus: effect of SDS (sodium dodecylsulfate) surfactant on biosorption properties. Chem Eng J 158:474–481

13. Saleh TA, Gupta VK (2012) Photo-catalyzed degradation of hazardous dye methyl orange by use of a composite catalyst consisting of multi-walled carbon nanotubes and titanium dioxide. J Colloid Interface Sci 371:101–106

14. Gupta VK, Agarwal S, Saleh TA (2011) Synthesis and characterization of alumina-coated carbon nanotubes and their application for lead removal. J Hazard Mater 185:17–23

15. Gupta VK, Srivastava SK, Mohan D, Sharma S (1998) Design parameters for fixed bed reactors of activated carbon developed from fertilizer waste for the removal of some heavy metal ions. Waste Manag 17:517–522

16. Jain AK, Gupta VK (2003) A comparative study of adsorbents prepared from industrial wastes for removal of dyes. Sep Sci Technol 38:463–481

17. Saleh TA, Gupta VK (2012) Column with CNT/magnesium oxide composite for lead(II) removal from water. Environ Sci Pollut Res 19:1224–1228

18. Guler UA, Sarioglu M (2013) Single and binary biosorption of Cu (II), Ni (II) and methylene blue by raw and pretreated Spirogyra sp.: equilibrium and kinetic modeling. J Environ Chem Eng 1:369–377

19. Mona S, Kaushik A, Kaushik CP (2011) Biosorption of reactive dye by waste biomass of Nostoc linckia. Ecol Eng 37:1589–1594

20. Irem S, Khan QM, Islam E, Hashmat AJ, Ul Haq MA, Afzal M, Mustafa T (2013) Enhanced removal of reactive navy blue dye using powdered orange waste. Ecol Eng 58:399–405

21. Vučurović VM, Razmovski RN, Tekić MN (2012) Methylene blue (cationic dye) adsorption onto sugar beet pulp: equilibrium isotherm and kinetic studies. J Taiwan Inst Chem Eng 43:108–111

22. Deepaa K, Chandran P, Khan SS (2013) Bioremoval of Direct Red from aqueous solution by Pseudomonas putida and its adsorption isotherms and kinetics. Ecol Eng 58:207–213

23. Javadian H, Ahmadi M, Ghiasvand M, Kahrizi S, Katal R (2013) Removal of Cr(VI) by modified brown algae Sargassum bevanom from aqueous solution and industrial wastewater. J Taiwan Inst Chem Eng 44:977–989

24. Daneshvar E, Kousha M, Sohrabi MS, Khataee A, Converti A (2012) Biosorption of three acid dyes by the brown macroalga Stoechospermum marginatum: isotherm, kinetic and thermodynamic studies. Chem Eng J 195–196:297–306

25. Mittal A, Mittal J, Malviya A, Gupta VK (2010) Removal and recovery of Chrysoidine Y from aqueous solutions by waste materials. J Colloid Interface Sci 344:497–507

26. Barka N, Ouzaouit K, Abdennouri M, El Makhfouk M (2013) Dried prickly pear cactus (Opuntia ficus indica) cladodes as a low-cost and eco-friendly biosorbent for dyes removal from aqueous solutions. J Taiwan Inst Chem Eng 44:52–60

27. Barka N, Abdennouri M, Makhfouk MEL (2011) Removal of methylene blue and eriochrome black T from aqueous solutions by biosorption on Scolymus hispanicus L.: kinetics, equilibrium and thermodynamics. J Taiwan Inst Chem Eng 42:320–326

28. Ofomaja AE, Ukpebor EE, Uzoekwe SA (2011) Biosorption of Methyl violet onto palm kernel fiber: diffusion studies and multistage process design to minimize biosorbent mass and contact time. Biomass Bioenergy 35:4112–4123

29. Deniz F, Karaman S, Saygideger SD (2011) Biosorption of a model basic dye onto Pinus brutia Ten.: evaluating of equilibrium, kinetic and thermodynamic data. Desalination 270:199–205

30. Gupta VK, Nayak A (2012) Cadmium removal and recovery from aqueous solutions by novel adsorbents prepared from orange peel and Fe2O3 nanoparticles. Chem Eng J 180:81–90

31. Ncibi MC, Mahjou B, Seffen M (2007) Kinetic and equilibrium studies of methylene blue biosorption by Posidonia oceanica (L.) fibres. J Hazard Mater B139:280–285

32. Suyamboo BK, Srikrishnaperumal R (2014) Biosorption of crystal violet onto cyperus rotundus in batch system: kinetic and equilibrium modeling. Desalin Water Treat 52:4492–4507

33. Belala Z, Jeguirim M, Belhachemi M, Addoun F, Trouvé G (2011) Biosorption of basic dye from aqueous solutions by Date Stones and Palm-Trees Waste: kinetic, equilibrium and thermodynamic studies. Desalination 27:180–187

34. Oboh IO, Aluyor EO, Audu TOK (2011) Application of Luffa Cylindrica in natural form as biosorbent to removal of divalent metals from aqueous solutions -kinetic and equilibrium study. Waste Water-Treat Reutil 195–212

35. Wang LG, Yan GB (2011) Adsorptive removal of direct yellow 161 dye from aqueous solution using bamboo charcoals activated with different chemicals. Desalination 274:81–90

36. Mall ID, Srivastava VC, Kumar GVA, Mishra IM (2006) Characterization and utilization of mesoporous fertilizer plant waste carbon for adsorptive removal of dyes from aqueous solution. Colloids Surf A278:175–187

37. Tanobe VOA, Sydenstricker THD, Munaro M, Amico SC (2005) A comprehensive characterization of chemically treated Brazilian sponge-gourds (Luffa cylindrica). Polym Test 24:474–482

38. Wang Y, Shen XY (2012) Optimum Plasma Surface Treatment of Luffa Fibers. J Macromol Sci Part B: Phys 51:662–670

39. González-García P, Centeno TA, Urones-Garrote E, Ávila-Brande D, Otero-Díaz LC (2013) Microstructure and surface properties of lignocellulosic-based activated carbons. Appl Surf Sci 265:731–737

40. El-Haddad M, Regti A, Slimani R, Lazar S (2014) Assessment of the biosorption kinetic and thermodynamic for the removal of safranin dye from aqueous solutions using calcined mussel shells. J Ind Eng Chem 20:717–724

41. Javadian H, Sorkhrodi FZ, Koutenaei BB (2014) Experimental investigation on enhancing aqueous cadmium removal via nanostructure composite of modified hexagonal type mesoporous silica with polyaniline/polypyrrole nanoparticles. J Ind Eng Chem 20:3678–3688

42. Crini G, Peindy HN, Gimbert F, Robert C (2007) Removal of C.I. Basic Green 4 (malachite green) from aqueous solutions by adsorption using cyclodextrin based adsorbent: kinetic and equilibrium studies. Sep Purif Technol 53:97–110

43. Saeed A, Iqbal M, Zafar IS (2009) Immobilization of Trichoderma viride for enhanced methylene blue biosorption: batch and column studies. J Hazard Mater 168:406–415

44. Han R, Wang Y, Han P, Shi J, Yang J, Lu Y (2006) Removal of methylene blue from aqueous solution by chaff in batch mode. J Hazard Mater B137:550–557

45. Sadaf S, Bhatti HN (2014) Batch and fixed bed column studies for the removal of Indosol Yellow BG dye by peanut husk. J Taiwan Inst Chem Eng 45:541–553

46. Deniz F, Saygideger SD (2010) Equilibrium, kinetic and thermodynamic studies of Acid Orange 52 dye biosorption by Paulownia tomentosa Steud. Leaf powder as a low-cost natural biosorbent. Bioresour Technol 101:5137–5143

47. Mouni L, Merabet D, Bouzaza A, Belkhiri L (2011) Adsorption of Pb(II) from aqueous solutions using activated carbon developed from Apricot stone. Desalination 276:148–153

48. Aksu Z, Karabayir G (2008) Comparison of biosorption properties of different kinds of fungi for the removal of Gryfalan Black RL metal-complex dye. Bioresour Technol 9:7730–7741

49. Çolak F, Atar N, Olgun A (2009) Biosorption of acidic dyes from aqueous solution by Paenibacillus macerans: kinetic, thermodynamic and equilibrium studies. Chem Eng J 150:122–130

50. Khan AA, Ahmad R, KhanA Mondal PK (2013) Preparation of unsaturated polyester Ce(IV) phosphate by plastic waste bottles and its application for removal of Malachite green dye from water samples. Arabian J Chem 6:361–368

51. Aksu Z, Akın AB (2010) Comparison of Remazol Black B biosorptive properties of live and treated activated sludge. Chem Eng J 165:184–193

52. Reddy DHK, Ramana DKV, Seshaiah K, Reddy AVR (2011) Biosorption of Ni (II) from aqueous phase by Moringa oleifera bark, a low cost biosorbant. Desalination 268:150–157

53. Vijayaraghavan K, Padmesh TVN, Palanivelu K, Velan M (2006) Biosorption of nickel (II) ions onto Sargassum wightii: application of two-parameter and three-parameter isotherm models. J Hazard Mater B133:304–308

54. Ghasemi M, Khosroshahy MZ, Abbasabadi AB, Ghasemi N, Javadian H, Fattahi M (2015) Microwave-assisted functionalization of Rosa Canina-L fruits activated carbon with tetraethylenepentamine and its adsorption behavior toward Ni(II) in aqueous solution: kinetic, equilibrium and thermodynamic studies. Powder Technol 274:362–371

55. Javadian H, Ghorbani F, Tayebi HA, Asl SMH (2015) Study of the adsorption of Cd (II) from aqueous solution using zeolite-based geopolymer, synthesized from coal fly ash; kinetic, isotherm and thermodynamic studies. Arabian J Chem 8:837–849

56. Wang L (2012) Application of activated carbon de rived from 'waste' bamboo culms for the adsorption of azo disperse dye: kinetic, equilibrium and thermodynamic studies. J Environ Manag 102:79–87

57. Demir H, Top A, Balköse D, Ülkü S (2008) Dye adsorption behavior of Luffa cylindrica fiber. J Hazard Mater 153:389–394

58. Gürses A, Hassani A, Kıranşan M, Açışlı ö, Karaca S (2014) Removal of methylene blue from aqueous solution using by untreated lignite as potential low-cost adsorbent: kinetic, thermodynamic and equilibrium approach. J Water Process Eng 2:10–21

59. Abdelwahab O (2008) Evaluation of the use of loofa activated carbons as potential adsorbents for aqueous solutions containing dye. Desalination 222:357–367

60. Altınışık A, Gür E, Seki Y (2010) A natural sorbent, Luffa cylindrica for the removal of a model basic dye. J Hazard Mater 179:658–664

61. Albadarin AB, Mangwandi C (2015) Mechanisms of Alizarin Red S and Methylene blue biosorption onto olive stone by-product: isotherm study in single and binary systems. J Environ Manag 164:86–93

62. Sarat Chandra T, Mudliar SN, Vidyashankar S, Mukherji S, Sarada R, Krishnamurthi S, Chauhan VS (2015) Defatted algal biomass as a non-conventional low-cost adsorbent: surface characterization and methylene blue adsorption characteristics. Bioresour Technol 184:395–404

63. Ghaedi M, Kokhdan MN (2015) Removal of methylene blue from aqueous solution by wood millet carbon optimization using response surface methodology. Spectrochim Acta A 136:141–148

64. Ghaedi M, Golestani Nasab A, Khodadoust S, Rajabi M, Azizian S (2014) Application of activated carbon as adsorbents for efficient removal of methylene blue: kinetics and equilibrium study. J Ind Eng Chem 20:2317–2324

65. Lagergren S (1898) Zur theorie der sogenannten adsorption gelöster stoffe. Kungliga Svenska Vetenskapsakademiens Handlingar 24:1–39

66. Javadian H, Koutenaei BB, Shekarian E, Sorkhrodi FZ, Khatti R, Toosi MR (2014)Application of functionalize d nano HMS type mesoporous silica with N-(2-ami noethyl)-3-a minopropyl methyldimeth oxysilan e as a suitable adsorbent for removal of Pb(II) from aqueous media and industrial wastewater. J Saudi Chem Soc xxx xxx–xxx

67. Javadian H, Taghavi M (2014) Application of novel Polypyrrole/thiol-functionalized zeolite Beta/MCM-41 type mesoporous silica nanocomposite for adsorption of Hg2+ from aqueous solution and industrial wastewater: kinetic, isotherm and thermodynamic studies. Appl Surf Sci 289:487–494

68. Javadian H, Vahedian P, Toosi MR (2013) Adsorption characteristics of Ni(II) from aqueous solution and industrial wastewater onto Polyaniline/HMS nanocomposite powder. Appl Surf Sci 284:13–22

69. Weber WJ, Morris JC (1963) Kinetics of adsorption on carbon from solution. J Sanit Eng Div Am Soc Civ Eng 89:31–59

70. Akar E, Aylinşik A, Seki Y (2013) Using of activated carbon produced from spent tea leaves for the removal of malachite green from aqueous solution. Ecol Eng 52:19–27

71. ChenY Zhai SR, Liu N, Song Y, An QD, Song XW (2013) Dye removal of activated carbons prepared from NaOH-pretreated rice husks by low-temperature solution-processed carbonization and H3PO4 activation. Bioresour Technol 144:401–409

72. Cherifi H, Bentahar F, Hanini S (2013) Kinetic studies on the adsorption of methylene blue onto vegetal fiber activated carbons. Appl Surf Sci 282:52–59

73. Han X, Wang W, Ma X (2011) Adsorption characteristics of methylene blue onto low cost biomass material lotus leaf. Chem Eng J 171:1–8

74. Hu Z, Chen H, Ji F, Yuan S (2010) Removal of Congo Red from aqueous solution by cattail root. J Hazard Mater 173:292–297

75. Rehman MSU, Kim I, Han JI (2012) Adsorption of methylene blue dye from aqueous solution by sugar extracted spent rice biomass. Carbohydr Polym 90:1314–1322

76. Albadarin AB, Mangwandi C, Al-Muhtaseb AH, Walker GM, Allen SJ, Ahmad MNM (2012) Kinetic and thermodynamics of chromium ions adsorption onto low-cost dolomite adsorbent. Chem Eng J 179:193–202

77. Han R, Zou W, Yu W, Cheng S, Wang Y, Shi J (2007) Biosorption of methylene blue from aqueous solution by fallen phoenix tree's leaves. J Hazard Mater 141:156–162

78. Oladoja NA, Aboluwoye CO, Akinkugbe AO (2009) Evaluation of loofah as a sorbent in the decolorization of basic dye contaminated aqueous system. Ind Eng Chem Res 48:2786–2794

12

Amphoteric gellan gum-based terpolymer–montmorillonite composite: synthesis, swelling, and dye adsorption studies

Sirajo Abubakar Zauro[1] · B. Vishalakshi[1]

Abstract A terpolymer gel, Gellan gum-graft-poly(2-acrylamido-2-methyl-1-propanesulfonic acid-co-dimethylaminopropyl methacrylamide) and its composite with the clay, Montmorillonite, was prepared by free-radical polymerization and crosslinking reactions in solution. The terpolymer gel and the clay composite were characterized using FTIR, TGA, SEM, and X-ray diffraction techniques. Swelling studies were carried out in different pH and salt solutions. The gel showed maximum swelling capacity in alkaline medium, while the composite showed higher swelling in neutral medium. The swelling of the gel and the composite followed second kinetics model and water transport is found to be a less Fickian diffusion process. The terpolymer gel and the composite were evaluated for the adsorption of rhodamine B (RhB) and chromotrope 2R (C2R) dyes. Rhodamine B is found to be adsorbed to a higher extent than chromotrope 2R and the adsorption isotherm studies suggested that adsorption of both RhB and C2R on the terpolymer gel was best explained by Langmuir model, while the adsorption on the Composite fitted best into Freundlich model. Similarly, the adsorption kinetics data for both RhB and C2R dyes followed the second-order kinetics.

Keywords Gellan gum · 2-Acrylamido-2-methyl-1-propanesulfonic acid · Dimethylaminopropyl methacrylamide · Montmorillonite · Swelling · Dye adsorption

✉ B. Vishalakshi
vishalakshi2009@yahoo.com

[1] Department of Post-Graduate Studies and Research in Chemistry, Mangalore University, Mangalagangothri, Dakshina Kannada, Mangalore, Karnataka 574199, India

Introduction

The increasing demand for manufactured products worldwide and the use of synthetic dyes in various industries such as textile, leather, paper, rubber, plastic, cosmetic, etc. led to the proportionate release of a large quantity of effluent into the environment. In addition, this effluent contained non-biodegradable toxic and carcinogenic dye substances into the environment [1–5]. Rhodamine B (RhB) and chromotrope 2R (C2R) are synthetic dyes that are commonly used in leather, textile, and paper industries and cause various health hazards [6, 7].

The composites and nanocomposite of polymer–clay have been gaining increase attention by researchers globally due to the hybrid properties which they exhibit when compared with either the polymer or clay separately [8]. A wide range of polymer–clay composite/nanocomposite has been produced and used for a variety of applications such as water treatment [5, 8], dye adsorption [2, 9–11], etc.

Several physical and chemical methods like chemical precipitation, ion exchange, membrane separation, chemical reduction, chemical oxidation, advanced oxidation processes (AOPs), etc. [12] have been employed in the removal of toxic substances from the environment. However, these methods are ineffective in removing most of the dyes molecules and are time-consuming, not cost-effective, and sometimes generate large amount of sludge that are toxic to the biotic organisms in the environment. Hence, adsorption using biopolymer-based composites has been described as one of the effective and promising techniques for removal of pollutants due to its simplicity, inexpensiveness, etc. [13–18].

Several biopolymer-based hydrogels such as Gum gatti [19], Gur gum [14], Kappa carrageenan [20–22], chitosan

[23], and Guaran [24] were studied as adsorbents for the removal of dyes from aqueous solution. Casey and Wilson [25] reported the adsorption of Methylene blue (MB) dye on Chitosan-PVA composite films and direct relationship between film composition (Chitosan-PVA) with solution pH and the uptake of MB were observed. Similarly, Datskevich et al. [26] synthesized cationic starch and sodium alginate-based composite and studied the adsorption of Methyl orange and MB under different conditions. The adsorption of congo red on Chitosan/Montmorillonite composite has been studied by Wang and Wang [1]. Vasugi and Girija [4] reported the adsorption of reactive blue dye on hydroxyapatite-alginate composite.

The composite materials consisting of clay and a biopolymer are very effective in removal of dyes due to the availability of numerous functional groups on the biopolymer and the clay for binding with the dye molecules, rendering the materials useful as adsorbents.

The aim of the present study is to obtain a functional composite hydrogel consisting of clay, a biopolymer, and a synthetic polymer to be evaluated as an adsorbent for dyes. This has been achieved by polymerizing AMPS, DMAPMAm, and MBA in the presence of gellan gum and montmorillonite (MMT) clay in water and its effectiveness as an adsorbent for removal of dyes has been studied using chromotrope 2R and rhodamine B as model ionic dyes.

Materials and methods

Materials

Gellan gum (GG) was purchased from Sigma-Aldrich Chemicals Pvt Ltd., Bangalore, India. 2-Acrylamidomethyl-2-propane sulfonic acid (AMPS), dimethylaminopropyl methacrylamide (DMAPMAm), N, N-methylene-bis-acrylamide (MBA), and montmorillonite (MMT) were obtained from Sigma-Aldrich Chemie, GmbH, Germany. Ammonium peroxodisulphate (APS) was obtained from Spectro Chem Pvt. Ltd., Mumbai, India. Rhodamine B (RhB) was obtained from s.d. Fine Chemical Limited, Mumbai, India. Chromotrope 2R (C2R) was purchased from Loba Chemie Limited Mumbai, India. Acetone was obtained from Nice Chemicals Pvt Ltd., Kerala, India. Methanol was obtained from Himedia Laboratories Pvt Ltd., Mumbai, India. NaCl, KCl, FeCl$_3$, CaCl$_2$, and Na$_2$SO$_4$ were obtained from Merck Ltd., Mumbai, India. DMAPMAm was purified by passing through column containing alumina gel before use. All other reagents were used as received. Distilled water was used throughout the experiments.

Methods

Synthesis of GG-g-AMPS

GG-g-AMPS was prepared via free-radical polymerization process as follows: 0.15 g GG was dissolved in distilled and stirred overnight. To the resultant solution, varying amounts (0.1–0.30 g) of AMPS were added followed by APS (0.05) under continues stirring. The temperature was raised to 40 °C under continues stirring for 2 h. The gel was precipitated with acetone and washed with methanol several times, and dried in an oven at 50 °C for 24 h.

Synthesis of GG-g-poly(AMPS-co-DMAPMAm)

The graft copolymer GG-g-poly(AMPS-co-DMAPMAm) gel was synthesized based on the established methods reported by Nie et al. [27] with a little modification as follows: a known amount of GG (0.1 g) was dissolved in distilled water and stirred overnight at room temperature. A specified amount of AMPS (0.1–0.30 g) and DMAPMAm (0.15–0.50 g) were added to the above solution. To the mixture above, APS (0.05 g) and MBA (0.05 g) were added and stirred by raising the temperature to 60 °C slowly for 4 h maintaining the temperature at 60 °C until a gel-like solution was formed. It was then allowed to cool for an hour to complete the polymerization and added to excess acetone to remove un-reacted components. The gels obtained were then washed with 50% ethanol and placed in a hot oven at 50 °C until constant weight was obtained. The GG-g-poly(AMPS-co-DMAPMAm) gel formation was optimized. The percentage yield and grafting percentage (GP) were calculated by the following equation:

$$\%\text{Yield} = \frac{\text{Experimental yield}}{\text{Theoretical yield}} \times 100, \qquad (1)$$

$$\text{GP}(\%) = \frac{(w_1 - w_0)}{w_0} \times 100, \qquad (2)$$

where w_0 and w_1 are the weight of grafted gels and monomers, respectively.

Synthesis of GG-g-poly(AMPS-co-DMAPMAm)/MMT

The GG-g-poly(AMPS-co-DMAPMAm)/MMT composite hydrogel was made following the same procedure as in 2.2.2 with the addition of MMT (0.01–0.03 g) after adding DMAPMA under continuous stirring slowly during 1 h.

Characterization

The GG, GG-g-AMPS, GG-g-poly(AMPS-co-DMAPMA)-8, and GG-g-poly(AMPS-co-DMAPMA)/MMT-3 samples were characterized using FTIR, TGA, SEM, and XRD techniques. The FTIR were recorded using FTIR-Prestige-21, Shimadzu Japan, in the range of 4000–400 cm^{-1} wavenumber during 40 scans, with a resolution of 2 cm^{-1}. Thermograms were recorded using standard DSC-TGA (Q600 V20.9 model) Japan, by heating the samples in the ranges of 30–700 °C, under a nitrogen atmosphere at 10 °C/min. Surface morphology of the samples was obtained on gold coating JOEL JSM-6380LA analytical Scanning electron microscope (SEM) under magnification of 2000 at 20 kV. XRD pattern was recorded on X-ray diffractometer (Rigatu Miniflex 600-XRD instrument, USA) using Cu Ká radiation generated at 35 kV and 35 mA in the differential angle 2θ at a range of $0°$–$80°$ in steps of 0.020/s.

Swelling studies

Swelling experiments of the GG-g-poly(AMPS-co-DMAPMA)-8 and GG-g-poly(AMPS-co-DMAPMA)/MMT-3 samples were carried out in different media (pH and salts solution). A known amount of the samples were weighed and immersed into swelling media at room temperature. After specified interval of time, the samples were removed and the excess surface water was wiped away gently using blotting (tissue) paper and re-weighed. This procedure was repeated until equilibrium is reached. The data were reported as the mean of three different measurements. The effects of nature of different salts solution (0.1 M) on the swelling ratio were also studied in the same manner.

The swelling ratio (SR) and swelling equilibrium (S_{eq}) were calculated by the following equations:

$$\mathrm{SR}\left(\tfrac{g}{g}\right) = \frac{(W_t - W_o)}{W_o}, \tag{3}$$

$$S_{eq}\left(\tfrac{g}{g}\right) = \frac{(W_e - W_0)}{W_o}, \tag{4}$$

where W_0, W_t, and W_e are the weight of the gel at time $t = 0$, $t = t$, and at equilibrium, respectively [28].

Dyes adsorption studies

A known amount of GG-g-poly(AMPS-co-DMAPMAm)-8 and GG-g-poly(AMPS-co-DMAPMAm)/MMT-3 samples were left immersed in 100 mg/L solutions of RhB and C2R dyes. At different time intervals, 2.5 mL of the supernatant solution were withdrawn and the absorbance values were measured using UV–visible spectrophotometer (UV-1800 Shimadzu, Japan) λ_{max} of 554 and 510 nm for RhB and C2R, respectively. Calibration curves were used to convert the absorbance measured into concentration using standard solutions of 2, 4, 6, 8, and 10 mg/L of the dyes. Different initial concentrations (10, 30, 50, 70, and 100 mg/L) were used for equilibrium adsorption studies by immersing

varied amount of the adsorbent and allowed to stand for 14 h and the resultant solutions were decanted and the absorbance were recorded. The amount of dyes adsorbed at time t (q_t) and at equilibrium (q_e) in mg/g was calculated using the following equations [29, 30]:

$$q_t = \frac{(C_0 - C_t)}{M} \times V, \tag{5}$$

$$q_e = \frac{(C_0 - C_e)}{M} \times V, \tag{6}$$

where q_t and q_e are the amount of dyes adsorbed (mg/g) at time $t = t$ and at equilibrium, respectively. C_0, C_t, and C_e are dyes concentration (mg/L) at time $t = 0$, $t = t$, and at equilibrium, respectively, M is the weight of the gel (g) and V is the volume (L) of the dye solution.

Results and discussion

The composite hydrogels were prepared by crosslink copolymerization of AMPS, DMAPMAm, and MBA in water in the presence of GG. MMT was incorporated in situ in the copolymer network. During the polymerization reaction, the bi-functional MBA copolymerizes with AMPS and DMAPMAm to form a network, while GG takes part in the free-radical polymerization reaction by forming macroradicals [31]. Thus, a composite gel is formed by entrapment of MMT clay in the copolymer network. The free-radical reaction mechanism along with the formation of the gel network is shown in Scheme 1.

The composition of the gels/composites and the percentage yield is presented in Table 1. The optimized product [GG-g-poly(AMPS-co-DMAPMAm)-8] was used for composite formation and used as representative sample for swelling and dye adsorption studies. The grafting conditions were optimized by varying monomer (DMAPMAm and AMPS) contents and keeping all other parameters constant.

The GP increases as the AMPS content increases from 0.1 to 0.25 g, and decreases as AMPS content increases to 0.30 g (Table 1). For DMAPMAm, the GP follows a similar pattern as in AMPS. The decreases in GP as the content of monomers increases could be attributed to the less reactive side on the GG as its content remains constant, and hence, there are more molecules of DMAPMAm and AMPS than GG and this could lead to the formation of homopolymer and hence low yield.

FTIR

FTIR Spectra of GG, GG-g-AMPS, GG-g-poly(AMPS-co-DMAPMAm)-8, and GG-g-poly(AMPS-co-DMAPMAm)/MMT-3 composite gels are shown in Fig. 1. The spectrum

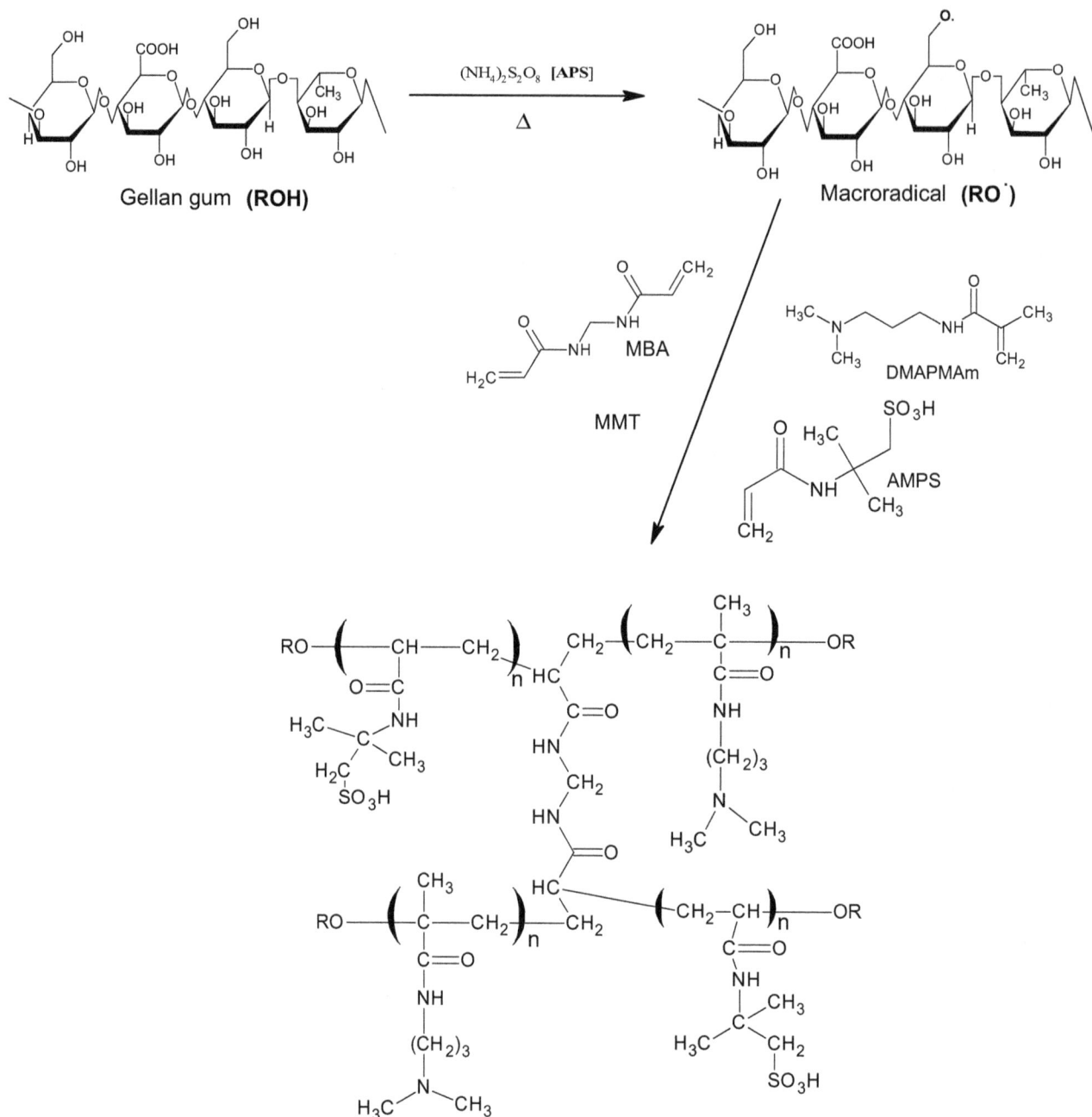

Scheme 1 Proposed scheme for the formation of GG-g-poly(AMPS-co-DMAPMAm) gel

(Fig. 1a) showed a broad absorption band at 3290 cm^{-1} which is due to stretching of O–H and a medium absorption peak at 2926 cm^{-1} corresponding to the C–H stretching of CH$_2$ groups. The absorption at 1605 cm^{-1} is related to the C=O stretching of COO$^-$ of the GG. The peak at 1016 cm^{-1} is assigned to C–O bond stretching frequencies [32]. Comparing the GG and GG-g-AMPS (Fig. 1b) spectra, new characteristic peaks were observed at 1643, 1438, and 923 cm^{-1} which are attributed to asymmetric stretching vibration of C=O, C–N stretching, and S–O stretching of the SO$_3$H, respectively [33]. The additional characteristic absorption bands of 1537 and 2771 cm^{-1} for N–H stretching and C–H stretching of –N(CH$_3$)$_2$ of DMAPMAm [25] were observed in the spectra of GG-g-poly(AMPS-co-DMAPMAm)-8 (Fig. 1c). Similarly, in addition to the peaks on the spectra (Fig. 1a–c), peaks at 1040, 814, and 621 cm^{-1} for Si–O–Si, Al–Al–OH, and Si–Al–OH [34], respectively, were observed on the spectrum of GG-g-poly (AMPS-co-DMAPMAm)/MMT-3 (Fig. 1d) indicating the entrapment of MMT on the gel matrices.

Table 1 Composition of hydrogels/composite and percentage yield

Gel	Code	GG (g)	AMPS (g)	DMAPMAm (g)	APS (g)	MBA (g)	MMT	GP (%)	Yield (%)
GG-g-poly(AMPS-co-DMAPMAm)	1	0.1	0.1	0.15	0.05	0.05		63.18	57.87
	2	0.1	0.1	0.20	0.05	0.05		64.08	60.32
	3	0.1	0.15	0.20	0.05	0.05		72.91	58.73
	4	0.1	0.15	0.25	0.05	0.05		73.81	69.80
	5	0.1	0.20	0.25	0.05	0.05		79.13	50.91
	6	0.1	0.20	0.30	0.05	0.05		79.72	62.32
	7	0.1	0.25	0.30	0.05	0.05		89.12	68.92
	8	0.1	0.25	0.40	0.05	0.05		89.82	79.56
	9	0.1	0.30	0.40	0.05	0.05		88.12	64.39
	10	0.1	0.30	0.50	0.05	0.05		87.65	63.87
Composite GG-g-poly(AMPS-co-DMAPMAm)/MMT	1	0.1	0.25	0.40	0.05	0.05	0.01	59.23	69.21
	2	0.1	0.25	0.40	0.05	0.05	0.02	59.87	67.12
	3	0.1	0.25	0.40	0.05	0.05	0.03	58.94	74.18

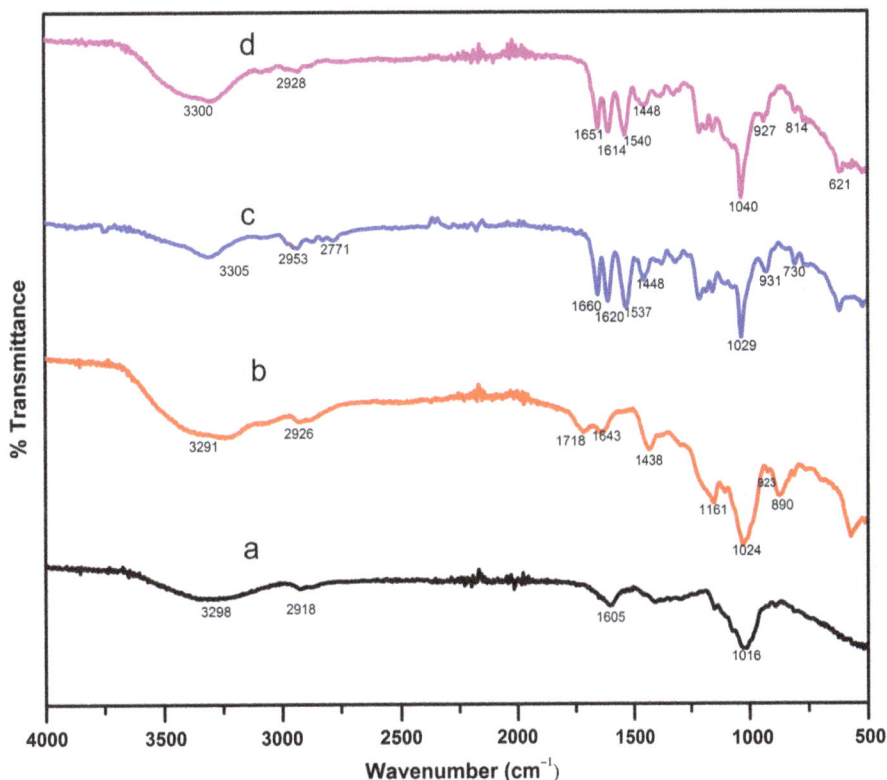

Fig. 1 FTIR spectra of **a** GG, **b** GG-g-APMS, **c** GG-g-poly(AMPS-co-DMAPMAm)-8, and **d** GG-g-poly(AMPS-co-DMAPMAm)/MMT-3

TGA

The thermograms of GG, GG-g-AMPS, GG-g-poly(AMPS-co-DMAPMAm)-8, and GG-g-poly(AMPS-co-DMAPMAm)/MMT-3 are presented in Fig. 2. GG (Fig. 2a) shows three degradation steps. The first step of degradation occurs between temperatures of 35–100 °C with the weight loss of 14%, and is attributed to the loss of moisture content in the polysaccharide. The second step of the decomposition occurs in the range of 210–260 °C with a major weight loss of 36% due to the breaking of the glycosidic linkage of the GG. The final decomposition of GG occurs around 270–540 °C with the weight loss of 32%. About 14% of the GG sample remains as residual matter at

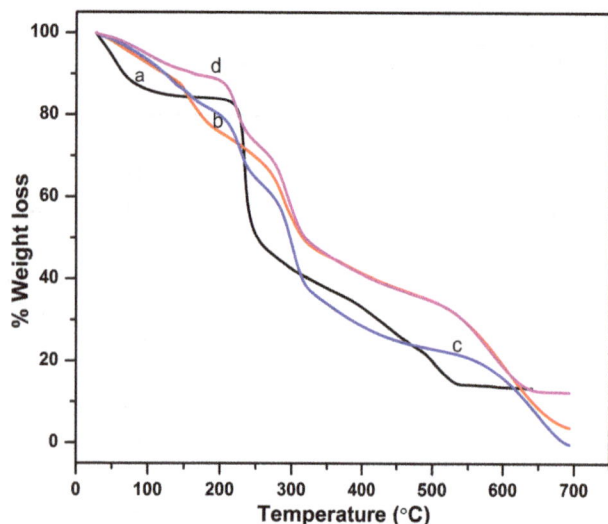

Fig. 2 Thermograms of **a** GG, **b** GG-g-APMS, **c** GG-g-poly(AMPS-co-DMAPMAm)-8, and **d** GG-g-poly(AMPS-co-DMAPMAm)/MMT-3

550 °C. For GG-g-AMPS (Fig. 2b), four decomposition steps occur. With the first step in the range of 30–140 °C with a weight loss of 13% due to loss of moisture in the sample and 10% weight loss between 182 and 265 °C, this might be due to the dissociation of GG from AMPS. Between 320 and 520 °C, a major weight loss (17%) occurred and is linked to the degradation of the polysaccharide and final step decomposition at 555 °C with 25% weight loss leaving 4% as the residue. The thermogram of GG-g-poly(AMPS-co-DMAPMAm)-8 (Fig. 2c) shows four degradation steps with the first step losing 16% of the weight between 35 and 160 °C due to the elimination of moisture from the system. The second steps occurred in the range of 324–510 °C with the loss of 18% which could be due to the breaking of the grafting between GG and the copolymers. The third degradation step results in weight loss of 28% between the temperature range of 324–510 °C. The complete degradation of the grafted gel occurred around 580 °C. For the GG -g-poly(AMPS-co-DMAPMAm)/MMT-3 (Fig. 2d), the degradation steps are similar to that of GG-g-AMPS and are in four stages. The first was the elimination of water molecule from the composite gel in the range of 33–180 °C. The second steps of the degradation occurred in the temperature range of 205–256 °C with a weight loss of 16% and could be due to the degradation of GG. The third step occurred between 317 and 535 °C with major weight loss of 20% and the final step of degradation occurred at 633 °C leaving around 13% as residual matter. The GG-g-AMPS, and GG-g-poly(AMPS-co-DMAPMAm)-8 showed low thermal stability compared GG and GG-g-poly(AMPS-co-DMAPMAm)/MMT-3. The high stability of GG-g-poly(AMPS-co-

DMAPMAm)/MMT-3 toward heat is attributed to the incorporation of MMT in the system [35].

SEM

The surface morphology of GG, GG-g-AMPS, GG-g-poly(AMPS-co-DMAPMAm)-3, and GG-g-poly(AMPS-co-DMAPMAm)/MMT-8 is presented in Fig. 3. It could be deduced from Fig. 3b that grafting of AMPS on the GG changes the fibrous homogeneous surface of GG (Fig. 3a) into heterogeneous. Likewise, Fig. 3c shows a very distinct crystalline-like morphology suggesting the grafting of GG on poly(AMPS-co-DMAPMAm)-8. On the other hand, exfoliating MMT in the system also shows a considerable change in the surface morphology producing cotton like accumulation with an irregular shape that appears fibrous (Fig. 3d).

XRD

The XRD patterns of MMT, GG, GG-g-AMPS, GG-g-poly(AMPS-co-DMAPMAm)-8, and GG-g-poly(AMPS-co-DMAPMAm)/MMT-3 samples are shown in Fig. 4. MMT diffractogram (Fig. 4a) shows a complete amorphous nature of the clay MMT. GG diffractogram (Fig. 4b) showed two major peaks at 2θ values of $21.31°$ (medium) and $30.47°$ (sharp). The exhibition of this sharp peak at lower 2θ value indicated the crystalline nature of the GG. In GG-g-AMPS sample (Fig. 4c); the disappearance of some peaks and appearing of new one at a 2θ value of $5.82°$ indicated the grafting of AMPS on GG. The appearance of many sharp peaks at low 2θ ($9.92°$, $14.12°$ and $16.32°$) values (Fig. 4d) indicated the incorporation of DMAPMAm on GG-g-AMPS and it further indicates the semicrystalline nature of the gel network [36]. The intercalation of MMT within the polymer gel network is evidence of the major shift of the peak from the 2θ value of $18.79°$ (Fig. 4d) to $22.18°$. Similarly, the increases of d-spacing for MMT to 11.4527 Å from the normal 9.8 Å [34] are another evidence to show the intercalation of MMT into the gel network. The decrease in the intensity of the peak around 2θ value of 9.1 and the disappearance of the peak at 2θ of $7.2°$ from Fig. 4d are further evidences to prove the intercalation of MMT into the gel matrix. The disappearance of a small broad peak at a 2θ value of $31.69°$ in Fig. 4e also showed the intercalation of MMT into the gel network which resulted in decreases in the degree of crystallinity [37].

Swelling responsive in different salt solution

The effect of different salt solution (Fig. 5) on the swelling ratio of GG-g-poly(AMPA-co-DMAPMAm)-8 and GG-g-poly(AMPA-co-DMAPMAm)/MMT-3 samples shows

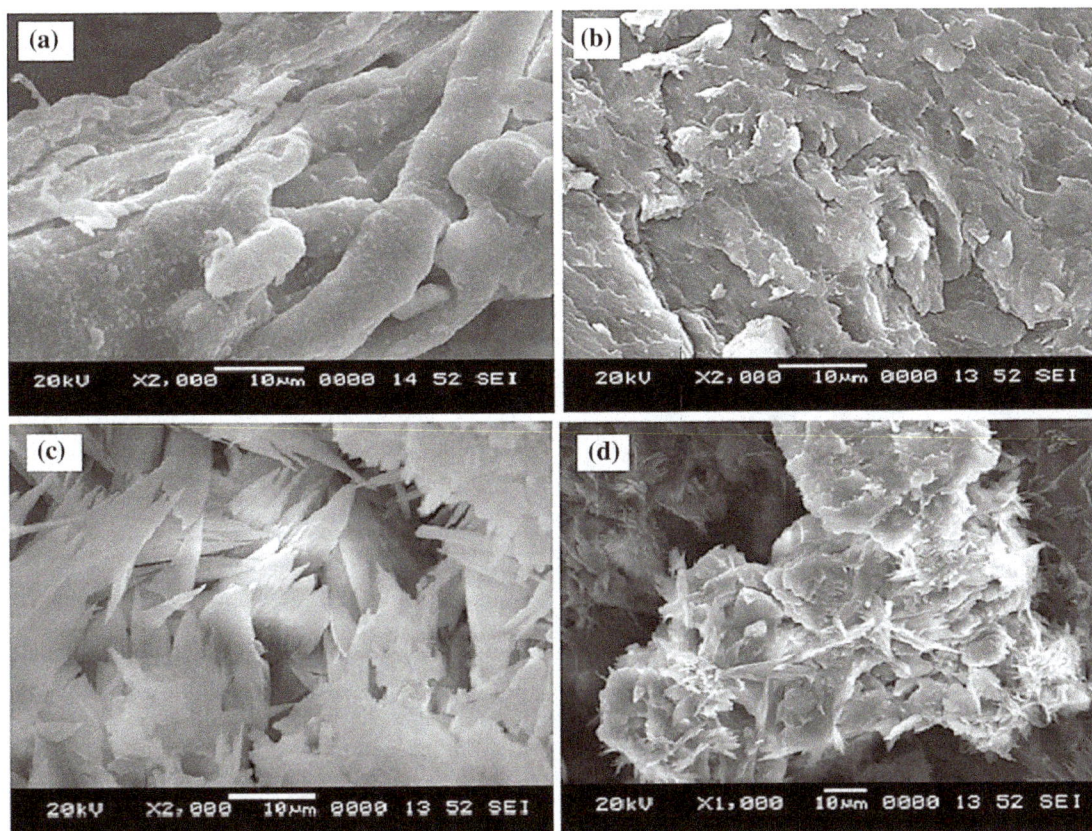

Fig. 3 SEM images of **a** GG, **b** GG-g-AMPS, **c** GG-g-poly(AMPS-co-DMAPMAm)-8, and **d** GG-g-poly(AMPS-co-DMAPMAm)/MMT-3

Fig. 4 XRD diffractograms of **a** MMT, **b** GG, **c** GG-g-AMPS, **d** GG-g-poly(AMPA-co-DMAPMAm)-8, and **e** GG-g-poly(AMPA-co-DMAPMAm)/MMT-3

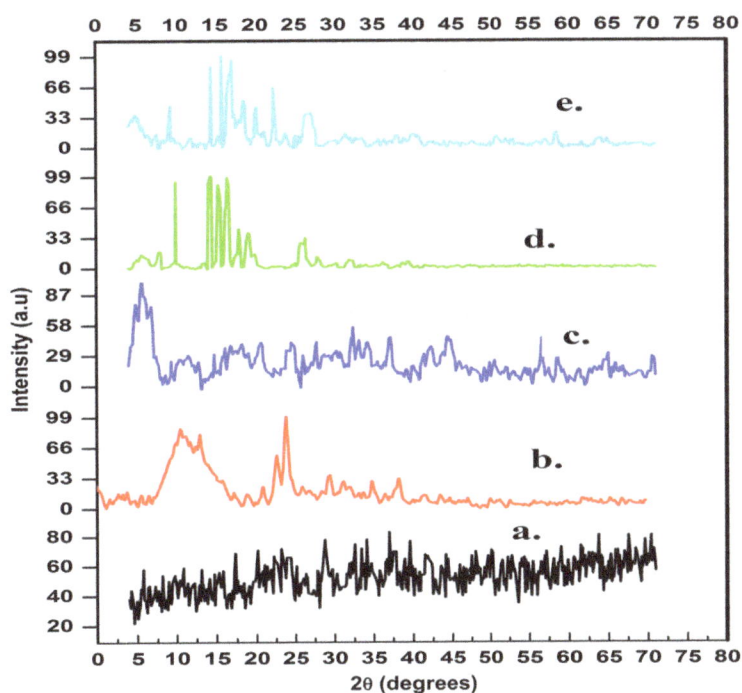

higher swelling capacity in NaCl solution compared to other salt solution with the least swelling response in FeCl$_3$. The swelling behavior is affected by many factors such as nature of cations (charge and radius of cations). The greater the charge of cations, the greater the cross linking degree which results in decrease in swelling [28].

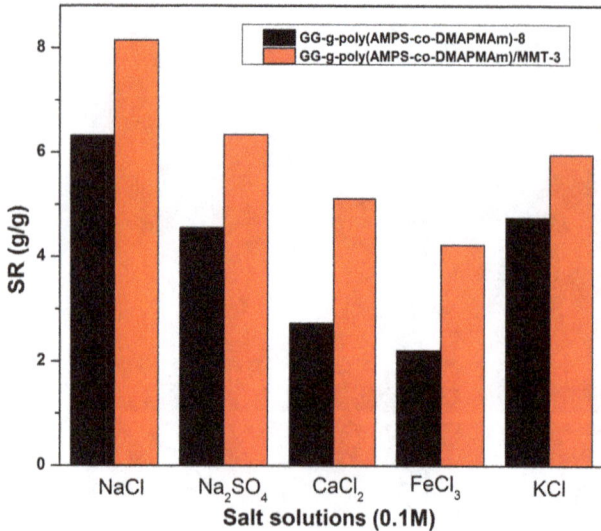

Fig. 5 Effects of different salt solution (0.1 M) on swelling ratio of GG-g-poly(AMPS-co-DMAPMAm)-8 and GG-g-poly(AMPS-co-DMAPMAm)/MMT-3

The swelling of the gel is due to osmotic pressure difference developed between the gel and the external salt solution due to the charge screening effect of the salt solution. The composite hydrogels exhibited salt sensitivity as reported in the literature [38, 39].

Effect of pH on the swelling ratio

The swelling ratios of the gel in different pH media were studied and the results reported in Fig. 6a. Figure 6 showed higher swelling ratio (SR) in basic medium (pH 9.0) with lower SR in acidic medium (pH 1.2) by GG-g-poly(AMPS-co-DMAPMAm)-8. The presence of free amino group on DMAPMAm leads to the formation of many hydrogen bonds in an alkaline medium which will restrict the relaxation of network chain. While in acidic medium, the free amino group is expected to ionize which may result in the breakage of hydrogen bond and generate electrostatic repulsion on the polymer chain [40]. In acidic medium, the attraction between SO_3^{2-} and quaternary ammonium group restricts the swelling [41]. Furthermore, the carbonyl group of the AMPS and DMAPMAm forms hydrogen bonding with the GG which also aids in reducing the swelling [42, 43].

Swelling kinetics

The mechanisms of swelling studies for the gel/composite were carried out based on the standard methods reported in the literature [44, 45]. The swelling parameters of the gel/composite were determined from the various plots (Figs. 6, 7). The parameters such as swelling equilibrium

ratio (S_{eq}) in gram of water per gram of gel/composite, initial swelling rate (R_i) in g of water/g gel/composite min^{-1}, swelling rate constant (K_s), in g gel/composite/g water min^{-1}), swelling exponent (n), and maximum equilibrium swelling (S_{max}) in g of water g^{-1} of sample were calculated using the dynamic swelling data obtained from various plots shown in Figs. 6 and 7. The swelling mechanism of the gel/composite was experimentally determined by employing a second-order kinetic equation as follows:

$$\frac{ds}{dt} = k_s(s_{eq} - s)^2, \tag{7}$$

where k_s and s_{eq} are the swelling rate constant and degree of swelling at equilibrium, respectively. The above equation on integration over the limit $S = S_0$ at $t = t_o$ and $S = S$ at $t = t$ gives

$$\frac{t}{SR} = \frac{1}{k_s s_{eq}^2} + \frac{1}{s_{eq}} t, \tag{8}$$

where $k_s s_{eq}^2$ is equal to R_i which is the initial swelling rate, s_{eq} is the equilibrium swelling, and k_s is the swelling rate constant. The plot of t/SR vs t (Figs. 6b, 7b) produced a linear straight line with a slope of $1/s_{eq}$ and intercept of $\frac{1}{k_s s_{eq}^2}$. This indicates that the second-order kinetics is followed by the swelling process. Furthermore, the calculated s_{eq} from the slope are in good agreement with the experimental value as shown in Table 2. The initial swelling rate (R_i) of the GG-g-poly(AMPS-co-DMAPMAm)/MMT-3 decreases drastically from acidic to basic medium. A similar finding was reported [45]. However, the swelling rate constant (k_s) increases as the pH values increase.

The mechanism of diffusion is one of the factors that govern the applicability of materials [42]. The absorption process involves the diffusion of water molecules into the free spaces of the materials which increase the segmental mobility and consequently result in an expansion of chain segment between crosslink and later result in swelling. The dynamics of water sorption process was studied using the simple empirical equation called power law equation which is used mostly in determining the mechanism of diffusion in the polymeric network [46]:

$$F = Kt^n. \tag{9}$$

The above equation can be rewritten in form of ln as follows:

$$\ln F = \ln K + n \ln t. \tag{10}$$

The values of K and n were calculated from the intercept and slope of $\ln F$ vs t plots (Figs. 6c, 7c) and tabulated in Table 2, where F, n, and K are swelling power, swelling exponent, and swelling rate constant, respectively.

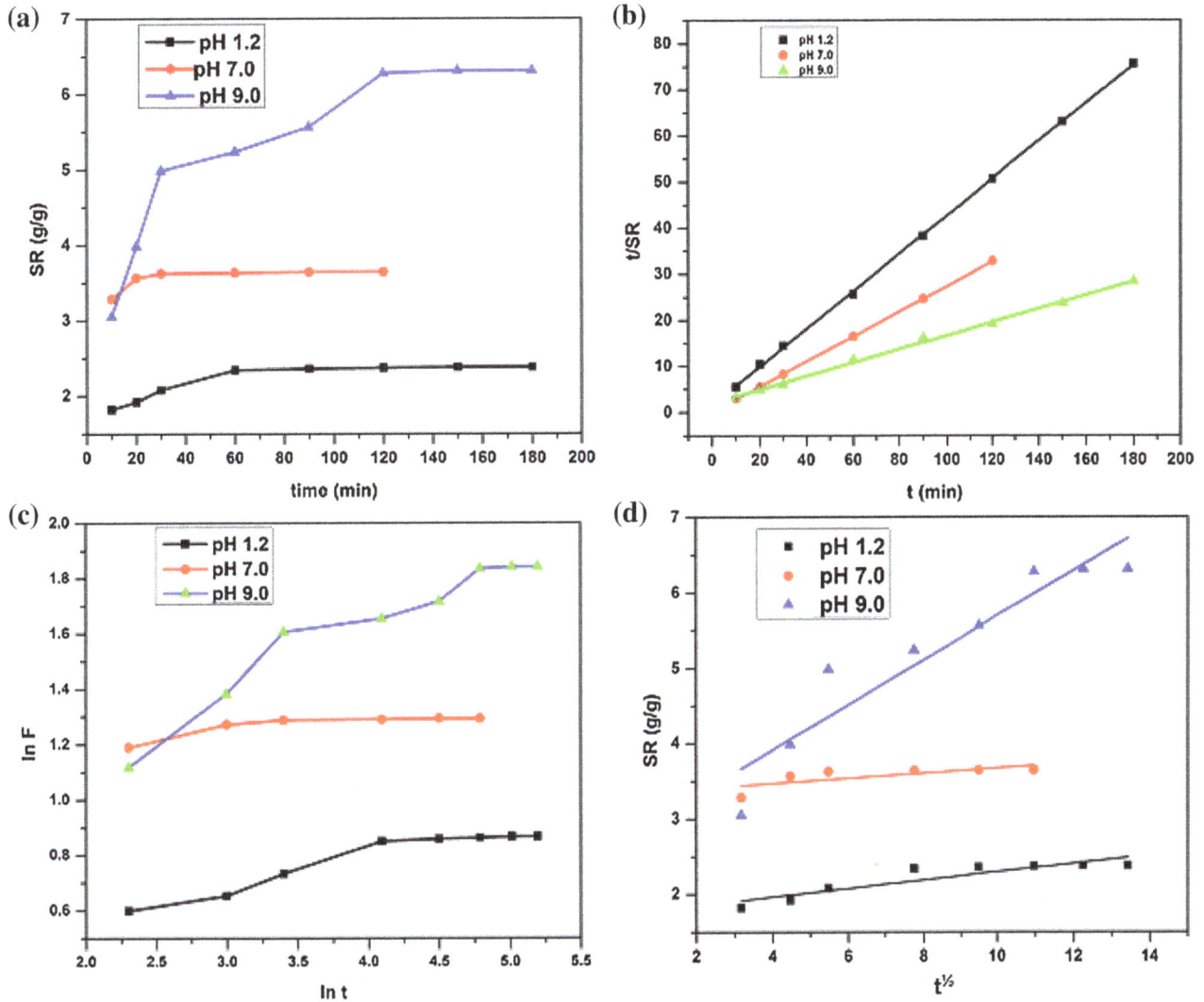

Fig. 6 Swelling curves for GG-g-poly(AMPS-co-DMAPMAm)-8 gel

Depending on the diffusion rate of the material relaxation, three different diffusion mechanisms are proposed [42–45, 47]:

1. Fickian diffusion in which the diffusion rate is less than the relaxation rate ($n = 0.50$);
2. Diffusion which is rapid compared the relaxation processes ($n = 1$); and
3. Non-Fickian or anomalous diffusion which occurs when the rate of diffusion and that of relaxation are comparable ($0.50 < n < 1$).

The Fickian diffusion, actually, refers to a situation where water penetration rate in the gels is less than the polymer chain relaxation rate. Therefore, $n = 0.5$ indicates a perfect Fickian process. Nevertheless, when the water penetration rate is much below the polymer chain relaxation rate, it is possible to record the n values below 0.5. This situation is still regarded as Fickian diffusion or "Less

Fickian diffusion" behavior [46]. In this study, the values of n (Table 2) are all below 0.5. Hence, it is said to follow a less Fickian diffusion mechanism.

Dye adsorption studies

The dye adsorption studies were carried out using rhodamine B and chromotrope 2R as model dyes and their structures are given in Fig. 8.

Effects of contact time on the adsorption capacity

The effect of contact time on the adsorption of dyes RhB and C2R on both GG-g-poly(AMPS-co-DMAPMAm)-8 and GG-g-poly(AMPS-co-Dmapmam)/MMT-3 (Fig. 9) showed an increase in the adsorption capacity slowly with increase in time. RhB showed higher adsorption compared to C2R; this could be attributed to the presence of many

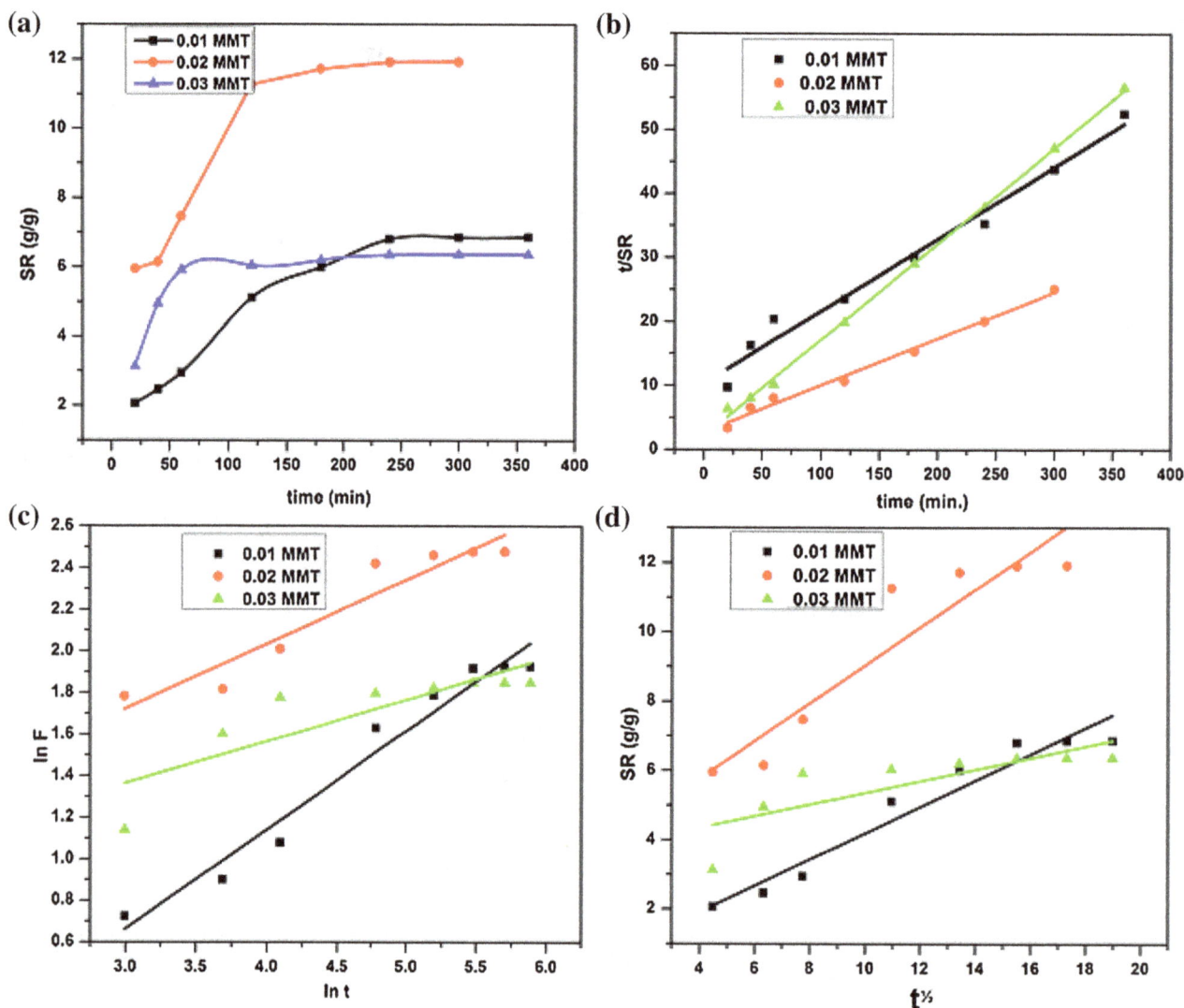

Fig. 7 Swelling curves for GG-g-poly(AMPS-co-DMAPMAm)/MMT-3

Table 2 Kinetics swelling and diffusion parameters of GG-g-poly(AMPS-co-DMAPMAm)-8 and GG-g-poly(AMPS-co-DMAPMAm/MMT-3

PH	$S_{eq}(g/g)$	R_i (g of water/g of sample)	K_s (g/g)	N	K	S_{max}
GG-g-poly(AMPS-co-DMAPMAm)-8						
1.2	2.45	1.69	0.099	0.10	1.03	2.38
7.0	3.68	0.20	0.379	0.04	7.48	3.65
9.0	6.85	2.07	0.010	0.24	2.40	6.32
GG-g-poly(AMPS-co-DMAPMAm/MMT-3						
1.2	8.93	10.33	0.0012	0.48	3.733	6.86
7.0	13.70	2.71	0.0020	0.31	4.383	11.93
9.0	6.67	2.12	0.0106	0.2	3.679	6.36

negative (acidic groups) site on the adsorbent which could result in the formation of an electrostatic attraction with the positive (basic groups) part of the adsorbate. The adsorption of C2R on both the adsorbents is proceeded at slower rate especially on the GG-g-poly(AMPS-co-DMAPMAm)/MMT-3. This could be due to the repulsion between the anionic (basic site) groups on both C2R and the composites [48].

Fig. 8 Structures of the dyes used: **a** rhodamine B and **b** chromotrope 2R

(a)

(b)

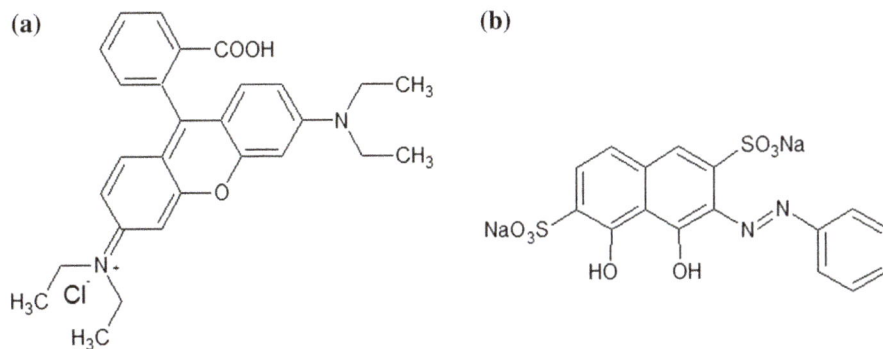

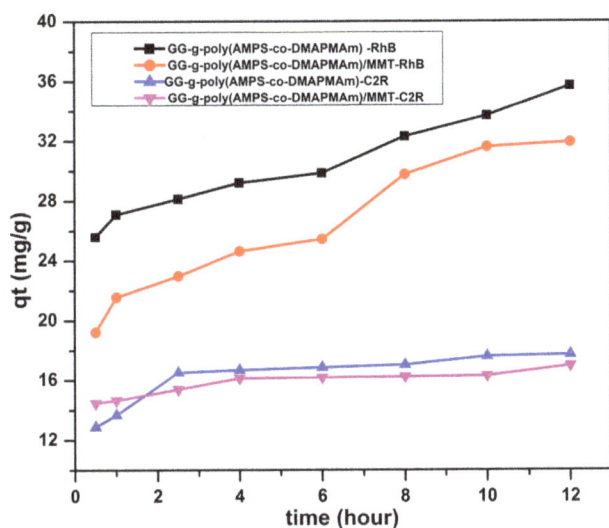

Fig. 9 Amount of dyes (rhodamine B and chromotrope 2R) adsorbed (mg/g) on GG-g-poly(AMPS-co-DMAPMAm)-8 gel and GG-g-poly (AMPS-co-DMAPMAm)/MMT-3 composite over time

Adsorption isotherm

Adsorption isotherms usually describe the performance of adsorbents in adsorption processes by describing the surface interaction between the adsorbent and adsorbate [49]. There are various isotherm models used to describe the adsorption processes. In this study, the two most common used adsorption isotherms, namely, Langmuir isotherm [50, 51] and Freundlich isotherm [52], are employed.

The Langmuir isotherm is a model which quantitatively describes equilibrium monolayer adsorbate formation on the surface of the adsorbent, and is expressed as follows:

$$\frac{C_e}{q_e} = \frac{1}{q_m} \times C_e + \frac{1}{K_L q_m}, \tag{11}$$

where C_e and q_e are the equilibrium concentration of dye (mg/L) and the amount of dye adsorbed (mg/g), respectively, q_m, is the maximum adsorption corresponding to complete monolayer coverage on the surface (mg/g), K_L is the Langmuir constant which is related to the energy of adsorption (L/mg). K_L and q_m are determined from the

intercept and slope of the linear plot of C_e/q_e versus C_e (Figs. 10a, b, 11a, b) and presented in Table 3. The essential feature of the Langmuir isotherm can be represented in terms of separation factor (dimensionless equilibrium parameter) R_L [53, 54], which can be expressed as follows:

$$R_L = \frac{1}{1 + K_L C_o}, \tag{12}$$

where C_0 is the initial concentrations of dyes, K_L is the constant related to the energy of adsorption (Langmuir Constant). RL value indicates the favorability nature of adsorption. If $R_L > 1$, the adsorption is unfavorable; if $R_L = 1$, the adsorption is linear; if $0 < R_L < 1$, the adsorption is favorable; and if $R_L = 0$, then the adsorption is irreversible. From the data reported in Table 3, the R_L is greater than 0 but less than 1 indicating the favorability of Langmuir isotherm for the adsorption of RhB and C2R. Similarly, comparing the q_m calculated (33.33 and 16.13 mg/g) with the experimental q_m (35.7 and 17.73 mg/g), respectively, for RhB and C2R on GG-g-poly(AMPS-co-DMAPMAm)-8. This indicated the formation of a monolayer of RhB and C2R on the surfaces of GG-g-poly(AMPS-co-DMAPMAm)-8.

The Freundlich adsorption isotherm is based on the assumption that encompasses the heterogeneity of the surface and the adsorption capacity related to the equilibrium concentration of the adsorbate. The Freundlich isotherm is commonly expressed as follows:

$$\ln q_e = \ln k_f + \frac{1}{n}\ln C_e, \tag{13}$$

where q_e and C_e are the amount of dyes adsorbed (mg/g) and the equilibrium concentration of dyes (mg/L), respectively, K_f and n are Freundlich adsorption isotherm constants that represent the adsorption capacity and the degree of nonlinearity between the dye concentration and the adsorption, respectively. The values of K_f and n were calculated from the intercept and slope of the plot between ln q_e and ln C_e (Figs. 10c, d, 11c, d) and are presented in Table 3. The value of n indicates whether the adsorption is favorable or otherwise. If it lies within the range of 1–10,

(a)

(b)

(c)

(d)

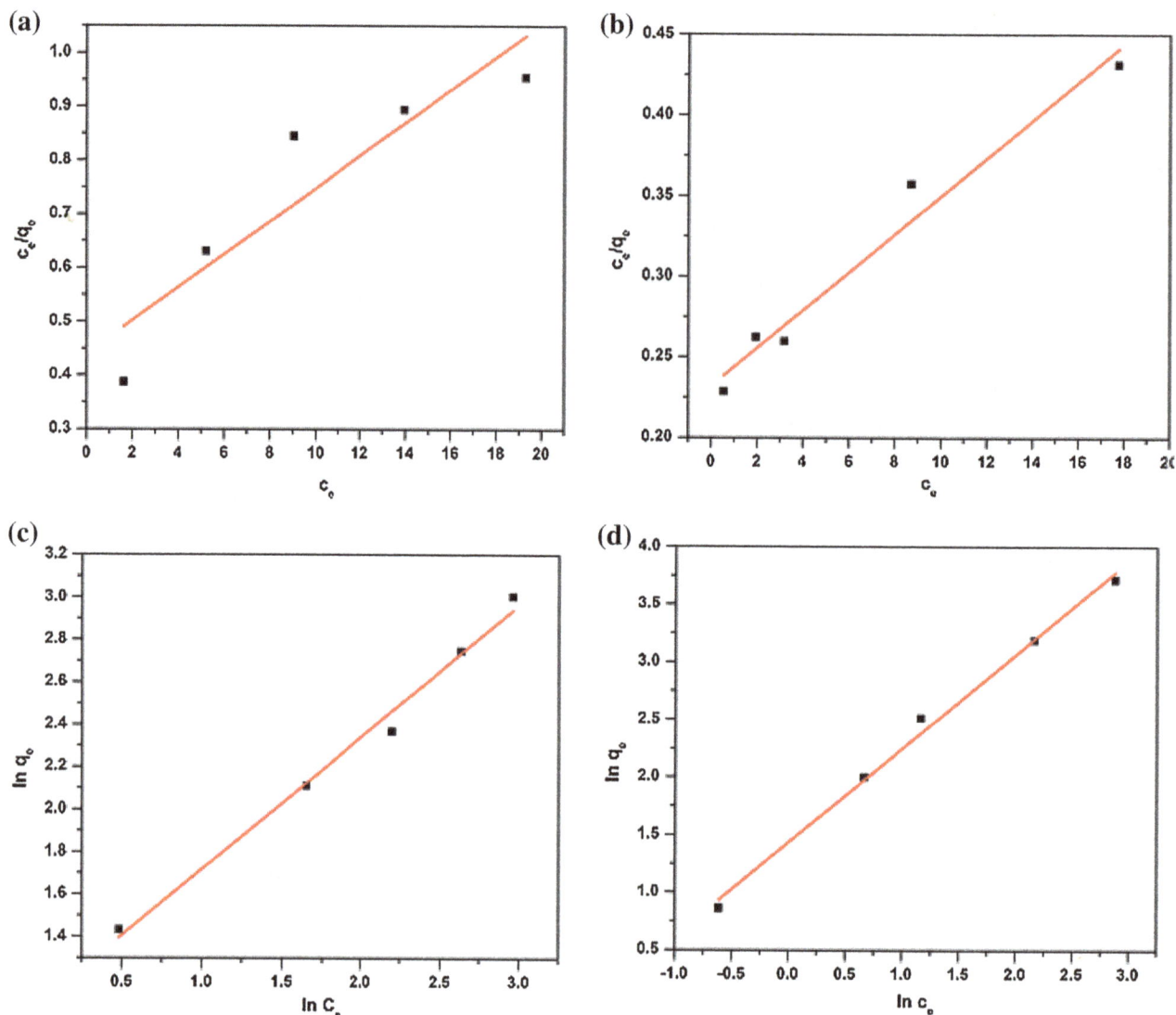

Fig. 10 Adsorption isotherms for rhodamine B dye. **a** Langmuir isotherm for GG-g-poly(AMPS-co-DMAPMAm)-8 gel, **b** Langmuir isotherm for GG-g-poly(AMPS-co-DMAPMAm)/MMT-3 composite gel, **c** Freundlich isotherm for GG-g-poly(AMPS-co-DMAPMAm)-8 gel, and **d** Freundlich isotherm for GG-g-poly(AMPS-co-DMAP-MAm)/MMT-3 composite

then the adsorption is considered favorable. In this case, the value of *n* lies between 1.23 and 4.83 which shows a favorable adsorption. Similarly, the R^2 values for the adsorption of RhB and C2R on GG-g-poly(AMPS-co-DMAPMAm)/MMT-3 are 0.994 and 0.996, respectively, which are higher when compared with 0.989 and 0.990, respectively, for RhB and C2R on GG-g-poly(AMPS-co-DMAPMAm)-8. Hence, we can say that adsorption of RhB and C2R on GG-g-poly(AMPS-co-DMAPMAm)/MMT-3 base fits into the Freundlich model.

Kinetic studies

The adsorption capacity of RhB and C2R dyes as a function of time by the adsorbents [GG-g-poly(AMPS-co-DMAPMAm)-8 and GG-g-poly(AMPS-co-DMAPMAm)/

MMT-3] is shown in Fig. 9. The rate of adsorption of the dye uptake was little slow especially with respect to C2R compared to RhB adsorption. The maximum adsorption observed in C2R was 17.72 and 16.99 mg/g, respectively, for GG-g-poly(AMPS-co-DMAPMAm)-8 and GG-g-poly (AMPS-co-DMAPMAm)/MMT-3 after 12 h. While higher adsorption capacity of 35.70 and 31.20 mg/g of RhB was recorded for GG-g-poly(AMPS-co-DMAPMAm)-8 and GG-g-poly(AMPS-co-DMAPMAm)/MMT-3, respectively, at 12 h. The adsorption of RhB on different adsorbents has been reported [7, 49, 55, 56].

To investigate the mechanism of adsorption, the adsorption data obtained in this work were subjected to various kinetics models. The models employed in this work are Lagergren's pseudo-first-order [55, 57] and pseudo-second-order kinetic models.

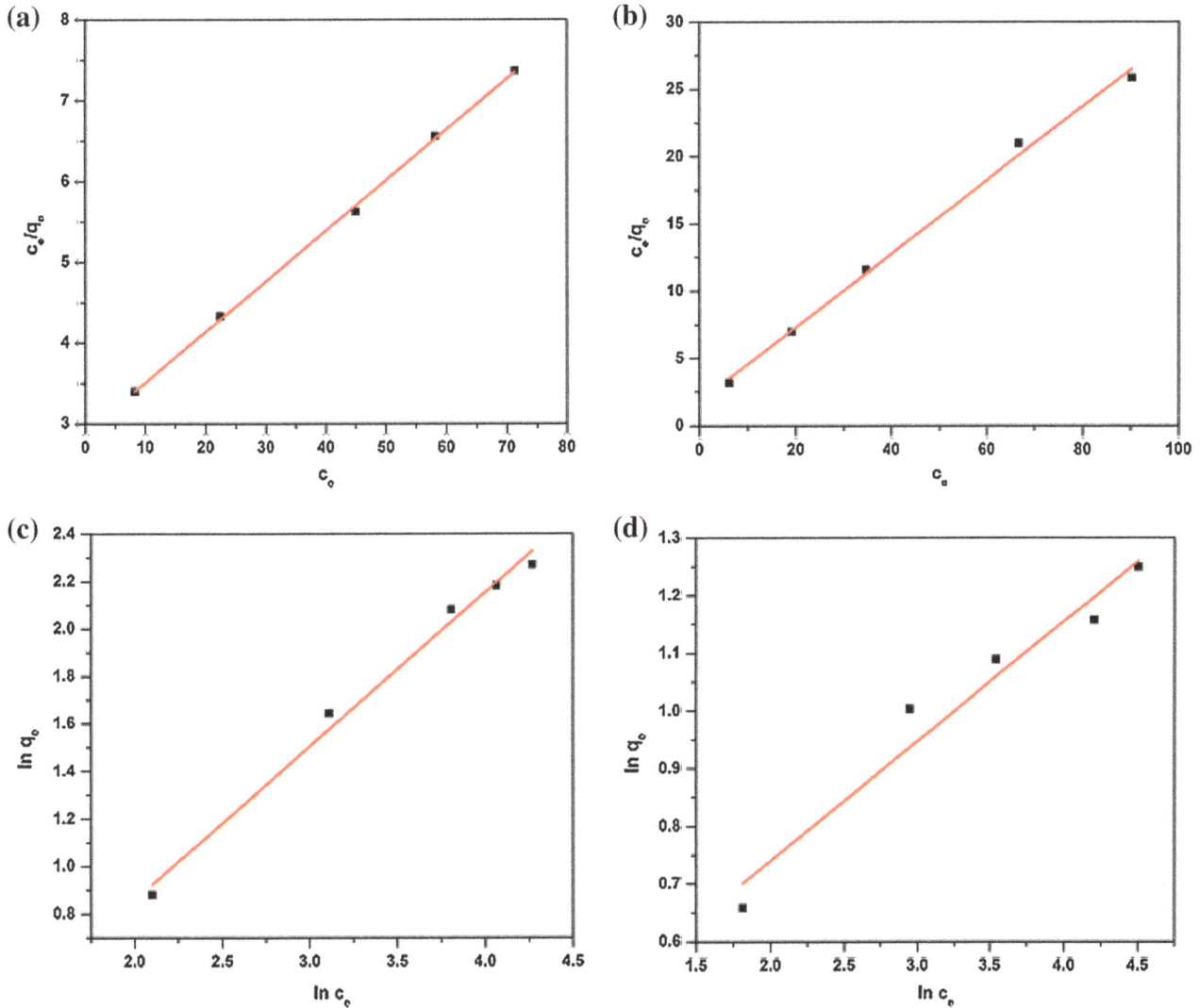

Fig. 11 Adsorption isotherms for chromotrope 2R dye. **a** Langmuir isotherm for GG-g-poly(AMPS-co-DMAPMAm)-8 gel, **b** Langmuir isotherm for GG-g-poly(AMPS-co-DMAPMAm)/MMT-3 composite gel, **c** Freundlich isotherm for GG-g-poly(AMPS-co-DMAPMAm)-8 gel, and **d** Freundlich isotherm for GG-g-poly(AMPS-co-DMAP-MAm)/MMT-3 composite

Pseudo-first-order kinetics

The Lagergren's pseudo-first-order kinetic model is based on assumption that the rate of adsorption of adsorbate with time is directly proportional to difference in equilibrium concentration and concentration with time and this can be represented as follows:

$$\log(q_e - q_t) = \log q_e - \frac{K_1}{2.303}t, \tag{14}$$

where q_e and q_t are the amount of dyes adsorbed (mg/g) at equilibrium and at time t, respectively. K_1 is the rate constant (h^{-1}) for the pseudo-first-order kinetics and t is the time (h) of adsorption. The value of K_1 and R^2 were calculated from the plot of $\log(q_e - q_t)$ versus t and tabulated in Table 4. The $q_{e\ cal}$ for the first-order rate equation was found to be in sharp disagreement with the $q_{e\ exp}$ in all cases. Furthermore, the values of correlation coefficients are low, which is an indication of bad quality linearization. Hence, the adsorption cannot be said to be of first order. It has been suggested that the differences in experimental and theoretical q_e values are that there is a time lag due to external resistance controlling at the beginning of the adsorption [57].

Pseudo-second-order kinetics

The kinetics studies of RhB and C2R adsorption on the adsorbents was carried out using the initial dyes concentration of 100 mg/L in all cases. The pseudo-second-order rate expression of Ho and McKay [58] was adopted in this study and the best model that fit the adsorption was

Table 3 Isotherm model parameters for the adsorption of rhodamine B and chromotrope 2R dyes on GG-g-poly(AMPS-co-DMAPMAm)-8 and GG-g-poly(AMPS-co-DMAPMAm)/MMT-3

Dyes		RhB		C2R	
Adsorbent		GG-g-poly(AMPS-co-DMAPMAm)-8	GG-g-poly(AMPS-co-DMAPMAm)/MMT-3	GG-g-poly(AMPS-co-DMAPMAm)-8	GG-g-poly(AMPS-co-DMAPMAm)/MMT-3
	C_o (mg/L)	10–100	10–100	10–100	10–100
Langmuir model	R_L	0.43–0.88	0.797–0.975	0.045–0.320	0.19–0.165
	K_L	0.013	0.0026	0.212	0.509
	q_m (mg/g)	33.33	90.91	16.13	3.66
	R^2	0.841	0.969	0.999	0.996
Freundlich model	k_f	10.91	2.76	1.21	0.89
	n	1.61	1.23	1.54	4.83
	R^2	0.989	0.994	0.990	0.996

R^2 linear regression correlation co-efficient

Table 4 Pseudo-first-order and pseudo-second-order kinetics model data for the adsorption of rhodamine B and chromotrope 2R dyes on GG-g-poly(AMPS-co-DMAPMAm)-8 and GG-g-poly(AMPS-co-DMAPMAm)/MMT-3

Dye		RhB		C2R	
Adsorbent		GG-g-poly(AMPS-co-DMAPMAm)-8	GG-g-poly(AMPS-co-DMAPMAm)/MMT-3	GG-g-poly(AMPS-co-DMAPMAm)-8	GG-g-poly(AMPS-co-DMAPMAm)/MMT-3
	$Q_{e\ exp}$ (mg/g)	35.70	31.92	17.73	17.56
Pseudo-first-order kinetic model	$q_{e\ cal}$ (mg/g)	45.42	15.34	5.84	2.88
	K_1 (hr^{-1})	0.15	0.198	0.32	0.124
	R^2	0.935	0.856	0.875	0.777
Pseudo-second-order kinetic model	q_{ecal} (mg/g).	37.04	34.48	18.18	17.24
	K_2 (g/mghr^{-1})	0.016	0.028	0.016	0.01
	R^2	0.991	0.984	0.999	0.996

R^2 linear regression correlation co-efficient

selected based on the values of the linear regression correlation co-efficient (R^2). The pseudo-second-order equation is given as follows:

$$t/q = \frac{1}{k_2 q_e^2} + \frac{1}{q_e} \times t, \qquad (15)$$

where k_2 is the adsorption rate constant for pseudo-second-order kinetics (gmg^{-1} h^{-1}), and qe is the adsorption capacity calculated from pseudo-second-order kinetic model (mgg^{-1}), q_e is the equilibrium adsorption (mgg^{-1}), and t is the adsorption time (h). The linear form of the pseudo-second-order kinetic model is given in Fig. 12. The values of k_2, and q_e were calculated from the slope and intercept of the linear plot of t/q vs t. The values of R^2 were

higher than those in the pseudo-first-order model and they approach unity in all cases. Hence, the adsorption is of the second-order kinetics. In addition, the $q_{e\ cal}$ values are in agreement with the $q_{e\ exp}$. Therefore, the experimental results support the assumption behind the model that the rate-limiting step in the adsorption of dyes are chemisorptions involving valence forces through the exchange of electrons between adsorbent and dyes [7]. A similar finding was reported in the literature [59, 60].

Desorption studies

The re-usability of GG-g-poly(AMPS-co-DMAPMAm)-8 gel and GG-g-poly(AMPS-co-DMAPMAm)/MMT-3

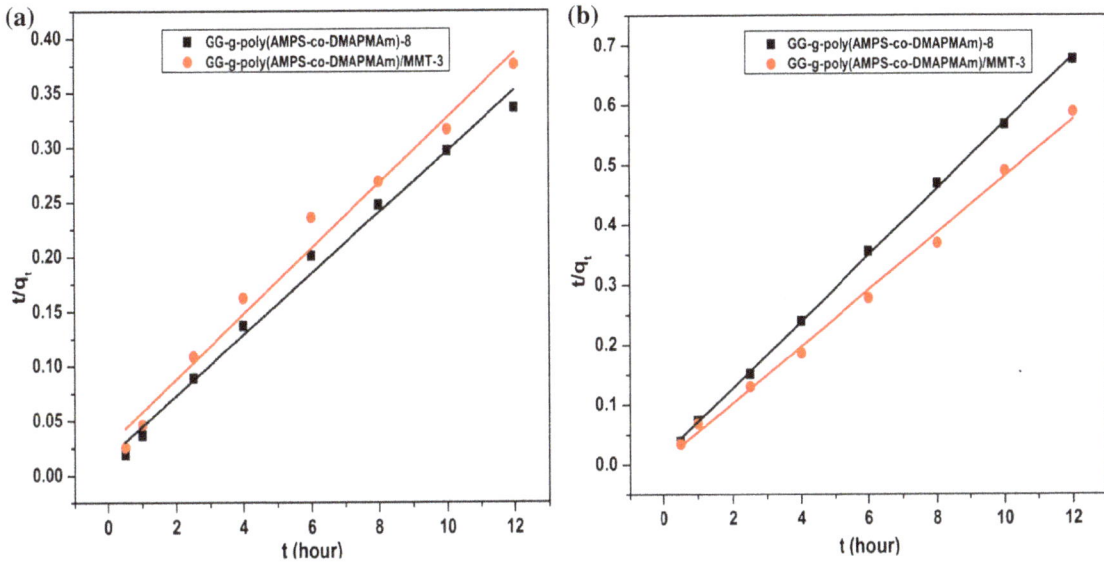

Fig. 12 Pseudo-second-order kinetics for the adsorption of **a** rhodamine B and **b** Chromotrope 2R Dyes on GG-g-poly(AMPS-co-DMAPMAm)-8 and GG-g-poly(AMPS-co-DMAPMAm)/MMT-3

Table 5 Desorption capacity (%) of GG-g-poly(AMPS-co-DMAPMAm)-8 and GG-g-poly(AMPS-co-DMAPMAm)/MMT-3 for RhB and C2R under pH 1.2 and pH 13.0

PH	Desorption (%)			
	GG-g-poly(AMPS-co-DMAPMAm)-8		GG-g-poly(AMPS-co-DMAPMAm)/MMT-3	
	RhB	C2R	RhB	C2R
First cycle				
1.2	91.57	48.50	94.25	48.34
13.0	83.18	41.85	91.69	43.92
Second cycle				
1.2	89.23	42.23	83.17	44.03
13.0	74.12	38.11	64.98	39.62

Table 6 Comparison of adsorption capacity of dyes onto different adsorbents

Adsorbate	Adsorbent	Q_e (mg/g)	References
RhB	GG-g-poly(AMPS-co-DMAPMAm)-8	35.70	Present work
RhB	GG-g-poly(AMPS-co-DMAPMAm)/MMT-3	31.92	Present work
C2R	GG-g-poly(AMPS-co-DMAPMAm)-8	17.73	Present work
C2R	GG-g-poly(AMPS-co-DMAPMAm)/MMT-3	17.56	Present work
RhB	Acid activated mango leaf powder	3.85	[7]
RhB	Palm shell-based activated carbon	2.92	[49]
RhB	Coffee powder	4.018	[59]
Methyl orange	Gellan gum-graft-poly(DMAEMA)	25.8	[31]
C2R	Carbons modified with lanthanum	164	[60]
Methylene blue	Poly(acrylic acid-co-acrylamide)	1313	[61]
Methylene blue and Direct blue	Polyacrylamide/chitosan	6.744	[63]
RhB	Rice husk activated carbon	275.2	[59]
Crystal violet	Car/poly(AAm-co-Na-AA)-MMT	46.15	[64]
Crystal violet	CarAlg/MMT	88.8	[23]

suspending about 15 mg of the adsorbent in 25 mL of solution (pH 1.2 and pH 13.0) allowed to stand for 8 h at room temperature. The solutions were filtered and diluted appropriately and the absorbance of the desorbed RhB and C2R were measured using UV–Vis spectrophotometer. The amounts of the two dyes were calculated using Eq. (6). The percentage of desorption for the dyes was calculated using the following equation [61]:

$$\text{Desorption}(\%) = \frac{\text{Amount of dye desorbed}}{\text{Amount of dye adsorbed}} \times 100. \quad (16)$$

It was clear from Table 5 that RhB can be efficiently removed by regenerating the adsorbent at least twice under pH 1.2 and 13.0, whereas desorption of C2R under the same conditions is low. In all cases, the % desorption is high at pH 1.2 and this could be attributed to the reaction between H^+ and a lone pair of electrons in both the secondary amine groups of the adsorbent and the tertiary nitrogen of RhB. Hence, more of RhB will go into solution [62].

The adsorption capacity of GG-g-poly(AMPS-co-DMAPMAm)-8 gel and GG-g-poly(AMPS-co-DMAP-MAm)/MMT-3 on RhB and C2R is compared with other previously developed adsorbents [7, 23, 31, 49, 59–61, 63, 64] towards different adsorbates and presented in Table 6.

Conclusion

In this work, an amphoteric terpolymer consisting of gel GG, AMPS, and DMAPMAm, and its clay composite containing MMT have successfully been synthesized and studied. The GG-g-poly(AMPS-co-DMAPMAm)-8 gel showed maximum swelling at pH of 9.0 and GG-g-poly (AMPS-co-DMAPMAm)/MMT-3 composite shows maximum swelling at neutral pH. The adsorption of RhB and C2R dyes on the gels/composite is higher for RhB compared to C2R. The adsorption kinetic studies revealed a second-order adsorption process. Furthermore, adsorption of RhB and C2R on to GG-g-poly(AMPS-co-DMAP-MAm)-8 and GG-g-poly(AMPS-co-DMAPMAm)/MMT-3 was found to best fit into the Langmuir and Freundlich models, respectively.

Acknowledgements One of the authors SAZ thanks the Government of India for providing the scholarship under the Indian Council for Cultural Relations (ICCR) and Usmanu Danfodiyo University, Sokoto, Nigeria for the study leave.

References

1. Wang L, Wang A (2007) Adsorption characteristics of Congo Red onto the chitosan/montmorillonite nanocomposite. J Hazard Mater 147:979–985
2. Luo P, Zhao Y, Zhang B, Liu J, Yang Y, Liu J (2010) Study on the adsorption of Neutral Red from aqueous solution onto halloysite nanotubes. Water Res 44:1489–1497
3. Jiang R, Fu Y, Zhu H, Yao J, Xiao L (2012) Removal of methyl orange from aqueous solutions by magnetic maghemite/chitosan nanocomposite films: adsorption kinetics and equilibrium. J Appl Polym Sci 125:540–549
4. Ganesan V, Girija EK (2015) Investigations on textile dye adsorption onto hydroxyapatite-alginate nanocomposite prepared by a modified method. Cellul Chem Technol 49:87–91
5. Gupta SK, Nayunigari MK, Misra R, Ansari FA, Dionysiou DD, Maity A, Bux F (2016) Synthesis and performance evaluation of a new polymeric composite for the treatment of textile wastewater. Ind Eng Chem Res 55:13–20
6. Chemwatch, Chromotrope 2R (2010) Material safety data sheet sc-214716. Available at:http://www.chemwatch.net/solutions/msds/chromotrope2r. Accessed on 12 May 2016
7. Khan TA, Sharma S, Ali I (2011) Adsorption of rhodamine B dye from aqueous solution onto acid activated mango (*Magnifera indica*) leaf powder: equilibrium, kinetic and thermodynamic studies. J Toxicol Environ Health Sci 3:286–297
8. Sebastian S, Mayadevi S, Beevi BS, Mandal S (2014) Layered clay-alginate composites for the adsorption of anionic dyes: a biocompatible solution for water/wastewater treatment. J Water Res Prot 6:177–184
9. Abd El-Latif MM, Ibrahim AM, El-Kady MF (2010) Adsorption equilibrium, kinetics and thermodynamics of methylene blue from aqueous solutions using biopolymer oak sawdust composite. J Am Sci 6:267–283
10. El Haddad M, Mamouni R, Saffaj N, Lazar S (2012) Adsorptive removal of basic dye rhodamine B from aqueous media onto animal bone meal as new low cost adsorbent. Glob J Hum Soc Sci Geog Environ Geo-Sci 12:19–29
11. Patil MR, Shrivastava VS (2015) Adsorption removal of carcinogenic acid violet19 dye from aqueous solution by polyaniline-Fe₂O₃ magnetic nano-composite. J Mater Environ Sci 6:11–21
12. Casey LS, Wilson LD (2015) Investigation of Chitosan-pva composite films and their adsorption properties. J Geosci Environ Prot 3:78–84
13. Datskevich EV, Prikhod'ko RV, Stolyarova IV, Lozovskii AV, Goncharuk VV (2010) Poly(acrylamide-sepiolite) composite hydrogels: preparation, swelling and dye adsorption properties. Russian J Appl Chem 83(10):1785–1793
14. Santos SCR, Boaventura RAR (2016) Adsorption of cationic and anionic azo dyes on sepiolite clay: Equilibrium and kinetic studies in batch model. J Environ Chem Eng 4:1473–1483
15. Mahida VP, Patel MP (2016) Superabsorbent amphoteric nanohydrogels: synthesis, characterization and dyes adsorption studies. Chin Chem Lett 27:471–474
16. Maity J, Ray SK (2016) Enhanced adsorption of Cr(VI) from water by guar gum based composite hydrogels. Int J Biol Macromol 89:246–255
17. Robati D, Mirza B, Ghazisaeidi R, Rajabi M, Moradi O, Tyagi I, Agarwal S, Gupta VK (2016) Adsorption behavior of methylene blue dye on nanocomposite multi-walled carbon nanotube functionalized thiol (MWCNT-SH) as new adsorbent. J Mol Liq 216:830–835
18. Postai DL, Demarchi CA, Zanatta F, Melo DCC, Rodrigues CA (2016) Adsorption of rhodamine B and methylene blue dyes

using waste of seeds of Aleurites moluccana, a low cost adsorbent. Alexandria Eng J 55:1713–1723

19. Shabbir M, Rather LJ, Bukhari MN, Shahid M, Khan MA, Mohammad F (2016) An eco-friendly dyeing of woolen yarn by Terminalia chebula extract with evaluations of kinetic and adsorption characteristics. J Adv Res 7:473–482

20. Reynel-Avila HE, Mendoza-Castillo DI, Bonilla-Petriciolet A (2016) Relevance of anionic dye properties on water decolorization performance using bone char: adsorption kinetics, isotherms and breakthrough curves. J Mol Liq 219:425–434

21. Sharma K, Kaith BS, Kumar V, Kalia S, Kumar V, Swart HC (2014) Water retention and dye adsorption behaviour of Gg-cl-poly(acrylic acid-aniline) based conductive hydrogels. Geoderma 232–234:45–55

22. Mahdavinia GR, Massoumi B, Jalili K, Kiani G (2012) Effect of sodium montmorillonite nanoclay on the water absorbency and cationic dye removal of carrageenan-based nanocomposite superabsorbents. J Polym Res 19(9):9947

23. Mahdavinia GR, Aghaie H, Sheykhloie H, Vardini MT, Etemadi H (2013) Synthesis of CarAlg/MMt nanocomposite hydrogels and adsorption of cationic crystal violet. Carbohydr Polym 98 (1):358–365

24. Mahdavinia GR, Baghban A, Zorofi S, Massoudi A (2014) Kappa-carrageenan biopolymer-based nanocomposite hydrogel and adsorption of methylene blue cationic dye from water. J Mater Environ Sci 5(2):330–337

25. Auta M, Hameed BH (2014) Chitosan–clay composite as highly effective and low-cost adsorbent for batch and fixed-bed adsorption of methylene blue. Chem Eng J 237:352–361

26. Sharma R, Kalia S, Kaith BS, Pathania D, Kumar A, Thakur P (2015) Guaran-based biodegradable and conducting interpenetrating polymer network composite hydrogels for adsorptive removal of methylene blue dye. Polym Degrad Stab 122:52–65

27. Nie X, Adalati A, Du J, Liu H, Xu S, Wang J (2014) Preparation of amphoteric nanocomposite hydrogels based on exfoliation of montmorillonite via in-situ intercalative polymerization of hydrophilic cationic and anionic monomers. Appl Clay Sci 97–98:132–137

28. Atta S, Khaliq S, Islam A, Javeria I, Jamil T, Athar MM, Shafiq MI, Ghaffar A (2015) Injectable biopolymer based hydrogels for drug delivery applications. Int J Biol Macromol 80:240–245

29. Mohsen M, Maziad NA, Gomaa E, Aly EH, Mohammed R (2015) Characterization of some hydrogels used in water purification: correlation of swelling and free-volume properties. Open J Org Polym Mater 5:79–88

30. Chen X (2015) Modeling of experimental adsorption isotherm data. Information 6:14–22

31. Karthika JS, Vishalakshi B (2015) Novel stimuli responsive gellan gum-graft-poly(DMAEMA) hydrogel as adsorbent for anionic dye. Int J Biol Macromol 81:544–655

32. Zhang S, Guan Y, Fu GQ, Chen BY, Peng F, Yao C, Sun R (2014) Organic/inorganic superabsorbent hydrogels based on xylan and montmorillonite. J Nanomater 2014:1–11

33. Wilpiszewska K, Spychaj T, Pa´zdzioch W (2016) Carboxymethyl starch/montmorillonite composite microparticles: Properties and controlled release of isoproturon. Carbohydr Polym 136:101–106

34. Guinier A (1994) Imperfect crystals and amorphous bodies. Freeman & Co., San Francisco

35. Verma SK, Pandey VS, Behari MYK (2015) Gellan gum-g-N-vinyl-2-pyrrolidone: synthesis, swelling, metal ion uptake and flocculation behavior. Int J Biol Macromol 72:1292–1300

36. Speakman AA (2013) Introduction to X-ray powder diffraction data analysis. Center for Materials Science and Engineering at MIT, USA

37. Binitha N, Suraja V, Yaakob Z, Sugunan S (2011) Synthesis of polyaniline-montmorillonite nanocomposites using H_2O_2 as the oxidant. Sains Malaysiana 40:215–219

38. Mahdavinia GR, Pourjavadi A, Hosseinzadeh H, Zohourian MJ (2004) Modified chitosan 4. Superabsorbent hydrogels from poly (acrylic acid-co-acrylamide) grafted chitosan with salt-and pH-responsiveness properties. Eur Polym J 40(7):1399–1407

39. Mahdavinia GR, Rahmani Z, Karami S, Pourjavadi A (2014) Magnetic/pH-sensitive κ-carrageenan/sodium alginate hydrogel nanocomposite beads: preparation, swelling behavior, and drug delivery. J Biomater Sci Polym Ed 25(17):1891–1906

40. Wang B, Xu X, Wang Z, Cheng S, Zhang X, Zhou R (2008) Synthesis and properties of pH and temperature sensitive P (NIPAAm-co-DMAEMA) hydrogels. Colloids Surf B 64:34–41

41. Shukla NB, Rattan S, Madras G (2012) Swelling and dye-adsorption characteristics of amphoteric superabsorbent polymer. Ind Eng Chem Res 51:14941–14948

42. Murthy PSK, Mohan YM, Sreeramulu J, Raju KM (2006) Semi-IPNs of starch and poly(acrylamide-co-sodium methacrylate): preparation, swelling and diffusion characteristics evaluation. React Funct Polym 66:1482–1493

43. Mahdavinia GR, Mosallanezhad A (2016) Facile and green rout to prepare magnetic and chitosan-crosslinked κ-carrageenan bionanocomposites for removal of methylene blue. J Water Process Eng 10:143–155

44. Katime I, Mendizábal E (2010) Swelling properties of new hydrogels based on the dimethyl amino ethyl acrylate methyl chloride quaternary salt with acrylic acid and 2-methylene butane-1,4-dioic acid monomers in aqueous solutions. Mater Sci Appl 1:162–167

45. Hiremath JN, Vishalakshi B (2012) Effects of crosslinking on swelling behavior of IPN hydrogels of guar gum and polyacrylamide. Der Pharma Chemica 4:946–955

46. Ganji F, Vasheghani-Farahani S, Vasheghani-Farahani E (2010) Theoretical description of hydrogel swelling: a review. Iranian Polym J 19:375–398

47. Mithun U, Vishalakshi B (2014) Swelling kinetics of a pH-sensitive polyelectrolyte complex of polyacrylamide-g-alginate and chitosan. Int J ChemTech Res 6:3579–3588

48. Shukla NB, Madras G (2013) Adsorption of anionic dyes on a reversibly swelling cationic superabsorbent polymer. J Appl Polym Sci 127(3):2251–2258

49. Mohammadi M, Hassani AJ, Mohamed A, Najafpour GD (2010) Removal of rhodamine B from aqueous solution using palm shell-based activated carbon: adsorption and kinetic studies. J Chem Eng Data 55:5777–5785

50. Dada AO, Olalekan AP, Olatunya AO, Dada O (2012) Langmuir, Freundlich, Temkin and Dubinin–Radushkevich isotherms studies of equilibrium sorption of Zn^{2+} unto phosphoric acid modified rice. IOSR J Appl Chem 3:38–45

51. Bas N, Yakar A, Bayramgil NP (2014) Removal of cobalt ions from aqueous solutions by using poly(N, N-dimethylaminopropyl methacrylamide/itaconic acid) hydrogels. J Appl Polym Sci 131:1–12

52. Zhao Z, Wang X, Zhao C, Zhu X, Du S (2010) Adsorption and desorption of antimony acetate on sodium montmorillonite. J Colloid Interface Sci 345:154–159

53. Krušić MK, Milosavljević N, Debeljković A, Üzüm ÖB, Karadağ E (2012) Removal of Pb^{2+} ions from water by poly(acrylamide-co-sodium methacrylate) hydrogels. Water Air Soil Pollut 223:4355–4368

54. Garba ZN, Bello I, Galadima A, Lawal AY (2016) Optimization of adsorption conditions using central composite design for the removal of copper (II) and lead (II) by defatted papaya seed. Karbala Int J Modern Sci 2:20–28

55. Al-Rashed SM, Al-Gaid AA (2012) Kinetics and thermodynamic studies on the adsorption behavior of rhodamine B dye on Duolite C-20 resin. J Saudi Chem Soc 16:209–215
56. Shen K, Gondal MA (2013) Removal of hazardous Rhodamine dye from water by adsorption on to exhausted coffee ground. J Saudi Chem Soc. doi:10.1016/j.jscs.2013.11.005 **(article in press)**
57. Hu X, Wang J, Liua Y, Li X, Zenga G, Baoc Z, Zenga X, Chena A, Longa F (2011) Adsorption of chromium (VI) by ethylene-diamine-modified cross-linked magnetic chitosan resin: Isotherms, kinetics and thermodynamics. J Hazard Mater 185:306–314
58. Ho YS, McKAY G (1998) A comparison of chemisorptions kinetics models applied to pollutant removal of various sorbents. Trans IchemE 76(4):332–340
59. Ding L, Zou B, Gao W, Liu Q, Wang Z, Guo Y, Wang X, Liu Y (2014) Adsorption of rhodamine-B from aqueous solution using treated rice husk-based activated carbon. Colloids Surf A 446:1–7
60. Goscianska J, Ptaszkowska M, Pietrzak R (2015) Equilibrium and kinetic studies of chromotrope 2R adsorption onto ordered mesoporous carbons modified with lanthanum. Chem Eng J 270:140–149
61. Mekewi MA, Madkour TK, Darwish AS, Hashish YM (2015) Does poly(acrylic acid-co-acrylamide) hydrogel be the pluperfet choiceness in treatment of dyeing wastewater? "From simple copolymer to gigantic aqua-waste remover". J Ind Eng Chem 30:359–371
62. Song W, Gao B, Xu X, Xing L, Han S, Duan P, Song W, Jia R (2016) Adsorption-desorption behavior of magnetic amine/Fe_3O_4 functionalized biopolymer resin towards anionic dyes from wastewater. Bioresour Technol 210:123–130
63. Dragan ES, Perju MM, Dinu MV (2012) Preparation and characterization of IPN composite hydrogels based on polyacrylamide and chitosan and their interactions with ionic dyes. Carbohydr Polym 88:270–281
64. Mahdavinia GR, Massoumi B, Jalili K, Kiani G (2012) Effect of sodium montmorillonite nanoclay on the water absorbency and cationic dye removal of carrageenan-based nanocomposite superabsorbents. J Polym Res 19(9):9947

13

Phthalocyanine green aluminum pigment prepared by inorganic acid radical/radical polymerization for waterborne textile applications

Benjamin Tawiah[1,2] · Benjamin K. Asinyo[2] · William Badoe[2] · Liping Zhang[1] · Shaohai Fu[1]

Abstract Polymer-encapsulated phthalocyanine green aluminum pigment was prepared via inorganic acid radical/radical polymerization route, and its properties were investigated by FT-IR, TGA, XPS, SEM, and TEM. SEM and TEM images showed that the aluminum pigment was encapsulated by a thin film of polymer which ensured good anti-corrosive performance in alkaline (pH 12) and acidic (pH 1) mediums. XPS results showed significant chemical shifts, and increase in binding energies to higher levels after raw aluminum pigment was phosphate coated and colored by phthalocyanine green pigment. TGA results suggest a marginal reduction in its thermal stability. Major absorbance peaks, such as aluminum phosphate ($AlPO_4$), different monomer units and CH_2 stretching vibration of phthalocyanine green G were highlighted in the FTIR spectra of the colored aluminum matrix. The polymer-encapsulated aluminum pigment (PAP) had excellent UPF properties regardless of the coating thickness, but the handle of the fabric was affected when the coating thickness increased beyond 0.04 mm. The prepared pigment showed excellent rubbing and washing fastness, but its handle and color strength were compromised when the content of monomer ratio by 100 % weight of PGAP increased beyond 10 %, was applied on cotton fabrics. This research provides a simple but effective route for the preparation of polymer-encapsulated aluminum pigments for waterborne textile applications.

Keywords Aluminum pigment · Phthalocyanine green · Polymer encapsulation · Radical polymerization · Inorganic acid radical

Introduction

Colored aluminum pigments having colorful pigment adhered closely, uniformly and firmly on its surface are suitably used in paints, automotive metallic finish, printing inks, molded resins and in decoration finish of plastics [1–3]. The application of aluminum pigment has expanded into security services, including the military, due to its high emissive properties and the ability to reflect IR rays in the solar spectrum [4–7]. Recently, colored aluminum pigments have been used in plastic components of objects, such as boats or buoys, to make them visible to RADAR detection due to their ability to reflect in the electromagnetic radiation [5, 8]. These and many other functional properties of aluminum pigments have rekindled research interest lately.

Traditionally, colored aluminum pigments are prepared by physically mixing colorful pigments with silver white aluminum pigments which makes it difficult to achieve vivid color tones because, the achromatic tone inherent in aluminum pigment is usually emphasized [2, 9, 10]. These pigments are usually common in the automotive industry where volatile organic compounds (VOCs) are commonly used as the medium for their application. These pigments, however, are not suitable for textiles applications where water play a major role throughout the manufacturing process to the care phase, hence the need to develop

✉ Benjamin Tawiah
ben.tawiahduke@aol.com

[1] Key Laboratory of Eco-Textile, Jiangnan University, Ministry of Education, Wuxi 214122, Jiangsu, China

[2] Department of Industrial Art (Textiles), Kwame Nkrumah University of Science and Technology, Private Mail Bag, Kumasi, Ghana

waterborne colored aluminum pigment for textiles applications. To surmount these problems, several techniques have been proposed as a solution to obtaining waterborne colored aluminum pigment by depositing pigment onto the surface of an aluminum pigment with the aid of polymeric coatings [9, 11–13]. Other researchers have suggested a route where silica is coated onto aluminum pigment using sol-gel method followed by color deposition with the aid of surface modification agents [9, 10], but the issue of corrosion and color deformation still persisted after a long exposure to high alkaline waterborne systems. As a result, the use of colored inorganic flaky materials, such as mica flakes, has been suggested [13–15]; which gives a pearly color tone of their own. The use of popular techniques, such as metal organic chemical vapor deposition, physical vapor deposition, laser cladding, and thermal spraying, has also been thoroughly investigated [16–18]. Meanwhile, drawbacks such as low hiding power, poor metallic luster, and lack of color vividness have persisted because of the difficulty in depositing ample amount of colorful pigment onto the surface of aluminum pigment [7, 12].

Recently, the traditional wet chemical coating methods using silica-coated aluminum pigments (Al/SiO$_2$) or phosphates as precursor materials with dyes have been reported [19, 20], but problems such as poor corrosive stability, color fading, and complicated preparation process have still remained [1, 9, 21]. The use of inorganic acid radical for coating aluminum pigment has been reported to enhance its corrosion protection ability and ensure uniform coating [12, 22–24] hence its extensive application in corrosion chemistry. Meanwhile, the application of inorganic radical/radical polymerization for coloring silver white aluminum pigment with phthalocyanine green G has not been reported.

The main object of this research is to provide aluminum pigment colored to the highest chroma by depositing phthalocyanine green G pigment on the surface of aluminum pigment, thereby improving the chroma and avoiding the issue of color fading and exfoliation inherent in conventionally pigment colored aluminum pigments. To achieve these objectives, a colorful pigment (phthalocyanine green G) with inherently high tinctorial strength and excellent fastness to different solvents, heat, light, weathering [15, 25] was chosen and adsorbed onto aluminum pigment having adsorption layer of inorganic acid radical with the green colorful pigment adhered to said adsorption layer. The phthalocyanine green G colored aluminum pigment was then coated with a polymer using radical polymerization, and its properties were investigated. The application of both techniques overcome the limitations of classical polymerization and corrosion inhibition methods and provides an efficient route for preparing colored aluminum pigment having excellent properties.

Experimental

Materials

Aluminum pigment (particle size 50 μm) was purchased from Tianjiu Metal Materials Co., Ltd., Changsha, China. Allyloxy nonyl alcohol polyoxyethylene (10) ether sulfate (DNS-86), sodium dodecyl sulfate (SDS), N-I3-(aminoethy1)-y-aminopropylmethyldimethoxysilane (Silane Si–602) (structure is shown in chat 1b) were purchased from Qingxin Haner Chemical Technology Co. Ltd., China. Binder (DM-5218), thickener (DM-5268) and crosslinker (FWO-B) were purchased from Demei Chemical Company Ltd. Phthalocyanine green G (purity 97 %) (Structure shown in chat 1a) was purchased from Anping Guanda Pigment industry Co. Ltd., China. Other analytical grade chemicals, such as methylmethacrylate, 1,6-hexanediol diacrylate, styrene (ST), azobisisobutyronitrile (AIBN), isopropyl alcohol, polyphosphoric acid, surfynol 440, acetone, absolute ethanol, sodium hydroxide, and methanol, were supplied by Sinopharm Chemical Reagent Co. Ltd., China and were used without further purification. Deionized water was used for the entire experiments (see Fig. 1).

Pretreatment of aluminum

The aluminum pigment was washed with acetone and dried at 70 °C in vacuum oven for 6 h.

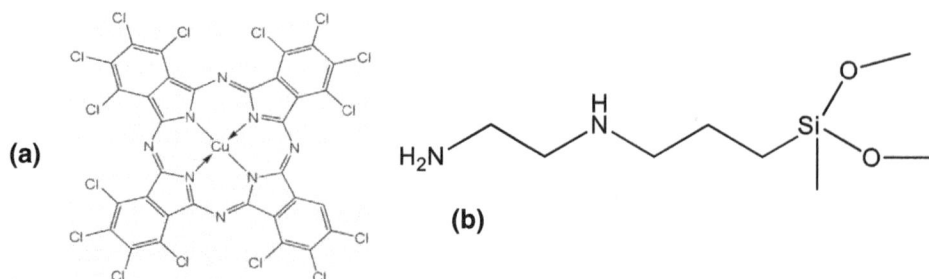

Fig. 1 Structure of (**a**) phthalocyanine green G (**b**) N-I3-(aminoethy1)-y-aminopropylmethyldimethoxysilane

Pigment dispersion

10 g SDS and DNS-86 in the ratio 1:1 based on 10–30 % weight of pigment (phthalocyanine green G) were mixed in 150 g deionized water. 1 g of glycerol and 6 g of N-|3-(aminoethy1)-y-aminopropylmethyldimethoxysilane were added and then subjected to mechanical stirring for 30 min at 450 r/m after which the pH was adjusted to 9. The pigment was further dispersed using bead mill Mini Zeta-03-E NETZSCH Grinding & Dispersing Machine (Germany) at 1800 r/m for 2 h to obtain finely dispersed phthalocyanine green G pigment with average particle size of 118 d nm. The dispersion was filtered through 0.5 μm pore-filtering sieve to remove broken pieces of glass beads.

Preparation of inorganic acid radical coating

12 g of aluminum pigment was dispersed in 80 g isopropyl alcohol using ultrasonicator JY98-3D, NingBo Scientz Biotechnology Co., Ltd, China at frequency of 28 kHz and power of 400 W for 10 min to obtain aluminum dispersion. 2 g of polyphosphoric acid (1 part by weight on the basis of 100 parts by weight of aluminum pigment) and 1 g of surfynol 440 were added, and the resultant mixture was kneaded for 1 h to adsorb the polyphosphoric radical to the surface of the aluminum pigment.

The aluminum pigment adsorbing inorganic acid radical prepared in step (see "Preparation of inorganic acid radical coating") was added to the dispersion of color pigment prepared in step (see "Pigment dispersion") to obtain aluminum pigment/color slurry. 20 g alcohol/water mixture was added to the slurry and stirred for 5 h. The slurry was subjected to solid–liquid extraction to obtain colored aluminum precipitates. The precipitates were dried at 60 °C for 6 h to obtain phthalocyanine green colored aluminum pigment (PGAP).

Radical polymerization

10 g of colored aluminum pigment was dispersed in 100 g toluene using ultrasonicator. The ultrasonicated colored aluminum pigment was transferred to a four-necked round bottom flask with a reflux condenser and a nitrogen gas inlet/out-let in a water bath equipped with thermometer (60 °C) with continuous stirring at 300 r/m. 0.3 g each of methyl methacrylate, 1,6-hexanediol diacrylate, styrene were added drop by drop over 1 h. 0.04 g of polymerization initiator (AIBN) was added thereunto drop by drop, and the temperature was raised to 85 °C with continuous stirring at 450 r/m for 20 h. The mixture was allowed to cool to room temperature and then centrifuged at 12,000 r/min for 10 min. The sediment was washed three times with methanol to remove the unreacted monomers and then

dried at room temperature under vacuum to obtain polymer-encapsulated PGAP.

Fabric coating

The coating paste was formulated on the following weight basis: polymer-encapsulated PGAP 22 %, glycerol 1 %, DM-5268 3 %, urea 0.3 %, FWO-B 4 %, DM-5218 23.7 %, and distilled water 50 %. These components were mixed to obtain a homogenous paste. Cotton fabrics obtained from Shandong Weiqiao Pioneering Group Co., Ltd, China were coated using rapid auto coat machine from Xiamen Rapid Company Ltd, China. The coated fabrics were dried in an oven at 60 °C for 30 min and then cured at 150 °C for 3 min.

Characterization

FT-IR

The phthalocyanine green aluminum pigment (PGAP) polymer-encapsulated PGAP, and phthalocyanine green G were characterized by FT-IR (NICOLET iS50, scan 400–4000 cm^{-1}). Smart iTX device equipped with AR-coated diamond crystal having interferometer speed of 1.0 cm/s. OMNIC^{TM} software was used to obtain the spectral images with a pixel size of 1.56×1.56 μm in four scans per pixel at a spectral resolution of 4 cm^{-1} in attenuated total reflectance (ATR) mode where 32 scans per sample were averaged to obtain the spectral images. A non-destructive sample preparation method was used where samples were placed into the diamond plate, and the Smart iTX pressure tower was adjusted to ensure a consistent contact between the plate and the sample. The ambient temperature and relative humidity for the spectrometer were set between 18 and 25 °C, and less than 40 %, respectively, after which the spectra were acquired in ATR mode.

Thermo-gravimetric (TGA) analyses

Thermo-gravimetric (TGA) analyses were performed using TG apparatus (TGA/SDTA851e, Mettler Toledo instrument co., LTD, Switzerland) with 10 °C/min ramp from 25 to 700 °C.

Anti-corrosion

Anti-corrosive stability of aluminum pigment was measured using the displacement method described by Zhang et al., [9]. 1 g of Al samples were immersed in 250 mL water with pH 1 and 12, respectively, at 25 ± 2 °C for 168 h. The amount of gas evolved was measured during the exposure time. The less the volume of gas evolved, the better the anti-corrosion performance.

Scanning electron microscopy (SEM)

Surface morphology of the samples was observed using Scanning Electron Microscope (Hitachi Model S-3200H) equipped with an STS X-Stream Imaging System.

Transmission electron microscopy (TEM)

TEM photograph was obtained using JEOL-1200EXII microscope (JEOL Ltd., Tokyo, Japan) operating at 80 keV equipped with a high-resolution Tietz F224 digital camera located below the imaging screen with a beam blanker acting as the camera shutter under a 120,000 magnification. Droplets of finely dispersed samples were dropped onto a TEM copper grid, and the solvent was evaporated at room temperature.

X-ray photoelectron spectroscopy

XPS was carried out on a RBD upgraded PHI-5000C ESCA system (Perkin Elmer) with Mg Kα radiation ($hv = 1253.6$ eV) and Al Kα radiation ($hv = 1486.6$ eV). The X-ray anode was run at 250 W, and the high voltage was kept at 14.0 kV with a detection angle at 54°. The pass energy was fixed at 23.5, 46.95 or 93.90 eV. The base pressure of the analyzer chamber was 5×10^{-8} Pa. The sample was pressed directly into a self-supported disk (10×10 mm) and mounted on a sample holder and then transferred into the analyzer chamber. The whole spectra [0–1100 (1200) eV] and the narrow spectra of all the elements with much high resolution were both recorded using RBD 147 interface (RBD Enterprises, USA) through the AugerScan 3.21 software. Binding energies were calibrated using the containment carbon (C1s = 284.6 eV) which is not included in the data. The data analysis was carried out using the RBD AugerScan 3.21 software provided by RBD Enterprises.

Pigment dispersion (milling)

The pigment was dispersed using bead mill Mini Zeta-03-E NETZSCH Grinding & Dispersing Machine (Germany) at 1800 r/m for 2 h.

Particle size

Particle size (D) and the particle size distribution were determined by dynamic light scattering method (DLS) with Nano ZS90 (Malvern Instruments Co., Ltd., England) at 25 °C.

Adhesive properties

The adhesion properties of the samples were evaluated by peel test. Peel test was carried out using the method described by Gao et al. [26]. The weight of aluminum pigments pulled off was measured by calculating the weight difference of the pre- and post-peeled tape using Eq. 3.

$$W_p = A_0 - A_1 \tag{1}$$

where W_p is the weight of aluminum pigments fallen off from coating during peel test, A_0 represents the original weight of the board after drying before peeling the adhesive tape off, and A_1 represents the weight of the board after the adhesive tape was peeled off.

Color performance

The colorimetric values of the samples in the CIE lab color space were measured on a Color-Eye automatic differential colorimeter (Xrite-8400, X-Rite Color Management Co., Ltd., USA) under illuminant D65 with the CIE 1964 Standard Observers, and the color strength (K/S) of the coated fabrics was determined using Kubelka–Munk equation.

$$\frac{K}{S} = \frac{(1 - R)^2}{2R} \tag{2}$$

where R defines the relationship between the spectral reflectance of the sample and its absorbance coefficient (K) and scattering coefficient (S).

The rubbing and washing fastness properties of the coated fabrics were evaluated according to the standard method GB/T 3921-2008 and GB/T3922-1995, respectively.

Coating thickness

Coating thickness was measured with Paramount thickness tester precision gauge according to the standard method ASTM D1777—96 (2011).

Ultraviolet protection factor (UPF)

Ultraviolet protection factor was measured using Cary 50 UV/Vis spectrophotometer (varian made in Australia). The UPF of the samples was evaluated according to the Australian/New Zealand Standard AS/NZS 4399:1996. UPF was calculated using Eq. (3).

$$\mathrm{UPF} = \frac{\int_{280}^{400} E_\lambda \times S_\lambda \times \mathrm{d}\lambda}{\int_{280}^{400} E_\lambda \times S_\lambda \times \mathrm{d}\lambda} \tag{3}$$

where S_λ is erythema action spectrum, E_λ is solar irradiance, d_λ is wavelength interval in nm, and T_λ is spectral transmittance of the specimen.

Fabric handle

The softness of coated cotton fabrics was evaluated according to the Kawabata Evaluation System for Fabrics (KES-FB) on a handle instrument (KES-FB2, Kato Giken Co., Ltd. Japan).

Results and discussion

Preparation of colored aluminum pigment

Coloration of aluminum pigment followed a three-step approach where finely dispersed copper phthalocyanine green G pigment reacted with the ammonium ion ($NH4^+$) present in Silane Si–602 at very high pressures during the milling process. During this process, the labile chlorine atoms underwent a nucleophilic substitution with the amine groups (NH_2) of Silane Si-602 to ensure strong coupling effect with the copper phthalocyanine green G pigment. Secondly, the aluminum pigment reacted with polyphosphoric acid radical to form aluminum phosphate ($AlPO4$) on its surface [27]. A mixture of the aluminum phosphate and the modified phthalocyanine green pigment created a link at the nitrogen monohydride (NH) region of the coupling age because, the ammonium ions combined with the organic radical to form aluminum nitrogen phosphate [Al-N(PO_3)] [28]. This reaction created a unique compound between the aluminum phosphate and the modified phthalocyanine green pigment leading to the formation of colored aluminum pigment as shown in scheme 1. To confirm this reaction, a high-resolution X-ray photoelectron spectroscopy (XPS) of "as received aluminum pigment" and phthalocyanine green aluminum pigment was taken, and the results are shown in Fig. 2. XPS is a surface-sensitive quantitative spectroscopic technique that measures the elemental compositions at parts-per-thousand range, empirical formula, chemical and electronic state of the elements that exist within a composite. The XPS spectra were obtained by irradiating samples with a beam of X-rays while simultaneously measuring the kinetic energy and number of electrons that escaped from the top 0 to 10 nm of the samples ("as received aluminum pigment and PGAP") analyzed.

It can be seen from Fig. 2 that "as received aluminum" pigment showed doublet peaks O 1s and Al 2p with maximum peak intensities of approximately 40,000 and 21,000 at the binding energies of 536.02 and 535.69 eV, respectively, compared to PGAP which showed peaks for compounds present in the matrix but at lower intensities

and different binding energies even for O 1s and Al 2p at approximately 34,000 and 4500 at binding energies of 533.97 and 537.25 eV. Besides these chemical shifts, copper phthalocyanine G; denoted as C.G 1s and phosphate compounds, also denoted as N(PO_3)2p3 were clearly shown in colored phthalocyanine green aluminum pigment (Fig. 2b) which are conspicuously missing in "as received aluminum pigment".

To ensure good adhesion property of phosphate/pigment coating on the aluminum pigment (metallic core), radical polymerization was done to encapsulate phthalocyanine green aluminum pigment with a thin film of polymer where the monomers tend to form dimers and trimers and then grow into a complex film of colorless polymer over the entire surface of PGAP with constant supply of heat over time.

Surface morphology

SEM and TEM are 3D surface topographic analysis tools capable of zooming to low magnifications to locate interesting areas of specimens to high magnifications down to nanometer surface features. These techniques were adopted to characterize the surface properties of the colored aluminum pigments and the images are shown in Fig. 3.

It can be seen that phthalocyanine green Al pigment (PGAP) (Fig. 3c, d) had a relatively thick layer of phosphate/pigment coatings compared to as received aluminum pigments (Fig. 3a, b). Also, it can be observed from (Fig. 3c) that colored phosphate on the surface of aluminum pigment was not very compact compared to the PAP (Fig. 3e, f). The polymer layer on the surface of PGAP gave PAP a relatively smoother surface suggesting a better barrier for corrosion resistance as a result of the radical polymerization. No significant difference can be seen in the TEM images of as received aluminum pigment (Fig. 3g), PGAP (Fig. 3h) and PAP (Fig. 3i), except what appears to be a smooth layer on the surface of PAP probably due to the encapsulated polymer layer.

FT-IR of colored aluminum pigment

FT-IR is a very useful surface characterization tool for single and multilayer films or matrix analysis. The spectra were acquired in ATR mode using Thermo Scientific Smart iTX device equipped with AR-coated diamond crystal plate.

Figure 4 shows the FT-IR spectra of PGAP and polymer-encapsulated PGAP and phthalocyanine green pigment. Minor broad shallow peak around 3432.37 cm^{-1} in phthalocyanine green pigment and PGAP though not obvious can be attributed to the O–H stretching vibration as

Scheme 1 Proposed reaction mechanism for preparation of phthalocyanine green aluminum pigment

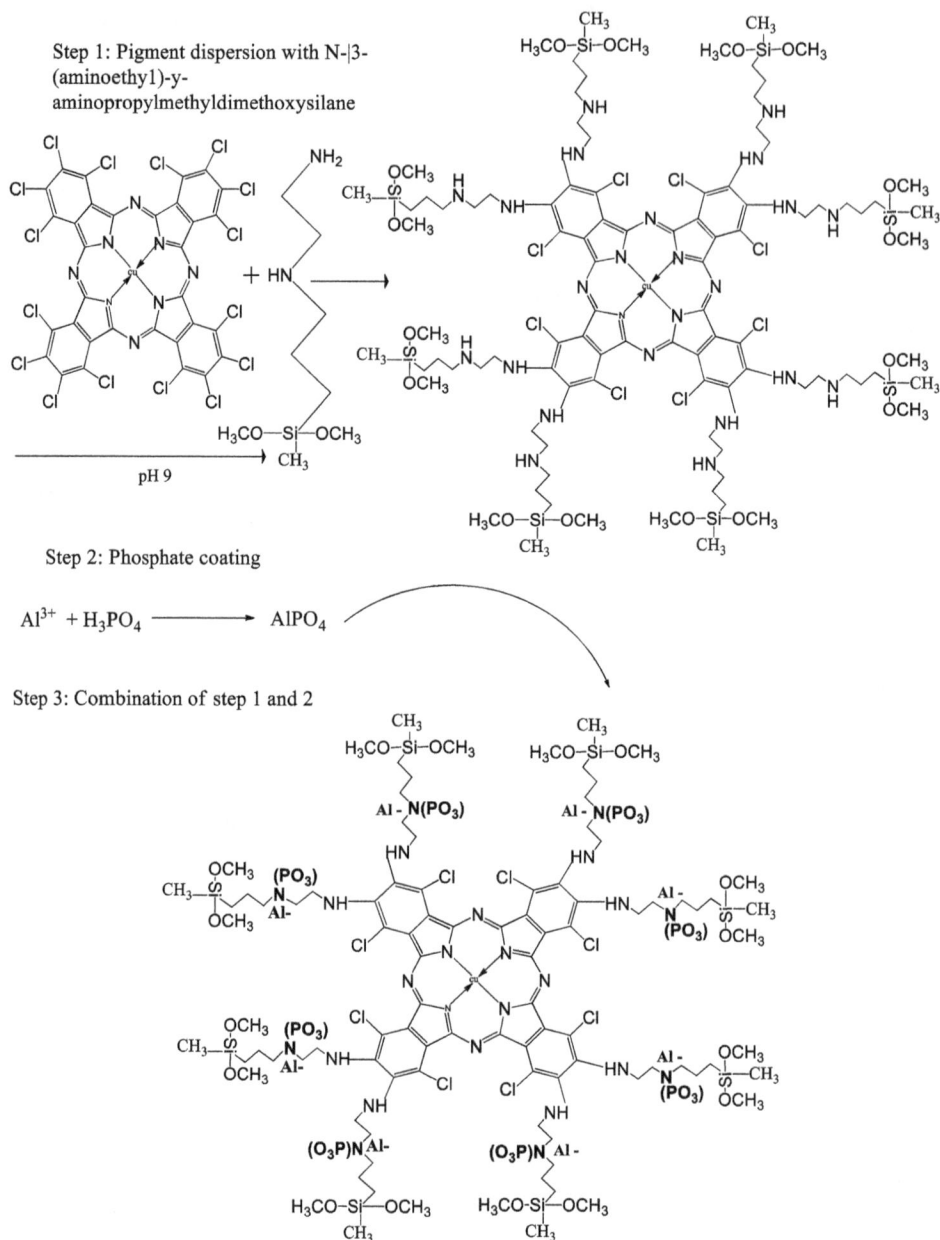

Step 1: Pigment dispersion with N-|3-(aminoethyl)-y-aminopropylmethyldimethoxysilane

Step 2: Phosphate coating

$$Al^{3+} + H_3PO_4 \longrightarrow AlPO_4$$

Step 3: Combination of step 1 and 2

a result of crystal water in phthalocyanine green but was not very prominent in polymer-encapsulated PAP (Fig. 4b). The minor peak recorded around 2926.79 cm^{-1} relating to C–H asymmetric stretching vibrations in phthalocyanine green pigment can be noticed in PGAP and polymer-encapsulated PGAP. The band observed in the FT-IR spectrum of PGAP and PAP around 1640.78, 1100, and 787.35 cm^{-1}, corresponds to the symmetric stretching mode of [PO4] [3] in AlPO$_4$ whereas the presence of peaks at 661.54 and 611.39 cm^{-1}, and the symmetric bending around 466.74 cm^{-1} indicates the presence of aluminum [29].

Characteristic absorbance band highlighted around 700.81 cm^{-1} could probably be due to out-of-plane ring

bending of C=C stretching vibrations in phenyl ring in styrene unit, whereas the C=O wag at around 508.16 cm^{-1} [30] in PAP (Fig. 4b) could be attributed to the presence of other monomer units. More so, the characteristic band around 948.36 cm^{-1} in PGAP and PAP can be attributed to the out-of-plane bending mode of N–Cu and C–Cl in phthalocyanine green pigment [30].

Adhesive properties

The weight of aluminum pigments peeled off from coated board during peel test (W_p) is used to describe the adhesion property of aluminum pigments in coatings [24, 26]. High W$_p$ values of aluminum pigments denote poor adhesion

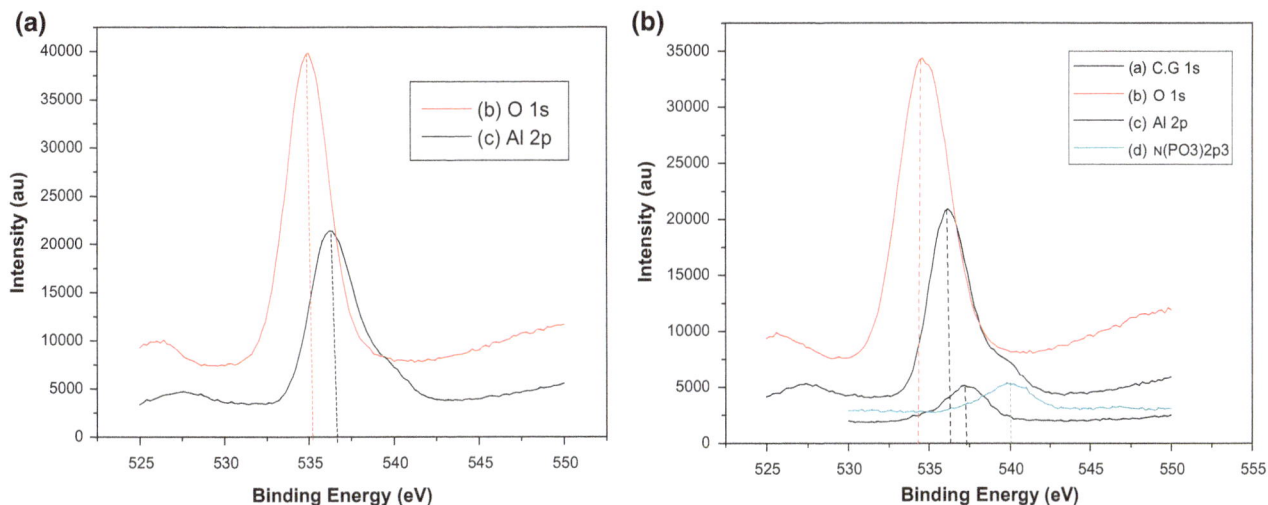

Fig. 2 High-resolution X-ray photoelectron spectra of (**a**) "as received aluminum pigment" (**b**) phthalocyanine green Al pigment (PGAP)

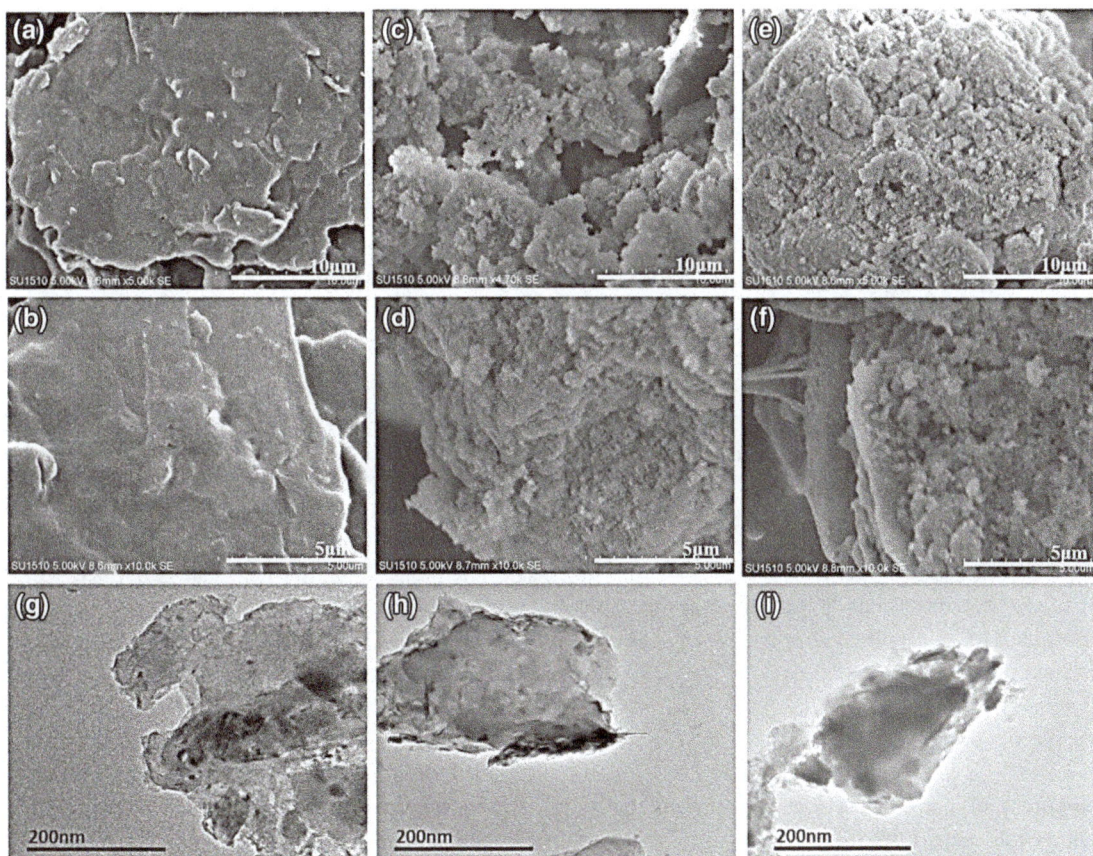

Fig. 3 SEM micrographs of (**a, b**) "as received Al pigment" (**c, d**) phthalocyanine green Al pigment (PGAP) (**e, f**) Polymer-encapsulated aluminum pigment (PAP) and TEM images of (**g**) as received aluminium pigment (**h**) phthalocyanine green Al pigment (PGAP) (**i**) polymer-encapsulated aluminum pigment (PAP)

performance. The adhesion efficiency of PGAP and PAP is shown in Fig. 5. PGAP had much poorer adhesion property (6.2 m g^{-2}) compared to PAP (0.5 m g^{-2}) probably due to the porous nature of the phosphate/pigment particles on the surface of the metallic core (aluminum pigment). The excellent adhesive property of PAP could be ascribed to the good adhesion force between the resin and the polymer film on the surface of PGAP. The polymer film on the

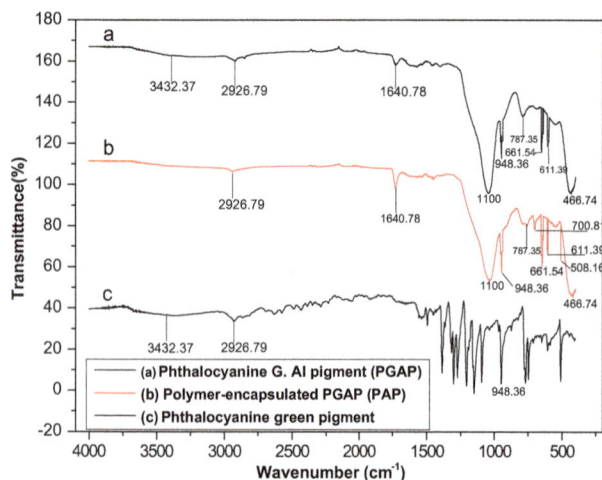

Fig. 4 FT-IR spectra of (a) phthalocyanine green Al pigment (PGAP) (b) polymer-encapsulated PGAP (PAP) (c) phthalocyanine green pigment

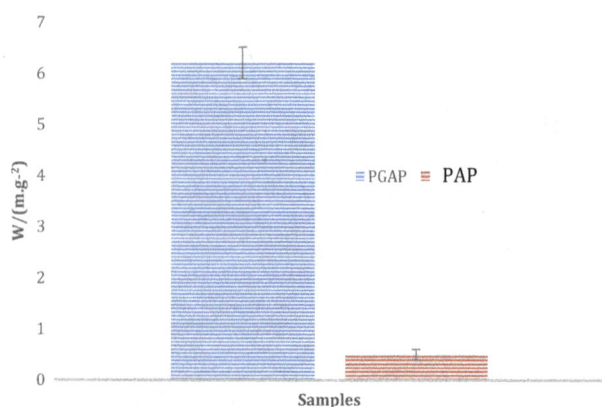

Fig. 5 Adhesion properties of (a) phthalocyanine green Al pigment (PGAP) (b) polymer-encapsulated aluminum pigment PGAP (PAP)

surface of PAP has, therefore, helped to minimize the problem of exfoliation usually experienced by conventionally colored aluminum pigments.

Anti-corrosive properties

The development of anti-corrosive and environmentally benign waterborne metallic colorants for wide variety of applications is very important. Metallic colorants especially for textiles applications must meet specific performance needs and as well satisfy the increasingly strict environmental regulation requirements. As a result, the anti-corrosive properties of the polymer-encapsulated Al pigment (PAP) meant for textiles applications were evaluated, and the results are shown in Fig. 6.

It can be seen that polymer-encapsulated phthalocyanine green Al pigment powder was highly stable in all

circumstances under various conditions compared to the as received Al pigments and PGAP. It can be seen from Fig. 6a and b that H_2 started to evolve for as received aluminum pigment just after it was introduced to acid and alkaline conditions unlike PGAP which saw evolution of gas after 24 h at minimal magnitudes similar to results obtained by Abd El-Ghaffar et al. [23]. Similar observations were made when the paste was subjected to accelerated aging test at 50 ± 2 °C, pH 10 (Fig. 6c). Under the same conditions, PAP was highly stable even after 168 h. The gassing phenomenon witnessed in PGAP could be attributed to the gradual penetration of acid and alkali solutions through the phosphate layer onto the metallic core after long exposure forming $AlCl_3$ and $NaAlO_2$.

Particle size distribution

The mean particle size of "as received Al Pigment", PGAP, and polymer-encapsulated PGAP was measured by dynamic light scattering (DLS), and the result is shown in Fig. 7. It can be seen that the particle size distribution of all the three samples had an insignificant difference with an average particle size of 1772.5, 1957, and 1998.3 d nm for as received Al pigment, PGAP, and PAP, respectively, with "as received Al pigment" bigger than PGAP by 184.5 d nm possibly due to phosphate treatment and subsequent coloration by phthalocyanine green G. PAP particle size was also bigger than PGAP by approximately 41.3 d nm. The minimal increase in size of PAP could be attributed to the encapsulated polymer-layer coating on the surface of PGAP.

Thermal properties of polymer-encapsulated PGAP

Most solid polymeric materials experience both physical and chemical changes when heat is applied which usually results in undesirable changes to the properties of the material [31, 32]. This often results in thermal degradation or even decomposition where the action of heat or elevated temperatures on a material causes a loss of physical, mechanical, or electrical properties or at worse generate harmful gases [32]. Figure 8 shows the thermal properties of polymer-encapsulated phthalocyanine green aluminum pigment (PGAP).

It can be seen that "as received Al pigment" showed a gentle weight loss of approximately 0.5 % between 380 and 700 °C due to the decomposition of adsorbed water and oxide layer on its surface [26]. PGAP also experienced a consistent weight loss starting from approximately 250 °C until 700   °C. This phenomenon could be attributed to the decomposition of water, absolute ethanol, and the pigment on the surface of the aluminum pigment (metallic core) [12, 33]. Polymer-encapsulated PGAP, however, recorded

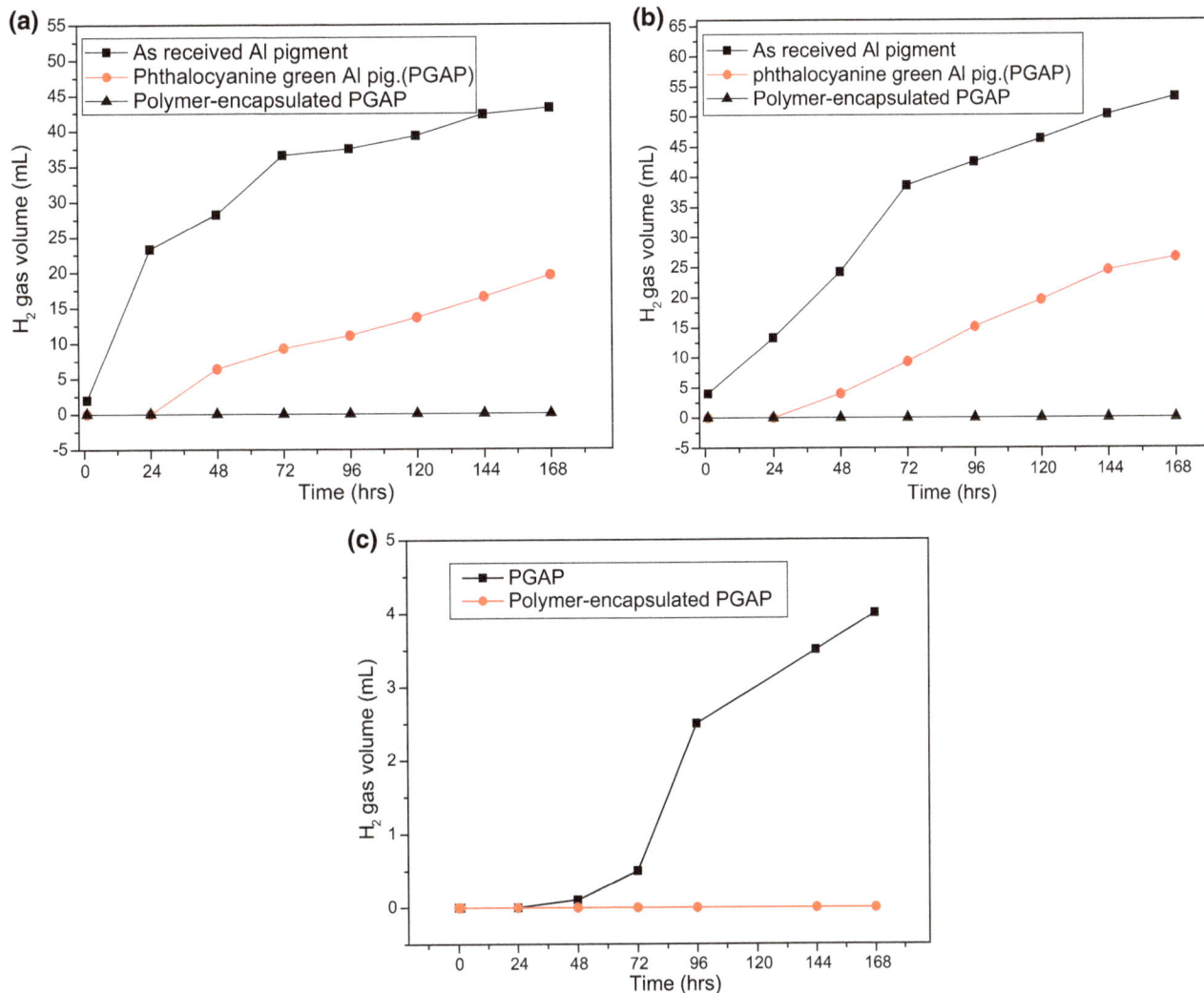

Fig. 6 Anti-corrosive stability of aluminum pigments in (a) acidic condition 25 ± 2 °C, pH 1 (b) alkaline condition 25 ± 2 °C, pH 12, (c) paste at 50 ± 2 °C, pH 10

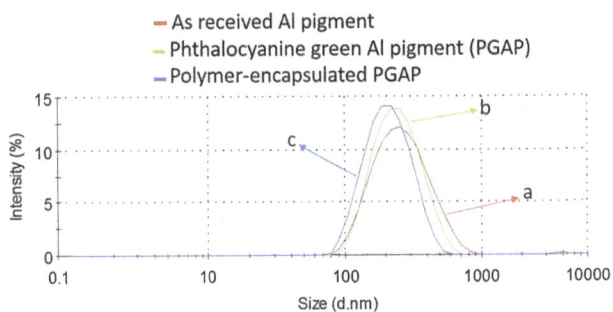

Fig. 7 Particle size distribution of (a) as received Al pigment (b) phthalocyanine green Al pigment (PGAP) (c) Polymer-encapsulated PGAP

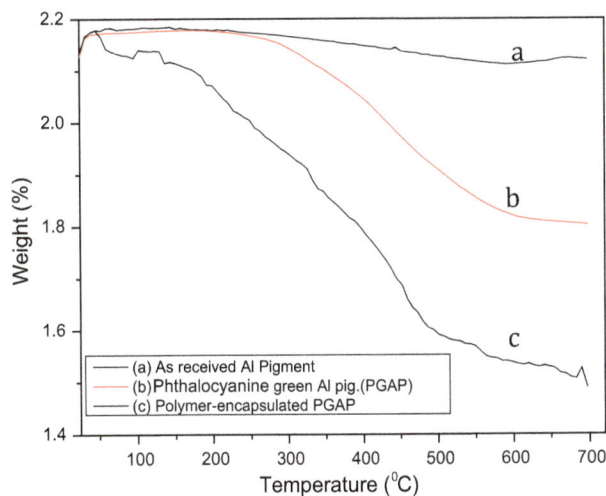

Fig. 8 Thermal properties of (a) "as received aluminum pigment" (b) phthalocyanine green aluminum pigment (PGAP) (c) polymer-encapsulated PGAP (PAP)

a much faster and higher rate of decomposition due to the reduced thermal stability of the encapsulated polymer at high temperature [34].

Table 1 Effect of polymer-encapsulated aluminum pigment (PAP) coating on color performance and handle of cotton fabrics

Polymer coating (%)	CT[a] (mm)	K/S value	λ_{max} (nm)	Rubbing fastness		Washing fastness		Handle
				Dry	Wet	Changing	Staining	
None	7.96	520		3	3	3–4	2–3	Softest
5	7.84	520		4	4	4–5	3–4	Softer
10	7.27	520		5	5	5	5	Soft
15	6.54	520		5	5	5	5	Poor
20	5.02	520		5	5	5	5	Poorer

CT coating thickness

[a] Coating was done on cotton fabric with thickness 0.21 mm

Color performance

Table 1 shows the effect of polymer coatings (monomer mass by weight of PGAP) on the color performance of polymer-encapsulated PGAP-coated fabrics. It can be seen that the K/S value decreased marginally as the coating thickness increased from 5 to 10 %. Beyond 10 %, however, the K/S value was adversely affected, but the λ_{max} remained fixed at 520 nm indicating that the hue was not affected except the considerable reduction in color strength (K/S). It can also be seen from Table 1 that PAP coated cotton fabric had good rubbing and washing fastness irrespective of the percentage polymer coating unlike PGAP which does not have polymer coating. The excellent rubbing and washing fastness could be attributed to the presence of polymer layer on the surface of PGAP even when the same amount of binder was used. The handle test indicates that the polymer coating had some negative effect on the handle of coated cotton fabrics because, as the percentage of monomer content increased, the handle was invariably affected as can be seen from Table 1.

Ultraviolet protection factor (UPF)

Regular and lengthy exposure to ultraviolet rays (UVA 400–320 nm, UVB 320–290 nm) can damage the immune system and cause skin diseases like erythema and skin burns, hence the need for UV protective clothing [34–36]. Cotton fabric was coated with polymer-encapsulated phthalocyanine green aluminum pigment, and the UPF was measured. Per the Australia/New Zealand standards AS/NZS 4399:1996, fabrics are considered to possess excellent UPF when they attain an average of 50 + UV rate; the greater the UPF, the better the UV protection.

It can be seen from Table 2 that UPF improved from 85.56 to 220.38 as coating thickness increased from 0.02 to 0.06 mm indicating that UPF of polymer-encapsulated phthalocyanine green aluminum pigment-coated cotton fabrics had excellent UV protection irrespective of the coating thickness similar to the results obtained in our

Table 2 UPF of polymer-encapsulated phthalocyanine green aluminum pigment coated cotton fabrics

C.T. (mm)[a]	UPF	UPF class	Handle
0	8.66	No class	Softest
0.02	85.56	Excellent	Softer
0.04	161.21	Excellent	Soft
0.06	220.38	Excellent	Poor

Thickness of cotton fabric 0.21 mm, Cotton fabric was used

[a] C.T. means coating thickness

previous findings [34]; meanwhile, handle of coated fabrics appeared compromised as the coating thickness increased beyond 0.04. The excellent UPF of polymer-encapsulated phthalocyanine green aluminum pigment-coated cotton fabrics could be attributed to the metallic core and excellent hiding power of the colored/polymer matrix.

Conclusion

Polymer-encapsulated phthalocyanine green aluminum pigment was prepared via combined inorganic acid radical/radical polymerization method and was characterized. The colored aluminum pigment was applied on cotton fabrics and its washing and rubbing fastness were evaluated according to the internationally accepted standards. The polymer-encapsulated phthalocyanine green aluminum pigment had good washing and rubbing fastness compared to non-encapsulated type, but the handle and K/S values were compromised as the encapsulation monomer content (by 100 % weight of PGAP) increased beyond 10 %. TGA results showed that its thermal stability was minimally compromised. Polymer-encapsulated PGAP had excellent UPF irrespective of the coating thickness.

Acknowledgments The authors are grateful to the National Natural Science Foundation of China (Grant No. 21346009), the Universities and Enterprises Prospective Joint Research Project of Jiangsu Province (BY2012050), Natural Science Foundation of Jiangsu Province (BK2012212) and a Project Funded by the Priority Academic

Program Development of Jiangsu Higher Education Institutions. We also thank Jiangnan University for supporting this research.

Compliance with ethical standards

Conflict of interest The authors declare no competing financial interest for this research work.

References

1. Sekar N (2013) 2—Optical effect pigments for technical textile applications. In: Gulrajani ML (ed) Advances in the dyeing and finishing of technical textiles. Woodhead Publishing, Oxford, pp 37–46
2. Zhang Y, Ye H, Liu H, Han K (2012) Preparation and characterization of colored aluminum pigments Al/SiO$_2$/Fe$_2$O$_3$ with double-layer structure. Powder Technol 229:206–213
3. Adams R (2008) Effect pigments fully explained. Focus Pigm 2008:1–2
4. Yuan L, Weng X, Xie J, Du W, Deng L (2013) Solvothermal synthesis and visible/infrared optical properties of Al/Fe3O4 core–shell magnetic composite pigments. J Alloy Compd 580:108–113
5. Yuan L, Weng X, Du W, Xie J, Deng L (2014) Optical and magnetic properties of Al/Fe$_3$O$_4$ core–shell low infrared emissivity pigments. J Alloy Compd 583:492–497
6. Wijewardane S, Goswami DY (2012) A review on surface control of thermal radiation by paints and coatings for new energy applications. Renew Sustain Energy Rev 16:1863–1873
7. Liu L, Han A, Ye M, Zhao M (2015) Synthesis and characterization of Al3 + doped LaFeO3 compounds: a novel inorganic pigments with high near-infrared reflectance. Sol Energy Mater Sol Cells 132:377–384
8. Choudhury AKR (2014) 2—Object appearance and colour. In: Choudhury AKR (ed) Principles of colour and appearance measurement. Woodhead Publishing, Oxford, pp 53–102()
9. Zhang Y, Ye H, Liu H (2012) Preparation and characterization of blue color aluminum pigments Al/SiO2/PB with double-layer structure. Powder Technol 217:614–618
10. Du B, Zhou SS, Li NL (2011) Research progress of coloring aluminum pigments by corrosion protection method. Procedia Environ Sci 10(Part A):807–813
11. Chen G, Zhu Z, Liu H, Wu Y, Zhu C (2013) Preparation of SiO$_2$ coated Ce$_2$S$_3$ red pigment with improved thermal stability. J Rare Earths 31:891–896
12. Pi P, Liu C, Wen X, Zheng L, Xu S, Cheng J (2015) Improved performance of aluminum pigments encapsulated in hybrid inorganic–organic films. Particuology 19:93–98
13. Zhou L, Huang SL, Kong JR, Zhou T, Zuo YJ (2013) Characterization of flaky aluminum pigments multi-coated by TiO$_2$ and SiO$_2$. Powder Technol 237:514–519
14. Liu H, Ye H, Zhang Y, Tang X (2008) Preparation and characterization of poly(trimethylolpropane triacrylate)/flaky aluminum composite particle by in situ polymerization. Dyes Pigm 79:236–241
15. Maile FJ, Pfaff G, Reynders P (2005) Effect pigments—past, present and future. Prog Org Coat 54:150–163
16. Hornig T, Lugscheider E, Seemann K (2004) 35—Vapour deposited coatings and thermal spraying. In: Totemeier WFGC (ed) Smithells metals reference book, 8th edn. Butterworth-Heinemann, Oxford, pp 1–16
17. Smallman RE, Ngan AHW (2014) Chapter 16—oxidation, corrosion and surface engineering. In: Smallman RE, Ngan AHW (eds) Modern physical metallurgy, 8th edn. Butterworth-Heinemann, Oxford, pp 617–657
18. Quintino L (2014) 1—Overview of coating technologies. In: Miranda R (ed) Surface modification by solid state processing. Woodhead Publishing, Amsterdam, pp 1–24
19. Nagano K, Mizoshita T (2008) Method of manufacturing aluminum flake pigment, aluminum flake pigment obtained by the manufacturing method and grinding media employed for the manufacturing method. Google Patents
20. Nagano K (2004) Aluminum flake pigment comprising aluminum flake as basic particle, method for producing the same, and coating ink using the same. Google Patents
21. Zhu H, Chen Z, Sheng Y, Thi TTL (2010) Flaky polyacrylic acid/aluminium composite particles prepared using in situ polymerization. Dyes Pigm 86:155–160
22. Millet F, Auvergne R, Caillol S, David G, Manseri A, Pébère N (2014) Improvement of corrosion protection of steel by incorporation of a new phosphonated fatty acid in a phosphorus-containing polymer coating obtained by UV curing. Prog Org Coat 77:285–291
23. Abd El-Ghaffar MA, Abdel-Wahab NA, Sanad MA, Sabaa MW (2015) High performance anti-corrosive powder coatings based on phosphate pigments containing poly(o-aminophenol). Progress Organ Coat 78:42–48
24. Gimeno MJ, Chamorro S, March R, Oró E, Pérez P, Gracenea J, Suay J (2014) Anticorrosive properties enhancement by means of phosphate pigments in an epoxy 2 k coating. Assessment by NSS and ACET. Prog Org Coat 77:2024–2030
25. Erk P, Hengelsberg H (2003) 119—Phthalocyanine dyes and pigments. In: Kadish K, Guilard R, Smith KM (eds) The porphyrin handbook. Academic, Amsterdam, pp 105–149
26. Gao AH, Pi PH, Wen XF, Zheng DF, Cai ZQ, Cheng J, Yang ZR (2012) Preparation and characterisation of aluminium pigments encapsulated by composite layer containing organic silane acrylate resin and SiO2. Pigm Resin Technol 41:149–155
27. Łuczka K, Grzmil B, Sreńscek-Nazzal J, Kowalczyk K (2013) Studies on obtaining of aluminium ammonium calcium phosphates. J Ind Eng Chem 19:1000–1007
28. Łuczka K, Grzmil B, Michalkiewicz B, Kowalczyk K (2015) Studies on obtaining of aluminium phosphates modified with ammonium, calcium and molybdenum. J Ind Eng Chem 23:257–264
29. Devamani RHP, Alagar M (2012) Synthesis and characterization of aluminium phosphate nanoparticles. Int J Appl Sci Eng Res 1:769–775
30. Zhang P, He J, Zhou X (2008) An FTIR standard addition method for quantification of bound styrene in its copolymers. Polym Testing 27:153–157
31. Van Krevelen DW, Te Nijenhuis K (2009) Chapter 21—thermal decomposition. In: Van Krevelen DW, Te Nijenhuis K (eds) Properties of polymers, fourth edn. Elsevier, Amsterdam, pp 763–777
32. Tsuchiya Y, Sumi K (1968) Thermal decomposition products of polyethylene. J Polym Sci A-1 Polym Chem 6:415–424
33. Wang H, Huang SL, Zuo YJ, Zhou T, Zhang LR (2011) Corrosion resistance of lamellar aluminium pigments coated by SiO$_2$ by sol-gel method. Corros Sci 53:161–167
34. Tawiah B, Narh C, Li M, Zhang L, Fu S (2015) Polymer-encapsulated colorful Al pigments with high NIR and UV reflectance and their application in textiles. Ind Eng Chem Res 54:11858–11865
35. Pinto da Silva L, Ferreira PJO, Duarte DJR, Miranda MS, Esteves da Silva JCG (2014) Structural, energetic, and UV–Vis spectral analysis of UVA filter 4-tert-Butyl-4′-methoxydibenzoylmethane. J Phys Chem A 118:1511–1518
36. Matito C, Agell N, Sanchez-Tena S, Torres JL, Cascante M (2011) Protective effect of structurally diverse grape procyanidin fractions against UV-induced cell damage and death. J Agric Food Chem 59:4489–4495

14

Influence of MEA and piperazine additives on the desulfurization ability of MDEA aqueous for natural gas purification

Shi Yunhai[1] · Liang Shan[1] · Li Wei[1] · Luo Dong[2] · Gatabazi Remy[1] · Du JianPeng[1]

Abstract The influence of monoethanolamine (MEA) and piperazine added into methyldiethanolamine (MDEA) aqueous solution on the desulfurization of natural gas was investigated by the method of equilibrium data determination in this paper. Four kinds of equilibrated systems, i.e. H_2S-NG-MEA-water, H_2S-NG-MDEA-water, H_2S-NG-(MEA-MDEA)-water, H_2S- NG-(MEA-MDEA-PZ)-water at the temperature ranging from 298.15 to 333.15 K were measured in a glass-jacketed gas absorption cell with a double-drive impeller device. The results show that the H_2S partial pressure increases with the increase of H_2S loading in liquid phase along an isotherm. The addition of MEA and PZ is beneficial for improving the desulfuration ability of MDEA. The ability of H_2S absorption for the four mixed alkanolamine systems is MEA > (MEA-MDEA-PZ) > (MEA-MDEA) > MDEA according to the order of size. The four equilibrium data can be well correlated with the Soave–Redlich–Kwong equation of state and electrolyte-NRTL activity coefficient model. The overall mean relative errors of total pressure and H_2S partial pressure between the calculated and experimental data of the four systems are 3.30 and 3.07 %, respectively. The experimental and calculated results are very useful for desulfuration and purification process of natural gas or other industrial gases.

Keywords Desulfuration · Hydrogen sulfide · Natural gas · Monoethanolamine · Methyldiethanolamine · Piperazine

List of symbols

Variables

wt.	Abbreviation of weight
vol.	Abbreviation of volume
NG	Abbreviation of natural gas
MEA	Abbreviation of monoethanolamine
MDEA	Abbreviation of methyldiethanolamine
PZ	Abbreviation of piperazine
n_{amines}	The total mole numbers of (MEA + MDEA + PZ), mol
L	H_2S loading in liquid phase, mol/mol
n	The mole number of each species, mol
V_{I_2}	The volume of I_2 standard solution consumed with titration, mL
c_{I_2}	The concentration of I_2 standard solution, mol/L
R	Universal gas constant, 8.3145 J/(mol K)
m	Weight of MEA, MDEA and PZ, g
V	Volume of gas sample, m^3
p_a	Atmosphere pressure, kPa
Δh	Reading difference of the glass U-tube manometer, kPa
p	The equilibrium total pressure, kPa
p_{read}	Fortin Barometer reading, kPa
t	Room temperature when testing, °C
M	Molecular weight, g/mol
N	Number of experimental points
H	Henry's constant of Eqs. 15, 16 and 18

✉ Shi Yunhai
shi_yunhai@sina.com; shi_yunhai@ecust.edu.cn; shi_yunhai@hotmail.com

[1] Research Centre of Chemical Engineering, East China University of Science and Technology, 130 Meilong Road, Box 368, Shanghai 200237, China

[2] Chongqing General Gas Purification Plant, Southwest Oil and Gas Field Branch Company, 44 Taohua Road, Chongqing 401220, China

$p_{exp.}$, $p_{lit.}$	Experimental and literature data of H_2S partial pressure, kPa
T	Absolute temperature, K
K	Equilibrium constants for R1-R6
a_i	Activity of component i
z_i	Valency of an ion i
x	Liquid phase mole fraction
y	Gas phase mole fraction
A_1, A_2, A_3, A_4	Parameters of Eq. 15
D	Dielectric constants
A, B	Parameters in Table 4
a, b	Parameters of Eq. 18
T_c	Critical temperature, K
p_c	Critical pressure, kPa
V_c	Critical volume, $m^3/kmol$
Z_c	Critical compressibility factor

Subscripts

I	Component i
J	W, 1, 2, 3, 4, 5
exp.	Experimental
lit.	Literature
cal.	Calculated

Superscipts

∞	Infinite dilution in pure water
ø	Reference state, standard state

Greek alphabet

ω	Parameter of Eq. 18
υ_i	Stoichiometric coefficient of component i
γ_i	Activity coefficient of component i
$\hat{\varphi}_i$	Fugacity coefficient of component i in a mixture

Introduction

The desulfuration of natural gas (NG), and the gas streams in petroleum refinery and chemical plant is of a great importance concerning energy efficiency and environment safety. The main method for these industries is the absorption of acid gases (mainly CO_2 and H_2S) by using aqueous alkanolamine solutions followed by the desorption from solutions by using steam stripping [1]. The monoethanolamine (MEA), methyldiethanolamine (MDEA) and their blends are the commonly used absorbents, and piperazine is often widely used as an additive.

MEA is the common gas treating alkanolamine solvent due to its high reactivity, low cost, ease of reclamation, and low solubility of hydrocarbons. The disadvantage of MEA is the large enthalpy of reaction with carbon dioxide, as well as the formation of stable carbamate which limits its absorption capability [2]. MDEA is difficult to react directly with CO_2 to form carbamate. That is to say, the

selectivity of MDEA absorption for H_2S is higher than that of MEA when H_2S and CO_2 are both present. Moreover, the regeneration cost for MDEA is lower than that of MEA [3]. A kind of solvent with aqueous blend alkanolamine by adding an additive is widely used to enhance the loading of acid gas. Piperazine is most commonly used as a chemical activator. It is reported that PZ is more effective than the other conventional activators. The major advantages of PZ are its high reaction rate, and high resistance of thermal and oxidative degradation. Besides, the blends of PZ and amines exhibit low amine volatility due to the non-ideality of the mixed amine solution [4, 5]. The advantages and disadvantages of MEA, MDEA and PZ have been summarized in literatures [6–10].

The gas–liquid equilibria data of H_2S in the aqueous MEA, MDEA and the blends of MEA and MDEA solution are reported in a lot of literatures with different concentration, temperature, H_2S loading and partial pressure. Lee et al. [11] measured the gas–liquid equilibrium of H_2S-MEA-H_2O system under the conditions of MEA concentration from 2.5 to 5.0 N, temperature at 298.15, 313.15, 333.15, 353.15, 373.15 and 393.15 K, and the H_2S partial pressure from 0.15 to 2317 kPa. Isaacs et al. [12] reported the solubilities of H_2S, CO_2 and their mixture in the 2.5 mol/L aqueous solution of MEA at 373.15 K and acid gases partial pressure from 0.03 kPa to 3.36 kPa. Jou et al. [13] determined the solubilities of H_2S and CO_2 dissolved in the aqueous MDEA solution under the conditions of temperature from 313.15 to 393.15 K and partial pressure of acid gas up to 6600 kPa. The experimental data were correlated with the procedure presented by Kent and Eisenberg. They [14, 15] also measured the solubilities of H_2S, CO_2, and $H_2S + CO_2$ in 35 %wt. aqueous MDEA solution at temperature from 313.15 to 373.15 K, and the experimental data were regressed by the Deshmukh-Mather correlation. Kuranov et al. [16] investigated the solubilities of single gas CO_2 and H_2S in the aqueous MDEA solution under the conditions of temperature from 313.15 to 413.15 K, and the total pressure up to 5 MPa. A mathematical model of taking into account contributions of chemical reaction and physical interaction was prosposed to correlate the experimental data. Kamps [17] reported the experimental data of solubilities of CO_2 and H_2S in 8 mol/kg aqueous MDEA solution under the conditions of temperature from 313.15 to 393.15 K, and the total pressure up to 7.6 MPa. Li et al. [18] investigated solubilities of H_2S in aqueous MEA and MDEA blend under the conditions of temperatures from 313.15 to 373.15 K, and at H_2S partial pressure up to 450 kPa.

Unlike the previous works, the NG was introduced as a makeup gas herein to actualize the industrial desulfuration process of natural gas, and the gas–liquid equilibrium of aqueous H_2S-(MEA-MDEA-PZ)-water solution was determined experimentally in this work. The gas–liquid equilibria data of the four systems of H$_2$S-NG-MEA-water, H$_2$S-NG-MDEA-water, H$_2$S-NG-(MEA-MDEA)-water and H$_2$S-NG-(MEA-MDEA-PZ)-water were measured in a homemade equilibrium apparatus under the conditions of the temperature from 298.15 to 333.15 K, and the H$_2$S partial pressure up to 60 kPa. And a thermodynamic model was used to correlate the experimental data.

Experimental section

Reagents and materials

Monoethanolamine (MEA, ≥99.0 %wt.), methyldiethanolamine (MDEA, ≥99.0 %wt.), piperazine (PZ, ≥99.0 %wt.), sodium thiosulfate (Na$_2$S$_2$O$_3$, ≥99.0 %wt.), soluble starch, kalium iodide (KI, ≥98.5 %wt.) and sodium sulfide (Na$_2$S, ≥98.0 %wt.) were purchased from Shanghai Ling Feng Chemical Reagent Co., Ltd., China. Iodine (I$_2$, ≥99.8 %wt.) was bought from Zhejiang Lingfu fine chemicals plant, China. Sulfuric aicd (H$_2$SO$_4$, ≥98.0 %wt.) and hydrochloric acid (HCl, ≥36 %wt.) were bought from Jiangsu Yonghua Fine Chemicals Co., Ltd., China. Zinc acetate [Zn(CH$_3$COO)$_2$, ≥99 %wt.] was supplied by Sinopharm Chemical Reagent Co., Ltd., China.

The natural gas in this work was obtained from laboratory natural gas pipeline which consists of 96.266 % CH$_4$ (vol, the same below), 1.770 % C$_2$H$_6$, 0.300 % C$_3$H$_8$, 0.062 % i-C$_4$H$_{10}$, 0.075 % n-C$_4$H$_{10}$, 0.125 % C$_5$H$_{12}$ and 1.442 % N$_2$.

Apparatus and experimental method

A static-analytic method was used to measure the gas–liquid equilibrium data of these systems stated above, and the experimental apparatus is shown in Fig. 1.

The apparatus in Fig. 1 consists of three parts. The first is a H$_2$S generator. The second one is an equilibrium cell, and the third one is the measurements of equilibrium temperature and pressure, as well as the samplings.

The H$_2$S generator is composed of a dropping funnel (1) containing 1 mol/L aqueous H$_2$SO$_4$ solution and the H$_2$S generator vessel (2) with 10 %wt. Na$_2$S solution. The equilibrium cell (9) is a glass-jacketed gas liquid absorber with two double drive impellers. The rotating speeds of the gas phase impeller (10-2) and liquid phase impeller (10-1) are controlled by its own direct current motor (12-1), (12-2), respectively. And the speed is displayed on the screen of revolution counter. The absorption temperature is measured by a mercurial thermometer (8) with a sensitivity of 0.1 °C, which is adjusted by the constant temperature circulating water (14), (15) with a water thermostat. The pressure difference was determined by a glass U-tube

manometer (18) with the minimum resolution of 0.1 mmHg (0.013 kPa). The absorption pressure equates the pressure difference plus the atmosphere pressure measured by a Fortin Barometer. Liquid sampling is undertaken with a 2 mL injection syringe connected with the liquid sampling valve (11), and analyzed by the methods of weighing and chemical iodine titration. Gas sampling is done quantitatively with a eudiometer (16).

Operation procedures The absorption alkanolamines agent is firstly added into the equilibrium cell (9) from the leveling bottle (5) and valve (6). Then the whole absorption unit including the pipelines is vacuumized and degassed by a vacuum air pump (21). Afterward the stopcock of dropping funnel (1) is opened and then let the aqueous sulfuric acid reacts with sodium sulfide to generate hydrogen sulfide. The gas of hydrogen sulfide is mixed with NG derived from the pipeline and its valve (4). The gas mixture is introduced into the equilibrium cell (9) and absorbed by the alkanolamines agent under a specified pressure and temperature. The system could be thought to reach the equilibration when the absorption time is about 1–1.5 h by preliminary test.

Analysis method The method of iodine quantity is used for determining the content of hydrogen sulfide in the liquid phase. About 1 mL liquid sample drawn from the equilibrium cell with a 2 mL injection syringe, and weighted by an electric analytical balance with accuracy of 0.0001 g. Then it is slowly injected underneath the liquid interface of a 250 mL volumetric flask containing 25 mL aqueous 0.1 mol/L zinc acetate. The injection syringe is washed with this aqueous zinc acetate for three times, and for another two times washed with pure water. All of the cleaning water should be collected and mixed with the aqueous zinc acetate. Afterwards, the pH value of the aqueous solution is adjusted to 6.5 ~ 7.0 with 0.01 mol/L HCl solution. Add appropriate amount of iodine standard solution into the liquid, sealed and preserved under a dark place for at least 5 min. The mixture is titrated with 0.01 mol/L sodium thiosulfate standard solution to the color of buff; successively by added starch indicator, continuing titrated with the sodium thiosulfate standard solution to the color of blue disappearing as the titration end point [19]. The reactions included in this procedure can be written as from Eqs. (1–3).

$$H_2S + Zn(CH_3COO)_2 \rightarrow ZnS + 2CH_3COOH \qquad (1)$$

$$ZnS + 2HCl + I_2 \rightarrow ZnCl_2 + S + 2HI \qquad (2)$$

$$I_2 + 2Na_2S_2O_3 \rightarrow 2NaI + Na_2S_4O_6 \qquad (3)$$

L value represents the molar loading quantity of hydrogen sulfide per molar alkanolamines in liquid aqueous solution, which reflects the absorption ability of absorbents for H$_2$S. It is calculated as

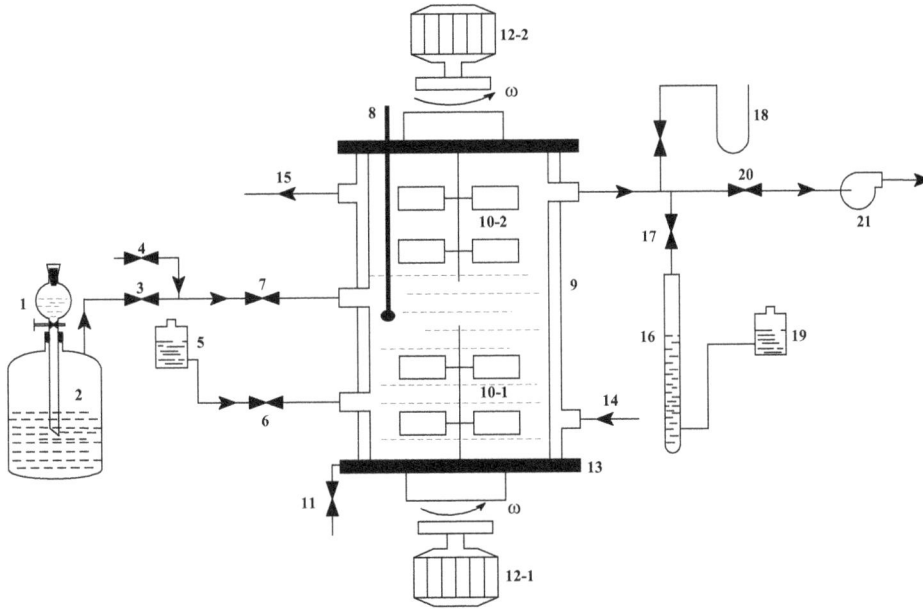

Fig. 1 Flow chart of the measurement equipment for gas–liquid equilibria *1* dropping funnel containing H_2SO_4, *2* H_2S generator vessel, *3* H_2S outlet valve, *4* NG inlet pipeline, *5* leveling bottle, *6* alkanolamines inlet pipeline and valve, *7* H_2S and NG mixture inlet valve, *8* mercurial thermometer, *9* glass-jacketed equilibrium cell, (*10-1*, *10-2*) electromagnet driving double agitator blades, *11* liquid

sampling valve, (*12-1*, *12-2*)-direct current motor, *13* two pieces of stainless steel flange, *14* constant temperature circulating water inlet, *15* constant temperature circulating water outlet, *16* eudiometer, *17* gas sampling valve, *18* glass U-tube manometer, *19* leveling bottle, *20* vacuum pump valve, *21* vacuum air pump

$$L(\text{mol H}_2\text{S/mol amine}) = \frac{n_{H_2S}}{n_{\text{amines}}}$$
$$= \frac{(V_{I_2}c_{I_2} - V_{Na_2S_2O_3}c_{Na_2S_2O_3}) \times 0.5 \times 10^{-3}}{n_{\text{amines}}} \quad (4)$$

where V_{I_2}, c_{I_2} are the volume consumed and molar concentration of iodine standard solution. $V_{Na_2S_2O_3}$, $c_{Na_2S_2O_3}$ are the volume titration demanded and molar concentration of sodium thiosulfate standard solution. n_{amines} is molar quantity of the mixed alkanolamines calculated by:

$$n_{\text{amines}} = \frac{m_{MEA}}{M_{MEA}} + \frac{m_{MDEA}}{M_{MDEA}} + \frac{m_{PZ}}{M_{PZ}} \quad (5)$$

In Eq. (5), m_i and M_i are the quality and molecular weight of the mixed alkanolamines component i.

The analysis method of gas phase: Firstly, 50 mL 0.1 mol/L zinc acetate solution is added into the leveling bottle (19). Keep the liquid interface of eudiometer (16) and leveling bottle (19), and record the initial scale value. Open the gas sampling valve (17) and let the gas phase into the eudiometer, then close the sampling valve and record the end reading. The difference of the ending and the initial readings is the gas sampling volume. Shake the eudiometer and let the sampling gas mix completely with the solution of zinc acetate. The reaction liquid is transferred to a clean 250 mL volumetric flask. Washing the eudiometer with

pure water two times, and the washed water is also added into the volumetric flask. Afterwards, an adequate quantity of iodine standard solution is added, and then sealed and preserved under a dark place for at least 5 min. Then, titration of sodium thiosulfate is adopted as described above for analyzing the containing of hydrogen sulfide. The Eq. (6) is used for calculation the molar fraction of hydrogen sulfide in the gas phase.

$$y_{H_2S} = \frac{(V_{I_2}c_{I_2} - V_{Na_2S_2O_3}c_{Na_2S_2O_3}) \times 0.5RT}{pV} \quad (6)$$

In Eq. (6), T and V are the equilibrium temperature and volume of the sampling gas phase. R is the universal gas constant, 8.3145 J/(mol K). p is the equilibrium total pressure, and calculated as

$$p = p_a + \Delta h \quad (7)$$

where Δh is the reading difference of the glass U-tube manometer (18), p_a is the atmosphere pressure measured by a Fortin Barometer. For Shanghai, it is calculated as

$$p_a = 0.9988 p_{\text{read}} \left(1 - \frac{1.634 \times 10^{-4}t}{1 + 1.818 \times 10^{-4}t} \times t \right) \quad (8)$$

In the Eq. (8), p_{read} is the reading value of the Fortin Barometer, and t is the room temperature during the experiments.

The reliability of the apparatus

In order to check the reliability of the apparatus and experimental method, gas–liquid equilibrium data of hydrogen sulfide dissolved in 2.5 mol/L MEA aqueous solution under 313.15 K and atmosphere pressure were measured and compared with the literature data [11]. The results are shown in Fig. 2.

The results shown in Fig. 2 indicate that the experimental value is agreed very well with the literature data, and the maximum relative error is less than 5 %. It shows that the apparatus is suitable for determination the gas–liquid equilibrium of hydrogen sulfide dissolved in alkanolamines solutions.

Correlation of experimental data with thermodynamics model

Hydrogen sulfide dissolved in aqueous alkanolamines is a system of electrolyte solution. Non-idealities of species in gas phase and liquid phase should be taken into account in the thermodynamic computation of multicomponent complex aqueous solution like H_2S-NG-(MEA-MDEA-PZ)-water system. In this work, the Soave–Redlich–Kwong (SRK) equation of state [20] is used to account for the non-ideality of gas phase, and the electrolyte-NRTL equation [21, 22] is adopted to describe that of the liquid phase. Herein, a brief description is made as follows.

Chemical equilibrium relationship of species in the liquid aqueous solution

The chemical equilibrium relationship of the species in the liquid phase can be written in the form of chemical dissociation [3, 23, 24] as follows.

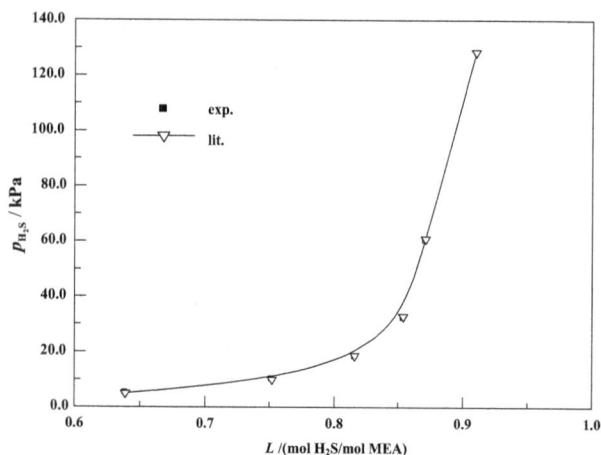

Fig. 2 The partial pressure of H_2S gas dissolved in 2.5 mol/L MEA aqueous solution compared with literature data at 313.15 K and atmosphere pressure

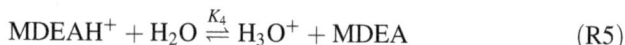

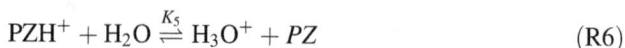

$$2H_2O \overset{K_W}{\rightleftharpoons} H_3O^+ + OH^- \qquad (R1)$$

$$H_2S + H_2O \overset{K_1}{\rightleftharpoons} H_3O^+ + HS^- \qquad (R2)$$

$$HS^- + H_2O \overset{K_2}{\rightleftharpoons} H_3O^+ + S^{2-} \qquad (R3)$$

$$MEAH^+ + H_2O \overset{K_3}{\rightleftharpoons} H_3O^+ + MEA \qquad (R4)$$

$$MDEAH^+ + H_2O \overset{K_4}{\rightleftharpoons} H_3O^+ + MDEA \qquad (R5)$$

$$PZH^+ + H_2O \overset{K_5}{\rightleftharpoons} H_3O^+ + PZ \qquad (R6)$$

The chemical equilibrium constants from Eqs. (R1–R6) can be written as:

$$K_j = \prod_i \hat{a}_{ji}^{\nu_{ji}} \quad (j = W, 1, 2, 3, 4, 5) \qquad (9)$$

where K_j is the chemical equilibrium constant of the above Eqs. (R1–R6), $\hat{a}_{ji}$ and ν_{ji} are the activity and stoichiometric coefficient of component i of the reaction j, respectively. The mass balance equations for the sulfur element and alkanolamine molecules can be expressed by

$$n_{H_2S}^0 = n_{H_2S} + n_{HS^-} + n_{S^{2-}} \qquad (10)$$

$$n_{MEA}^0 = n_{MEA} + n_{MEAH^+} \qquad (11)$$

$$n_{MDEA}^0 = n_{MDEA} + n_{MDEAH^+} \qquad (12)$$

$$n_{PZ}^0 = n_{PZ} + n_{PZH^+} \qquad (13)$$

Another restriction is the condition of liquid phase electroneutrality, and can be written as

$$\sum_i n_i z_i = 0 \qquad (14)$$

In Eq. (14), n_i and z_i are the molar quantity and charge number of ions (including ions of H_3O^+, HS^-, S^{2-}, $MEAH^+$, $MDEAH^+$, PZH^+, OH^+, etc.).

The relationship of equilibrium constant for Eq. (9) and Henry's constant of hydrogen sulfide with temperature can be expressed in Eq. (15), and the parameters are listed in Table 1.

$$\ln K_j \text{ or } \ln H_{H_2S}^p = A_1 + A_2/T + A_3 \ln T + A_4 T. \qquad (15)$$

Gas–liquid equilibria for molecules

The Henry's Law is adopted to express the gas–liquid equilibrium of H_2S:

$$p y_{H_2S} \overset{\wedge}{\varphi}_{H_2S} = \gamma_{H_2S}^* x_{H_2S} H_{H_2S}^\infty \exp\left[\frac{\nu_{H_2S}^\infty (p - p^\emptyset)}{RT}\right] \qquad (16)$$

where y_{H_2S}, x_{H_2S} are the molar fraction of hydrogen sulfide in gas phase and liquid phase. $\hat{\varphi}_{H_2S}$, $\gamma_{H_2S}^*$ are the fugacity

Table 1 The relationship of equilibrium constant for Eq. (9) and H$_2$S Henry's constant with temperature

Equation	A_1	A_2	A_3	A_4	References
Equilibrium constants					
R1	132.9	−13446.0	−22.48	0	[22]
R2	214.6	−12995.4	−33.55	0	[22]
R3	−32.0	−3338.0	0	0	[22]
R4	2.1211	−8189.38	0	−0.007484	[22]
R5	−56.2	−4044.8	7.848	0	[3]
R6	4.964	−9714.2	0	0	[25]
Henry's constant					
H$_2$S	358.138	−133236.8	−55.0511	0.059565	[22]

Table 2 Dielectric constants for MEA, MDEA, PZ and water

Solvent component	Equation	References
MEA	$D = 36.76 + 14836[1/T(K) − 1/273.15]$	[22]
MDEA	$D = 24.74 + 8989.3[1/T(K) − 1/273.15]$	[23]
PZ	$D = 4.719 − 1530[1/T(K) − 1/273.15]$	[29]
H$_2$O	$D = 78.65 + 31989[1/T(K) − 1/273.15]$	[23]

The dielectric constant of PZ is calculated as the method described in reference 29

Table 3 Binary interaction parameters used in the electrolyte-NRTL model ($\tau_{ij} = A + B/T$)

τ_{ij}, τ_{ji}	A	B	References
H$_2$O-MDEA	8.5092	−1573.9	[23]
MDEA-H$_2$O	−1.7141	−261.85	[23]
H$_2$O-PZ	3.66	−310	a
PZ-H$_2$O	6.46	−2648	a
H$_2$O-MEA	1.674	0	[23]
MEA-H$_2$O	0	−649.75	[23]
H$_2$O-H$_2$S	−3.674	1155.9	[23]
H$_2$S-H$_2$O	−3.674	1155.9	[23]
H$_2$O-MEAH$^+$, HS$^-$	6.844	501.83	[23]
MEAH$^+$, HS$^-$-H$_2$O	−3.560	−197.12	[23]
H$_2$O-MDEAH$^+$, HS$^-$	3.735	1036.04	[23]
MDEAH$^+$, HS$^-$-H$_2$O	−3.255	0	[23]
PZH$^+$, HS$^-$-H$_2$O	−3.79	0.98	a
H$_2$O-PZH$^+$, HS$^-$	9.07	0	a

a The parameters are fitted as the method described in reference 14

coefficient and activity factor of hydrogen sulfide in gas phase and liquid phase, respectively. $H_{H_2S}^\infty$, $v_{H_2S}^\infty$ and $p^\varnothing$ are the Henry's constant, molar volume of hydrogen sulfide and reference pressure under the condition of infinite dilute concentrations. For the components of solvent, like water, MEA, MDEA and PZ, the relationship of gas–liquid equilibrium can be written in

$$py_i\hat\varphi_i = \gamma_i x_i p_i^s \exp\left[\frac{v_i(p − p_i^s)}{RT}\right] \tag{17}$$

In Eq. (17), y_i, x_i are the molar fraction of component i in gas phase and liquid phase. $\hat\varphi_i$, γ_i are the fugacity

coefficient and activity factor of component i in gas phase and liquid phase, respectively. p_i^s, v_i are the saturated gas pressure and molar volume of component i under the equilibrium temperature T.

Activity coefficient of component i

The electrolyte-NRTL equation [25], which is composed of three contributions of excess Gibbs free energy counted by the Pitzer long-range interaction (PDH), corrected Born term (Born) and short-range solvation effect (NTRL), is used to calculate the activity coefficient of liquid phase

Table 4 The molecular properties of pure components

Component	Molecular weight	T_c/K	p_c/kPa	V_c/ (m^3/kmol)	Z_c
H$_2$S	34.08	373.2	8936.9	0.0986	0.284
H$_2$O	18.02	647.3	22090.0	0.0568	0.233
MEA	61.08	638.0	6870.0	0.2250	0.291
MDEA	119.16	677.8	3876.1	0.3932	0.192
PZ	86.14	364.85	5603.3	310.00	0.320

components. For similar weak electrolyte solutions, this model had been widely used to correlate the gas–liquid equilibrium in literatures [16, 22, 26–28]. The parameters, including relevant coefficients and interaction parameters in the model could be also obtained in the literatures. Table 2 lists the dielectric constant for solvent components, like MEA, MDEA, PZ and water. Table 3 collects the binary interaction parameters of components, which are used to calculate the activity coefficient of components by the electrolyte-NRTL model.

Fugacity coefficient for component i

The fugacity coefficients of components in the gas phase are calculated by the SRK equation of state. Table 4 lists the molecular properties of pure components of this gas–liquid equilibrium [19], which are used in the calculation with SRK equation of state.

Calculation procedure

The activity coefficient of components for liquid phase and fugacity coefficient for gas phase can be calculated by the electrolyte-NRTL equation and SRK equation of state, respectively. The total pressure for the equilibrated system can be calculated by the following Eq. (18), which ignored the partial pressure of other components excluding H$_2$S, solvent components and CH$_4$ in the gas phase.

$$p = \frac{\gamma^*_{H_2S} x_{H_2S} H^\infty_{H_2S}}{\hat{\varphi}_{H_2S}} \exp\left[\frac{v^\infty_{H_2S}(p - p^\emptyset)}{RT}\right] + \sum_i \frac{\gamma_i x_i p_i^s}{\hat{\varphi}_i} \exp\left[\frac{v_i(p - p_i^s)}{RT}\right] + p_{CH_4} \quad (18)$$

The Bubble point method was adopted to calculate the gas–liquid equilibria. Ordinarily, the known variables are the temperature T, concentration of hydrogen sulfide in the liquid phase x_{H_2S}, initial molar quantity of solvent components, like MEA, MDEA, PZ and water, by solving the equation set of Eqs. (9–14), as well as Eq. (16) and Eq. (17), the total pressure of the equilibrated system $p_{cal.}$ and molar fraction of hydrogen sulfide y_{H_2S} can be evaluated by the

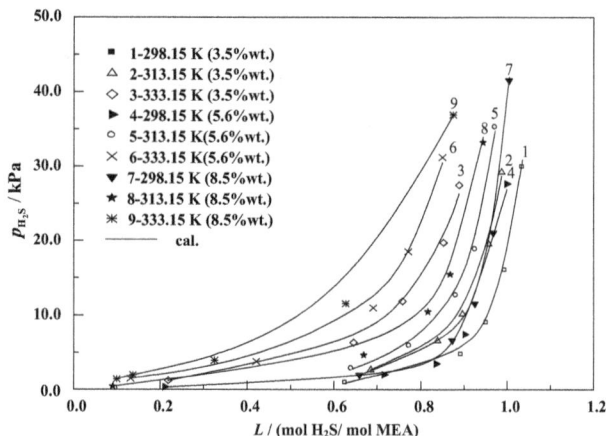

Fig. 3 Comparison of the calculated H$_2$S partial pressure with experimental data at various isotherms and different concentrations of aqueous MEA solution

objective function approaching to minimum, i.e., $OBJ =$

$$\sqrt{\left(\sum_j (p_{cal.} - p_{exp.})^2 + \sum_j (y_{H_2S,cal.} - y_{H_2S,exp.})^2\right)/2N} \rightarrow \min.$$

where N is the number of experimental points.

Results and discussions

Experimental data of H$_2$S dissolved in a single alkanolamine aqueous solution

Comparisons of the calculated partial pressures of hydrogen sulfide with the experimental data, which is dissolved in a different aqueous MEA solution at absorption isotherms of 298.15, 313.15 and 333.15 K, are shown in Fig. 3. As seen from Fig. 3, with the increasing of H$_2$S loading in liquid phase, the partial pressure of H$_2$S in the gas phase increases under any an isotherm. While, the increment of H$_2$S partial pressure shows a small value under the condition of relatively lesser H$_2$S loading along an isotherm at first; then it sharply increases at the higher H$_2$S loading along the same isotherm. For an example, at the isotherm of aqueous H$_2$S-NG-8.5 %wt. MEA-water solution at 333.15 K (Line 9 in Fig. 3), the H$_2$S partial pressure increases from 1.462 to 3.997 kPa by the difference value of 2.535 kPa when H$_2$S loading in the aqueous solution changes from 0.100 to 0.325 (the difference value of 0.225). But it increases rapidly from 11.578 to 36.914 kPa when H$_2$S loading varies from 0.627 to 0.874 (the difference value of 0.247). The reason is that the desulfuration of H$_2$S by aqueous alkanolamine solutions has the features of both chemical absorption and physical absorption. The chemical reaction of hydrogen sulfide with MEA is to be equilibrium with a larger H$_2$S loading in

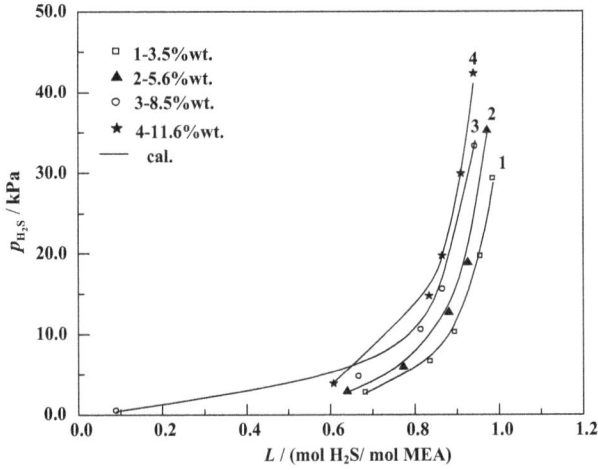

Fig. 4 Comparison of the calculated H_2S partial pressure with experimental value at 333.15 K under different isoconcentration of aqueous MEA solution

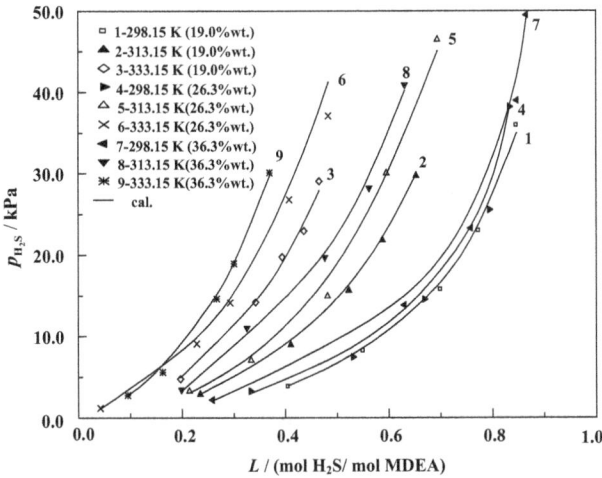

Fig. 6 Comparison of the calculated H_2S partial pressure with experimental value at 333.15 K under different isoconcentration of aqueous MDEA solution

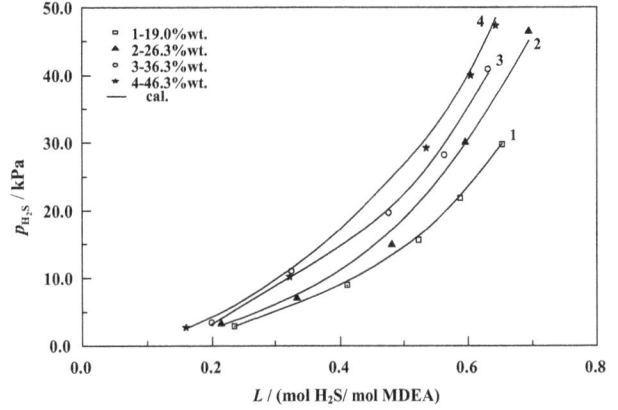

Fig. 5 Comparison of the calculated H_2S partial pressure with experimental data at various isotherms and different concentrations of aqueous MDEA solution

liquid. And then the chemical absorption is transferred into physical absorption, so the H_2S partial pressure in the gas phase increases theatrically sharply at the larger H_2S loading.

Meanwhile, the H_2S partial pressure increases with the rising of absorption temperature at the conditions of a constant H_2S loading and the same component concentrations of the aqueous absorption solution. The H_2S loading in liquid phase decreases with the rising of absorption temperature under the same H_2S partial pressure and the same component concentrations. These behaviors are the universal phenomena of the influence of temperature on the H_2S partial pressure under the conditions of the constant H_2S loading and the same component concentrations.

The average relative errors of total pressure and H_2S partial pressure between the theoretical values and the experimental data for the system of aqueous H_2S-NG-MEA-water are 3.01 and 3.26 %, respectively.

The H_2S partial pressure calculated by the electrolyte-NRTL model is compared with the experimental value at 333.15 K under different isoconcentration of aqueous MEA solution as shown in Fig. 4. The results show that the calculated values are in good consistent with the experimental ones. The H_2S partial pressure increases with the higher concentration of aqueous MEA solution at a constant H_2S loading; the loading of H_2S in the liquid phase decreases with higher concentration of the aqueous MEA solution at a constant H_2S partial pressure.

The relationship of H_2S partial pressure with H_2S loading in aqueous solution of the system consisting of H_2S-NG-MDEA-water is presented in Figs. 5 and 6. These two Figures have very similar features as the aqueous MEA solutions shown in Figs. 3 and 4, respectively. And the calculated results are also in agreement with the experimental data. Compared two H_2S absorption isotherms lines, for instance numbered as line 1, shown in Figs. 3 and 4, the H_2S loading is 1.02 in Fig. 3 under the condition of 20.0 kPa H_2S partial pressure and 298.15 K, which is larger than that of 0.742 in Fig. 4 at the same conditions. The result shows that the desulfuration ability of aqueous MEA solution is stronger than that of MDEA, although the concentration of aqueous MDEA is higher than that of MEA solution.

The average relative errors of total pressure and H_2S partial pressure between the calculated values and the experimental data for the system of aqueous H_2S-NG-MDEA-water are 3.46 and 2.91 %, respectively.

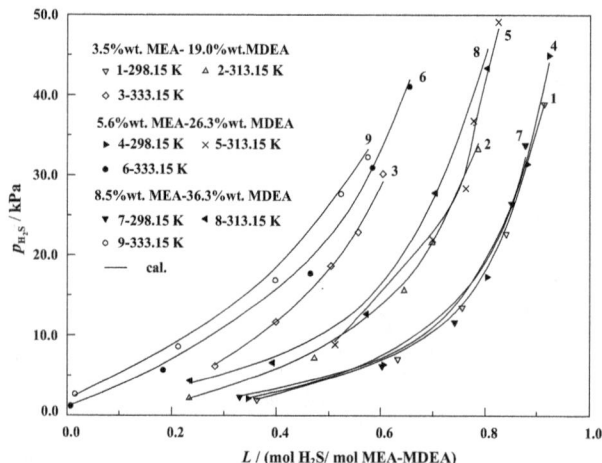

Fig. 7 Comparison of the calculated H_2S partial pressure with experimental data at different isotherms and different equi-composition of the mixed aqueous MEA-MDEA solution

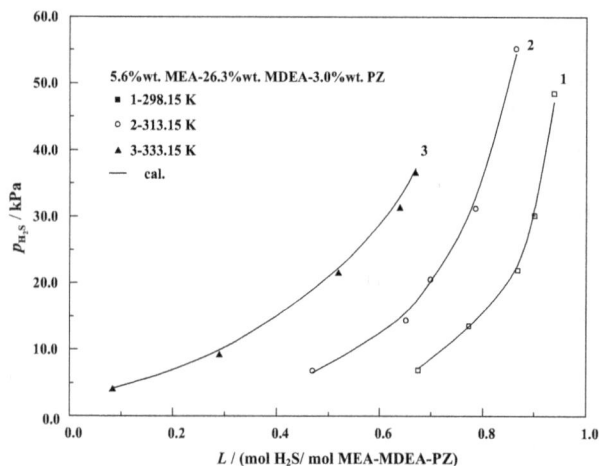

Fig. 8 Comparison of calculated H_2S partial pressure with experimental data at different isotherms for the aqueous 5.6 %wt. MEA-26.3 %wt. MDEA-3.0 %wt. PZ solution

Influence of MEA and piperazine added into the MDEA aqueous solutions

Comparison of calculated H_2S partial pressure with the experimental data of H_2S-NG-(MEA-MDEA)-water system at different isotherms and different equi-compositions is shown in Fig. 7. The blends of MEA and MDEA for H_2S absorption have the similar performance with the signal MEA or MDEA. It also can be seen in Fig. 7 that the total content of the mixed aqueous MEA-MDEA solution has little effect on the relationships between H_2S loading and H_2S partial pressure at lower temperatures, i.e., 298.15 and 313.15 K, but has significantly influence at higher temperature, i.e. 333.15 K. And the deviations become significant at high temperatures for the MEA-MDEA

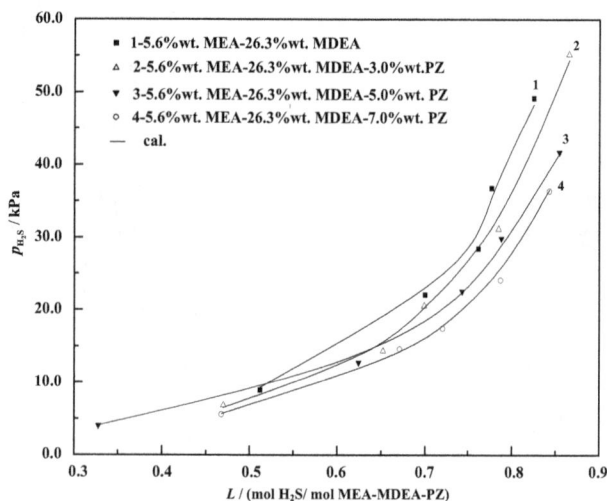

Fig. 9 Comparisons of calculated and experimental data of H_2S partial pressure for the mixed aqueous (MEA-MDEA-PZ) solutions with different PZ content at 313.15 K

solutions. The reason is that the MEA-MDEA solutions have a higher absorption capacity for H_2S, and the influences of total content of MEA-MDEA become more and more important with the increasing of temperature.

The average relative errors of total pressure and H_2S partial pressure between the calculated values and the experimental data for the system of aqueous H_2S-NG-(MEA-MDEA)-water are 3.12 and 3.00 %, respectively.

By comparing H_2S loading in liquid phase under a certain H_2S partial pressure and absorption temperature shown in Figs. 3, 5 and 7, the size order of H_2S absorption ability for the three systems, i.e. MEA, MDEA and the mixed MEA-MDEA, is MEA > (MEA-MDEA) > MDEA, which indicates that the addition of MEA into aqueous MDEA solution can improve the desulfuration ability of MDEA.

The relationship of the H_2S partial pressure with its loading in the liquid phase of the mixed aqueous (MEA-MDEA-PZ) solutions containing PZ is shown in Fig. 8. As discussed above, the H_2S partial pressure increases with the higher of H_2S loading along the isotherms shown in Fig. 8. The addition of the PZ into the mixed aqueous MEA-MDEA solution is beneficial for increasing its ability of desulfuration. The size order of H_2S absorption ability for the above four systems, i.e. MEA, MDEA, mixed (MEA-MDEA), and mixed (MEA-MDEA-PZ), is MEA > (MEA-MDEA-PZ) > (MEA-MDEA) > MDEA, which indicates that the addition of MEA and PZ into aqueous MDEA solutions can well enhance the desulfuration ability of MDEA.

The influence of PZ content on the relationship of H_2S partial with its loading in the liquid phase along the four iso-concentration curves of component MEA and MDEA at 313.15 K is shown in Fig. 9. As seen in Fig. 9, the more content of PZ, the greater H_2S loading in the liquid phase

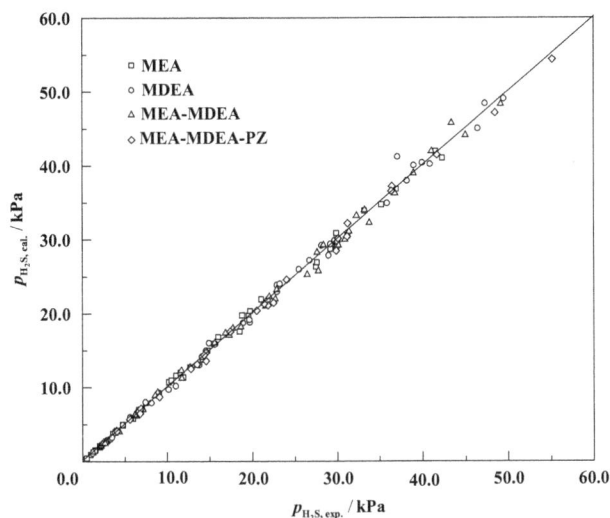

Fig. 10 Comparison of H_2S partial pressure calculated by the model with experimental data

under the condition of a constant H_2S partial pressure. And the more quantity of PZ containing, the lower H_2S partial pressure under the condition of a constant H_2S loading in the liquid phase. Especially, the four curves tend to be closer at the low H_2S loadings about from 0.3 to 0.6. The reason is that almost identical desulfuration ability for these four aqueous absorption systems at this range.

The average relative errors of total pressure and H_2S partial pressure between the calculated values and the experimental data for the system of aqueous H_2S-NG-(MEA-MDEA-PZ)-water are 3.47 and 3.08 %, respectively.

Comparison of calculated results by model and experimental data

The comparison of H_2S partial pressure calculated by the SRK equation of state and electrolyte-NRTL model with the experimental data for the four systems above is shown in Fig. 10. The overall average relative errors of total pressure and H_2S partial pressure are 3.30 and 3.07 %, respectively. It shows that the calculated results are very well consistent with the experimental data, and the selected models can well describe the non-idealities of the gas phase and the liquid phase containing aqueous weak electrolytes of hydrogen sulfide dissolved in a kind of mixed alkanolamine solution.

Conclusions

The gas–liquid equilibrium data for desulfuration of natural gas (NG) by using a mixed aqueous alkanolamine solution was investigated and the results were used for evaluating their ability of removing hydrogen sulfide. The apparatus

was a glass-jacketed gas absorption cell with a double-drive impeller device. Four gas–liquid equilibria systems, i.e. H_2S-NG-MEA-water, H_2S-NG-MDEA-water, H_2S-NG-(MEA- MDEA)- water, H_2S-NG-(MEA-MDEA-PZ)-water at the temperature ranging from 298.15 to 313.15 K were experimentally investigated in this work. The equilibrium data were correlated with SRK equation of state and electrolyte-NRTL activity coefficient model, which describe the non-idealities of the gas phase and the aqueous weak electrolyte solution. The results show that the H_2S partial pressure increases with the higher of H_2S loading along an isotherm. The quantity of H_2S loading in the aqueous phase decreases when the temperature increases under the condition of constant H_2S partial pressure. The addition of MEA and PZ is beneficial for improving the desulfuration ability of MDEA. The size order of H_2S absorption ability for the four systems, i.e. MEA, MDEA, the mixed (MEA-MDEA), and the mixed (MEA-MDEA-PZ), is MEA > (MEA-MDEA-PZ) > (MEA-MDEA) > MDEA. The four sets of gas–liquid equilibria data can be well correlated with the SRK equation of state and electrolyte-NRTL activity coefficient model. The overall average relative errors of total pressure and H_2S partial pressure between the calculated and experimental results of the above four absorption systems are 3.30 and 3.07 %, respectively. The experimental and calculated results are very useful for desulfuration and purification process of natural gas or other industrial gases.

References

1. Lemoine B, Li YG, Cadours R, Bouallou C, Richon D (2000) Partial vapor pressure of CO_2 and H_2S over aqueous methyl-diethanolamine solutions. Fluid Phase Equilib 172:261–277
2. Haghtalab A, Dehghani Tafti M (2007) Electrolyte UNIQUAC-NRF model to study the solubility of acid gases in alkanolamines. Ind Eng Chem Res 46:6053–6060
3. Vrachnos A, Kontogeorgis G, Voutsas E (2006) Thermodynamic modeling of acidic gas solubility in aqueous solutions of MEA, MDEA and MEA-MDEA blends[J]. Ind Eng Chem Res 245:5148–5154
4. Dash SK, Samanta AN, Bandyopadhyay SS (2011) Solubility of carbon dioxide in aqueous solution of 2-amino-2-methyl-1-propanol and piperazine. Fluid Phase Equilib 307:166–174
5. Haghtalab A, Izadi A (2014) Simultaneous measurement solubility of carbon dioxide + hydrogen sulfide into aqueous blends of alkanolamines at high pressure. Fluid Phase Equilib 375:181–190
6. Chakravarty T, Phukan UK, Weiland RH (1985) Reaction of acid gases with mixtures of amines. Chem Eng Prog 81:32–36
7. Bishnoi S, Rochelle GT (2000) Absorption of carbon dioxide into aqueous piperazine: reaction kinetics, mass transfer and solubility. Chem Eng Sci 55:5531–5543
8. Sun WC, Yong CB, Li MH (2005) Kinetics of the absorption of carbon dioxide into mixed aqueous solutions of 2-amino-2-methyl-1-propanol and piperazine. Chem Eng Sci 60:503–516
9. Bishnoi S, Rochelle GT (2002) Absorption of carbon dioxide in aqueous piperazine/methyldiethanolamine. AIChE J 48:2788–2799

10. Xu X, Cai ZY, Liang K (2010) A study on flue gas desulfurization using aqueous piperazine. Environ Chem 29:450–454 (**in Chinese**)
11. Lee JI, Otto FD, Mather AE (1976) Equilibrium in hydrogen sulfide-monoethanolamine-water system. J Chem Eng Data 21:207–208
12. Isaacs EE, Otto FD, Mather AE (1980) Solubility of mixtures of hydrogen sulfide and carbon dioxide in a monoethanolamine solution at low partial pressures. J Chem Eng Data 25:118–120
13. Jou FY, Mather AE, Otto FD (1982) Solubility of H_2S and CO_2 in aqueous methyldiethanolamine solutions. Ind Eng Chem Process Des Dev 21:539–544
14. Jou FY, Carroll JJ, Mather AE, Otto FD (1993) The solubility of carbon dioxide and hydrogen sulfide in a 35 wt% aqueous solution of methyldiethanolamine. Can J Chem Eng 71:264–268
15. Jou FY, Carroll JJ, Mather AE, Otto FD (1993) Solubility of mixtures of hydrogen sulfide and carbon dioxide in aqueous N-methldiethanolmine solutions. J Chem Eng Data 38:75–77
16. Kuranov G, Rumpf B, Smirnova NA, Maurer G (1996) Solubility of single gases carbon dioxide and hydrogen sulfide in aqueous solutions of N-methyldiethanolamine in the temperature range 313–413 K at pressure up to 5 MPa. Ind Eng Chem Res 35:1959–1966
17. Kamps ÁP-S, Balaban A, Jödecke M, Kuranov G, Smirnova NA, Maurer G (2001) Solubility of single gases carbon dioxide and hydrogen sulfide in aqueous solutions of N-methyldiethanolamine at temperatures from 313 to 393 K and pressures up to 7.6 MPa: new experimental data and model extension. Ind Eng Chem Res 40:696–706
18. Li M-H, Shen K-P (1993) Solubility of hydrogen sulfide in aqueous mixtures of monoethanolamine with N-methldiethanolamine. J Chem Eng Data 38:105–108
19. Perry RH, Green DW, Maloney JO (1997) Perry's chemical engineers' handbook, 7th edn. McGraw-hill, New York
20. Soave G (1972) Equilibrium constants from a modified Redlich-Kwong equation of state. Chem Eng Sci 27:1197–1203
21. Austgen DM, Rochelle GT, Peng X, Chen CC (1989) Model of vapor-liquid equilibria for aqueous acid gas-alkanolamine systems using the Electrolyte-NRTL equation. Ind Eng Chem Res 28:1060–1073
22. Austgen DM, Rochelle GT, Chen CC (1991) Model of vapor-liquid equilibria for aqueous acid gas-alkanolamine systems. 2. Representation of hydrogen sulfide and carbon dioxide solubility in aqueous MDEA and carbon dioxide solubility in aqueous mixtures of MDEA with MEA or DEA. Ind Eng Chem Res 30:543–555
23. Li Y, Mather AE (1997) Correlation and prediction of the solubility of CO_2 and H_2S in aqueous solutions of methyldiethanolamine. Ind Eng Chem Res 36:2760–2765
24. Bishnoi S, Rochelle GT (2002) Thermodynamics of piperazine/methyldiethanolamine/water/carbon dioxide. Ind Eng Chem Res 41:604–612
25. Mock B, Evans LB, Chen CC (1986) Thermodynamic representation of phase equilibria of mixed-solvent electrolyte systems. AIChE J 32:1655–1664
26. Hetzer HB, Robinson RA, Bates RG (1968) Dissociation constants of piperazinium ion and related thermodynamic quantities from 0° to 50°. J Phys Chem C 72:2081–2086
27. Vrachnos A, Voutsas E, Magoulas K, Lygeros A (2004) Thermodynamics of acid gas-MDEA-water systems. Ind Eng Chem Res 43:2798–2804
28. Masih HJ, Majid AA, Seyed SH, Medhi V, Naser SM (2005) Solubility of carbon dioxide in aqueous mixtures of N-methyldiethanolamine + piperazine + sulfolane. J Chem Eng Data 50:583–586
29. Derks PWJ, Dijkstra HBS, Hogendoorn JA, Versteeg GF (2005) Solubility of carbon dioxide in aqueous piperazine solutions. AIChE J 51:2311–2327

15

Photocatalytic degradation of tetracycline aqueous solutions by nanospherical α-Fe₂O₃ supported on 12-tungstosilicic acid as catalyst: using full factorial experimental design

Majid Saghi[1] · Kazem Mahanpoor[1]

Abstract In this paper, spherical α-Fe$_2$O$_3$ nanoparticles (NPs) were supported on the surface of 12-tungstosilicic acid (12-TSA·7H$_2$O) using two different solid-state dispersion (SSD) and forced hydrolysis and reflux condensation (FHRC) methods. Photocatalytic activity of supported α-Fe$_2$O$_3$ NPs (α-Fe$_2$O$_3$/12-TSA·7H$_2$O) for tetracycline (TC) degradation in aqueous solution was investigated using UV/H$_2$O$_2$ process and the results were compared with that of pure α-Fe$_2$O$_3$ NPs. α-Fe$_2$O$_3$ and 12-TSA·7H$_2$O were synthesized according to previous reports and all products were characterized by using FTIR, SEM, EDX and XRD. Design of experiments (DoEs) was utilized and photocatalytic degradation process was optimized using full factorial design. The experiments were designed considering four variables including pH, the initial concentration of TC, catalyst concentration and H$_2$O$_2$ concentration at three levels. TC concentration reduction in the medium was measured using UV/Vis spectroscopy at $\lambda_{max} = 357$ nm. The results of experiments indicated that supporting α-Fe$_2$O$_3$ NPs on the surface of 12-TSA·7H$_2$O through SSD and FHRC methods caused to improve the filtration, recovery and photocatalytic activity of NPs. Also, it was indicated that those NPs supported through SSD method, have better photocatalytic performance than those supported through FHRC method. The statistical analyses revealed that the maximum TC degradation (97.39%) is obtained under those conditions in which pH

and catalyst concentration variables are at maximum levels and the initial concentration of TC and H$_2$O$_2$ concentration variables are at minimum levels (pH 8, catalyst concentration = 150 ppm, initial concentration of TC = 30 ppm, H$_2$O$_2$ concentration = 0.1 ppm). A first order reaction with $k = 0.0098$ min^{-1} was observed for the photocatalytic degradation reaction.

Keywords Photocatalytic degradation · Tetracycline · α-Fe$_2$O$_3$ · 12-Tungstosilicic acid · α-Keggin

Introduction

From the perspective of green chemistry, degradation of chemical pollutants in wastewater has attracted a lot of attention. Antibiotics are one of the larger groups of these pollutants in wastewater released from pharmaceutical industries [1]. Besides, TC is one broad spectrum of antibiotics repeatedly detected in urban and industrial wastewaters, drinking water, surface water and groundwater [2–6]. The molecular structure of TC is shown in Fig. 1. Various techniques are used to degrade TC; one of these techniques is photocatalytic degradation [7]. NPs play an important role in heterogeneous photocatalysis. Metal oxide NPs, i.e., iron oxides, have a special position in the science and technologies because of having wide applications and unique properties [8]. α-Fe$_2$O$_3$ (hematite) which is the most common form of iron oxides, has the rhombohedral structure and it is an attractive compound because of its applications in data storage, gas sensor, magnets materials, pigment, catalysis and photocatalysis [9–14]. Various techniques including co-precipitation, sol–gel, thermal decomposition, Micelle synthesis, sonochemical synthesis, hydrothermal synthesis and FHRC have

✉ Kazem Mahanpoor
k-mahanpoor@iau-arak.ac.ir

Majid Saghi
m-saghi@iau-arak.ac.ir

[1] Department of Chemistry, Islamic Azad University, Arak Branch, Arāk, Iran

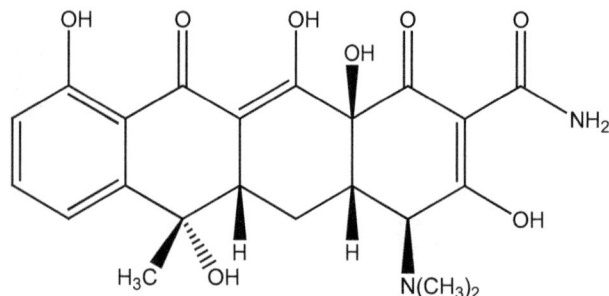

Fig. 1 Molecular structure of TC

been utilized to synthesize monodisperse α-Fe$_2$O$_3$ NPs.
[15–21]. Among various photocatalytic processes, water
and wastewater treatments are of the most important α-Fe$_2$O$_3$ NPs applications. In these processes, α-Fe$_2$O$_3$ NPs
could be used in the form of a fine powder or crystals
dispersed in water, but it is vital to know that filtering these
NPs following reaction is difficult and costly. To solve this
problem, researchers have examined methods for supporting α-Fe$_2$O$_3$ NPs on the surface of organic, inorganic or
organic/inorganic catalyst supports [22, 23]. Various
methods have been applied for supporting α-Fe$_2$O$_3$ NPs on
the surface of catalyst support. Utilizing any of these
methods depends on the chemical and physical properties
of catalyst and catalyst support as well as the purpose of the
process. One of these methods is SSD method in which
catalyst precursor and catalyst support are separately synthesized and then are mixed with specific weight ratio
using an appropriate solvent [24]. Then, during calcination,
the catalyst is both formed and thermally supported on the
surface of catalyst support. In another technique such as
FHRC, the catalyst support is added to the precursor
solution(s) during catalyst preparation (if it was stable in
reaction medium) and the catalyst is supported on the
surface of catalyst support while it is simultaneously
formed. In FHRC method, all steps related to the synthesis
of NPs were done on the surface of catalyst support and
"NP/catalyst support" was obtained after nucleation and
growth of NPs. Polyoxometalates (POMs) are a great class
of inorganic compounds as multi-core metal–oxygen
clusters [25]. If an atom named heteroatom (such as Si, P,
As, B, etc.) enters the molecular structure of POM in
addition to metal and oxygen, then heteropoly acids
(HPAs) will be obtained [26]. Thermodynamically, HPAs
have stable arrangements and maintain their crystal structure in aqueous and non-aqueous solutions. This class of
materials has various applications in catalysis [27], analytical chemistry [28], medicinal chemistry (anti-tumor,
anti-cancer, anti-bacteria, anti-microbial and anti-clotting)
[29–31], radioactive materials [32] and gas absorbents [33]
owing to their structural diversity and unique properties.
HPAs have different crystal structures of which α-, β-, γ-,

δ- and ε-Keggin, Wells–Dawson, Preysler, Stromberg and
Anderson–Evans are served as critical types.
12-tungstosilicic acid (hereafter, 12-TSA) is a HPA with
formula $H_4SiW_{12}O_{40}$ and α-Keggin crystal structure (see
Fig. 2). The central Si heteroatom is surrounded by a
tetrahedron whose oxygen vertices are each linked to one
of the four W_3O_{13} sets. Each W_3O_{13} set consists of three
W_3O_6 octahedrals linked in a triangular arrangement by
sharing edges and the four W_3O_{13} are linked together by
sharing corners [34]. So far, numerous experimental studies
have been done about supporting HPAs on the surface of
various organic and inorganic catalyst supports, but HPAs
have rarely been used as catalyst support [35–38].

12-TSA has suitable physical and chemical properties to
be used as a catalyst support. The pores existed on the
crystalline surface of 12-TSA provide a suitable condition
to support NPs [39]. To optimize a process like the photocatalytic degradation process, it is essential to study all
factors influencing the process. But studying the effects of
individual factors on the process is difficult and time-
consuming, especially if these factors are not independent
and they affect each other. Employing experimental design
could eliminate these problems because the interaction
effects of different factors could be attained using DoEs
only. Full factorial is an appropriate method for DoEs
because it could reduce the total number of experiments as
well as optimize the process by optimizing all the affecting
factors collectively, at a time [40]. The design could
determine the effect of each factor on the response as well
as how this effect varies with the change in level of other
factors.

Various crystal structures of α-Fe$_2$O$_3$ NPs including rod-
shape [21], spherical and elliptical forms [41] have been
synthesized and identified until now. In this work, spherical
α-Fe$_2$O$_3$ NPs are supported through two different SSD and
FHRC methods on the surface of 12-TSA·7H$_2$O (α-Fe$_2$O$_3$/
12-TSA·7H$_2$O). Then, the performance of pure and supported α-Fe$_2$O$_3$ NPs on the TC photocatalytic degradation
was investigated using full factorial experimental design.

Experimental

Material and apparatuses

All chemicals used in this work including sodium tungstate
dihydrate, sodium silicate, diethyl ether, iron (III) chloride
hexahydrate, urea, hydrogen peroxide (30% pure),
hydrochloric acid (37% pure), sulfuric acid (96% pure),
sodium hydroxide and ethanol were purchased from Merck
and were used without further purification. The required
TC was purchased from Razak pharmaceutical laboratory
(Tehran, Iran). Also, deionized water was used throughout

Fig. 2 α-Keggin structure of $[SiW_{12}O_{40}]^{4-}$

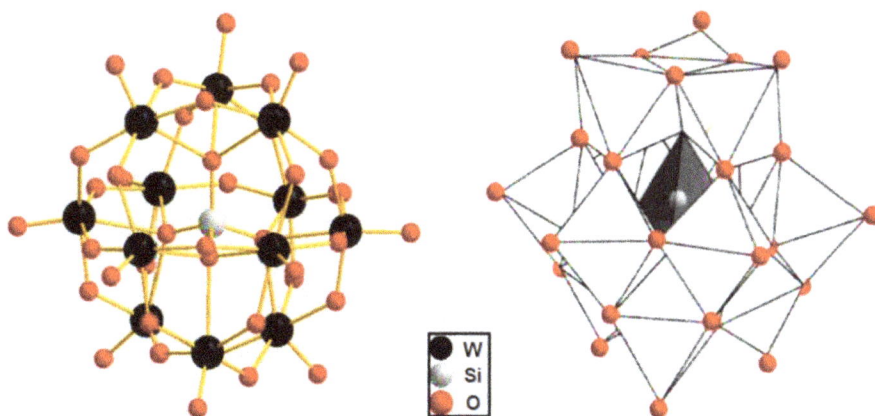

W
Si
O

the experiments. The Fourier transform infra-red (FTIR) spectra of products were recorded on a Perkin-Elmer spectrophotometer (Spectrum Two, model) in the range of $450-4000 \ cm^{-1}$. The shape, size and surface morphology of the synthesized 12-TSA·7H$_2$O and α-Fe$_2$O$_3$/12-TSA·7H$_2$O were examined using the obtained images of a Philips XL-30 scanning electron microscope (SEM). The X-ray diffraction (XRD) analysis of the samples was done using a DX27-mini diffractometer. BET surface area of materials was determined by N$_2$ adsorption–desorption method at 77 K, measured using a BELSORP-mini II instrument. The samples were degassed under vacuum at 473 K for 12 h before the BET measurement. All ultraviolet/visible (UV/Vis) absorption spectra were obtained using an Agilent 8453 spectrophotometer and the pH values were determined by a Metrohm pH meter model 827. Likewise, to separate the catalyst from samples, an ALC 4232 centrifuge was employed.

Synthesis of α-Fe$_2$O$_3$ NPs

The synthesis of α-Fe$_2$O$_3$ NPs was carried out according to Bharathi et al. [21]. Firstly, 100 ml iron (III) chloride hexahydrate 0.25 M which was considered as a source of Fe^{3+}, was poured into a flat-bottom flask. When Iron solution was agitated by stirrer, it was added drop by drop to it 100 ml urea 1 M (as a supplying agent of hydroxyl ions). The more gentle and regular adding urea, the smaller and more uniform-sized formed α-Fe$_2$O$_3$ particles will be. The obtained mixture was stirred for 30 min and then placed under the reflex at 90–95 °C for 12 h. Then, the precipitate after separation was washed with 100 ml deionized water because unreacted ions will be completely removed. The washed precipitate was dried at 70 °C for 2 h. Having fully dried, one light brown solid (iron hydroxide) was yielded. Finally, this solid remained at 300 °C for 1 h; hence the iron hydroxide particles will

transform to iron oxide. Consequently, a dark brown solid of α-Fe$_2$O$_3$ was obtained.

Synthesis of 12-TSA·7H$_2$O

12-TSA·7H$_2$O was synthesized according to literature procedure [42]. Firstly, 15 g sodium tungstate dihydrate was dissolved in 30 ml deionized water and then 1.16 g sodium silicate solution (density 1.375 g/ml) was added to it. The resulted mixture was heated up to about boiling point, and while it was stirred, 10 ml concentrated HCl was added to it during 30 min, smoothly. Then, the solution was naturally cooled down to RT and slight precipitate formed (silicic acid) in it was filtered. Again, 5 ml concentrated HCl was added to the solution and was transferred to separatory funnel after cooling it again down to RT. Then, 12 ml diethyl ether was added to it and well shaken. Therefore, three layers were formed inside separatory funnel, middle layer of which was yellow-colored. Bottom layer which was oily ether was separated and transferred into a beaker. To further extract, separatory funnel was further shaken again and the bottom layer was once more separated and transferred into the beaker. This extraction process was done so much that the yellow color of middle layer was fully faded. The extracted ether complex which was inside the beaker was transferred to another separatory funnel and then 16 ml HCl 25% (v/v) was added to it. Next, 4 ml diethyl ether was added to it, subsequently. The contents inside separatory funnel were shaken and bottom layer (ether) was transferred to the evaporating dish after separating. Evaporating dish was exposed to air and remained motionless to evaporate the solvent and form the 12-TSA·7H$_2$O crystals. Finally, 12-TSA·7H$_2$O formed crystals were placed at 70 °C for 2 h until it was completely dried. The chemical reaction occurred in the process of 12-TSA·7H$_2$O synthesis has been shown in (1) [42].

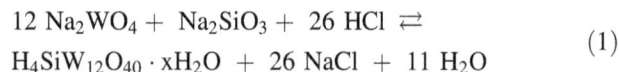

$$12\ Na_2WO_4 +\ Na_2SiO_3 +\ 26\ HCl\ \rightleftarrows$$
$$H_4SiW_{12}O_{40} \cdot xH_2O\ +\ 26\ NaCl\ +\ 11\ H_2O \tag{1}$$

Preparation of α-Fe$_2$O$_3$/12-TSA·7H$_2$O

SSD method

Firstly, the synthesized iron hydroxide (light brown solid) and 12-TSA·7H$_2$O catalyst support were mixed with weight ratio of 1:3 iron hydroxide/12-TSA·7H$_2$O (weight of catalyst support is three times of catalyst weight) using an agate pestle and mortar for 1 h. To have better mixture, ethanol was sprayed on the mixture until it becomes dough-form. During mixing, in the vaporization phase, ethanol is again added in order to keep the dough-form of the mixture. The resulted mixture was dried under air for 1 h and then was kept at 80 °C for 2 h. To do calcination and transform iron hydroxide particles fixed on the surface of 12-TSA·7H$_2$O into iron oxide (α-Fe$_2$O$_3$), the obtained solid was kept at 300 °C for 1 h.

FHRC method

Firstly, 50 ml iron (III) chloride hexahydrate 0.25 M was poured into a beaker. While it was agitated by stirrer, 3.5 g 12-TSA·7H$_2$O was gently added to it. The obtained mixture was stirred for 4–5 h. Then, stirring was stopped for 2 h until the solid within mixture was deposited. The solid accumulated at bottom of beaker was separated and transferred into one flat-bottom flask and the same 10 ml solution inside beaker was added to it. When mixture inside flat-bottom flask was being stirred, 50 ml urea 1 M was gradually added to it. The mixture was placed under reflux at 90–95 °C for 12 h. Then, the precipitate resulted after separation was washed with 100 ml ethanol/deionized water 1:1 solution because unreacted ions were completely removed. The washed precipitate was dried in the air for 2 h and then was kept at 80 °C for 2 h. In order to calcination, the obtained solid was kept at 300 °C for 1 h.

Full factorial experimental design

The photocatalytic efficiency of pure α-Fe$_2$O$_3$ NPs and α-Fe$_2$O$_3$/12-TSA·7H$_2$O prepared by SSD and FHRC methods on the TC degradation were investigated using DoE. The experiments were designed considering four variables including pH, the initial concentration of TC, catalyst concentration and H$_2$O$_2$ concentration at three levels. Experimental range and levels of variables are shown in Table 1. pH varied from 4 to 8 at three levels (4, 6 and 8), the initial concentration of TC from 30 to 70 ppm at three levels (30, 50 and 70 ppm), catalyst concentration from 50 to 150 ppm at

Table 1 Experimental range and levels of the variables

Variables	Range and levels		
	−1	0	+1
pH	4	6	8
Initial con. of TC (ppm)	30	50	70
Catalyst con. (ppm)	50	100	150
H$_2$O$_2$ con. (ppm)	0.1	0.3	0.5

three levels (50, 100 and 150 ppm) and H$_2$O$_2$ concentration from 0.1 to 0.5 ppm at three levels (0.1, 0.3 and 0.5 ppm). In Table 2, 19 experiments related to this factorial design and their experimental conditions have been listed. The removal efficiency of TC was a dependent response. In order to do DoEs, Minitab 16 version 16.2.0 statistical software was utilized. Also, analysis of variance (ANOVA) was run to analyze the results.

General procedure for photocatalytic degradation of TC

Figure 3 shows one schematic diagram of photocatalytic reactor used in the work. An MDF box was designed inside which a circular Pyrex reactor with 300 ml capacity was placed. On the upper section of the box, three mercury lamps (Philips 15 W) were built-in as UV light sources. The radiation is generated almost exclusively at 254 nm. These lamps were set up with the same intervals, so light was evenly radiated on the whole liquid surface inside the reactor. The liquid inside the reactor was agitated by magnetic stirrer and the air inside the box was conditioned by a fan (built-in at back of box). In order to carry out each experiment (according to Table 2), firstly 250 ml TC solution was made as specified concentration and poured inside the reactor. Then, at related pH, the specified amount of photocatalyst and H$_2$O$_2$ were added to the solution inside the reactor. In all experiments, pH adjustment was done via minimum use of H$_2$SO$_4$ and NaOH. Then, stirrer and UV lamps were immediately turned on to initiate the process. Sampling was done by a 5 ml syringe, every 10 min. To fully separate the catalyst from solution, the samples were centrifuged for 3 min with 3500 rpm speed. The TC concentration of the samples was determined using a UV/Vis spectrophotometer at $\lambda_{max} = 357$ nm. The percentage of initial concentration of pollutant decomposed by the photocatalytic process or the percent of photodegradation efficiency ($x\%$) as a function of time is given by

$$x\% = \frac{C_0 - C}{C_0} \times 100 \tag{2}$$

where C_0 and C are the concentration of TC (ppm) at $t = 0$ and t, respectively.

Table 2 Experimental conditions for photocatalytic process

Exp. no.	Variables			
	pH	Initial con. of TC (ppm)	Catalyst con. (ppm)	H$_2$O$_2$ con. (ppm)
1	−1	−1	−1	−1
2	+1	−1	+1	+1
3	−1	−1	+1	−1
4	−1	+1	+1	−1
5	+1	+1	+1	−1
6	0	0	0	0
7	+1	−1	+1	−1
8	−1	+1	−1	+1
9	+1	−1	−1	−1
10	+1	+1	+1	+1
11	−1	+1	−1	−1
12	+1	−1	−1	+1
13	+1	+1	−1	−1
14	−1	+1	+1	+1
15	+1	+1	−1	+1
16	0	0	0	0
17	−1	−1	+1	+1
18	0	0	0	0
19	−1	−1	−1	+1

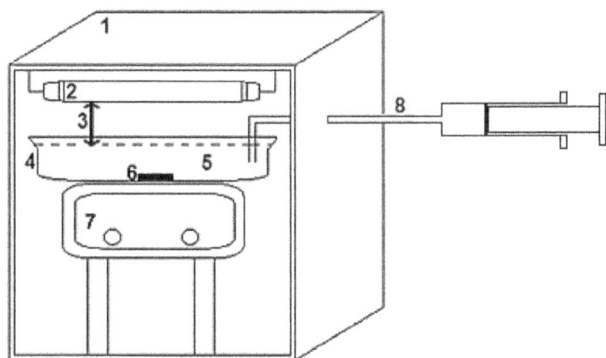

Fig. 3 Schematic diagram of photocatalytic reactor. *1* MDF box, 50 × 50 × 50 cm; *2* Mercury lamps, Philips 15 W; *3* The distance between surface of TC solution and lamps, 5 cm; *4* Reactor, 300 ml capacity; *5* TC solution, 250 ml; *6* Magnet; *7* Magnetic stirrer; *8* Sampling port

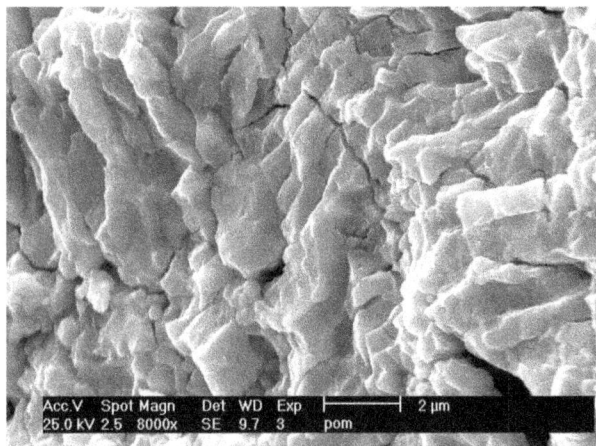

Fig. 4 SEM image of the synthesized 12-TSA·7H$_2$O

Results and discussion

Characterization

The synthesized 12-TSA·7H$_2$O

SEM image of the synthesized 12-TSA·7H$_2$O is shown in Fig. 4. Surface morphology of 12-TSA·7H$_2$O shows that this product has suitable structural properties and can be regarded as a catalyst support. In other words, the pores existed on the surface of this catalyst support provide a suitable condition to support α-Fe$_2$O$_3$ NPs. IR is a suitable method for the structural characterization of HPAs [26]. FTIR spectrum of the synthesized 12-TSA·7H$_2$O has been shown in Fig. 5a. There are four kinds of oxygen atoms in 12-TSA·7H$_2$O structure, 4 Si–O$_a$ in which one oxygen atom connects to Si, 12 W–O$_b$–W oxygen bridges (corner-sharing oxygen-bridge between different W$_3$O$_{13}$ groups), 12 W–O$_c$–W oxygen bridges (edge-sharing oxygen-bridge within W$_3$O$_{13}$ groups) and 12 W=O$_d$ terminal oxygen atoms. The symmetric and asymmetric stretching of the different kinds of W–O bonds are observed in the following spectral regions: Si–O_a bonds (1020 cm^{-1}), W = O_d bonds (1000–960 cm^{-1}), W–O_b–W bridges (890–850 cm^{-1}), W–O_c–W bridges (800–760 cm^{-1}) [43]. In Table 3, vibrational frequencies of the synthesized 12-TSA·7H$_2$O and equivalent values reported in previous studies [43, 44] have been listed. Comparing the vibrational frequencies reveals that 12-TSA·7H$_2$O has been well synthesized. XRD is one of the most important characterization tools used in solid state chemistry and materials science. Figure 6a shows the XRD pattern of the synthesized 12-TSA·7H$_2$O. This pattern indicates that the characteristic peaks corresponded to the 12-TSA were well appeared and it means that the synthesized 12-TSA·7H$_2$O crystals were well formed [44].

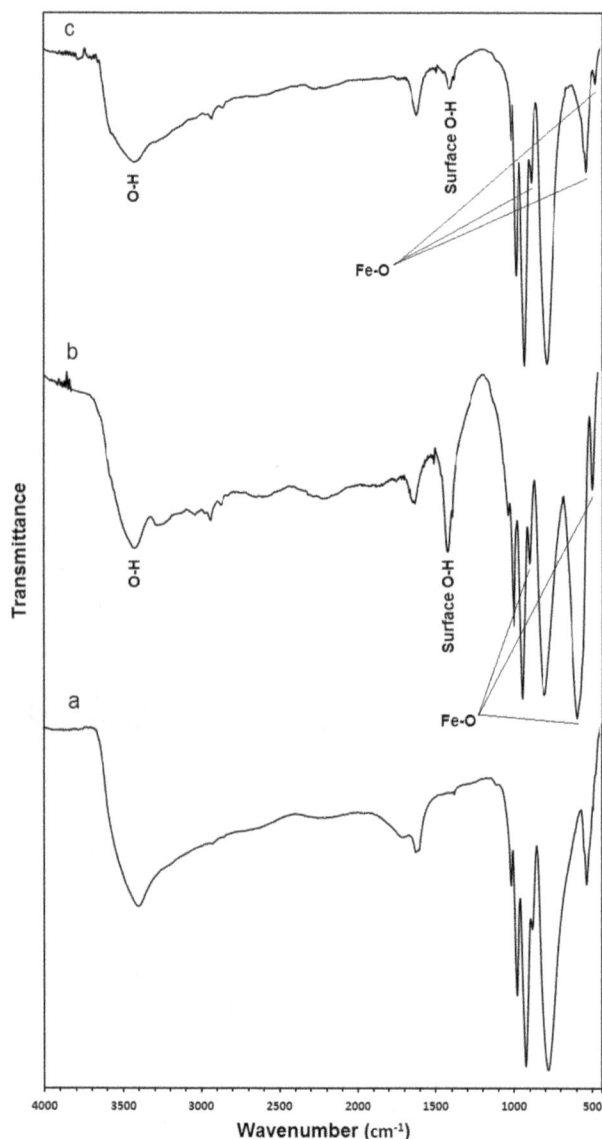

Fig. 5 FTIR spectra of the synthesized 12-TSA·7H$_2$O (*a*) and α-Fe$_2$O$_3$/12-TSA·7H$_2$O prepared by SSD (*b*) and FHRC (*c*) methods

The prepared α-Fe$_2$O$_3$/12-TSA·7H$_2$O

Figures 7 and 8 show SEM/EDX images of α-Fe$_2$O$_3$/12-TSA·7H$_2$O prepared by SSD and FHRC methods, respectively. These images indicate that in both methods, α-Fe$_2$O$_3$ particles were spherically supported on the surface of 12-TSA·7H$_2$O. The spheres in SSD method are bigger and have covered more area of 12-TSA·7H$_2$O than that of FHRC method. Possibly in SSD method, spherical α-Fe$_2$O$_3$ particles are adhered to each other and bigger spheres have formed while it did not occur in FHRC method and α-Fe$_2$O$_3$ particles were separately supported. It is assumed that the causes of this phenomena are as follows: (1) possibly, α-Fe$_2$O$_3$ synthesized particles by SSD method are smaller than that of FHRC method and this contributed to their adherence, (2) Supporting through SSD method is done in solid state and this increases the possibility of particles adhering to each other and forming bigger spheres and (3) supporting through FHRC method is done in liquid phase, so the particles could freely move and be separately fixed on the 12-TSA·7H$_2$O surface. In Fig. 5b, c, FTIR spectra of α-Fe$_2$O$_3$/12-TSA·7H$_2$O prepared by SSD and FHRC methods have been shown, respectively. It is clear that absorption peaks of 12-TSA·7H$_2$O have appeared without considerable change in the wavenumbers (only their intensities have been slightly changed). It means that in both methods, 12-TSA·7H$_2$O was stable and it had not been changed chemically during preparing α-Fe$_2$O$_3$/12-TSA·7H$_2$O. Also, absorption peaks of α-Fe$_2$O$_3$ have well appeared and are in agreement with results of Bharati et al. [21]. These absorption peaks which are related to stretching and bending modes of OH and Fe–O binding in FeOOH, in some cases overlapped with absorption peaks of 12-TSA·7H$_2$O. Comparing FTIR spectra reveals that absorption peaks of α-Fe$_2$O$_3$ related to SSD method are more intense than that of FHRC method. This partly confirms the results of SEM images. Hence in SSD method, surface of 12-TSA·7H$_2$O has been covered by more α-Fe$_2$O$_3$ particles. In Fig. 6b, c, XRD patterns of α-Fe$_2$O$_3$/12-TSA·7H$_2$O prepared by SSD and FHRC methods have been illustrated, respectively. In both of these patterns, characteristic peaks of 12-TSA·7H$_2$O have well appeared which indicates that 12-TSA·7H$_2$O was stable during the supporting process in both SSD and FHRC methods. In these patterns, the characteristic peaks of α-Fe$_2$O$_3$ which have also been marked have appeared and it is in agreement with results of Bharati et al. [21]. In XRD related to SSD method, intensity of 12-TSA·7H$_2$O and α-Fe$_2$O$_3$ characteristic peaks is lower and higher than that of FHRC method, respectively. This issue confirms the results of SEM and FTIR, so during supporting through SSD method, 12-TSA·7H$_2$O surface has been covered by the greater amount of α-Fe$_2$O$_3$ particles. The size of spherical α-Fe$_2$O$_3$

Table 3 Vibrational frequencies of the synthesized 12-TSA·7H$_2$O and equivalent values reported in previous reports

Number	The synthesized 12-TSA·7H$_2$O		[43, 44]
	Wavenumber (cm^{-1})	Transmittance %	
1	1019.04	13.29	1020 (weak)
2	980.68	8.81	981 (sharp)
3	924.31	5.92	928 (very sharp)
4	882.63	11.52	880 (medium)
5	780.28	5.77	785 (very sharp)
6	537.41	13.35	540 (medium)

Fig. 6 X-ray diffractogram of the synthesized 12-TSA·7H$_2$O (*a*), α-Fe$_2$O$_3$/12-TSA·7H$_2$O prepared by SSD (*b*) and FHRC (*c*) methods

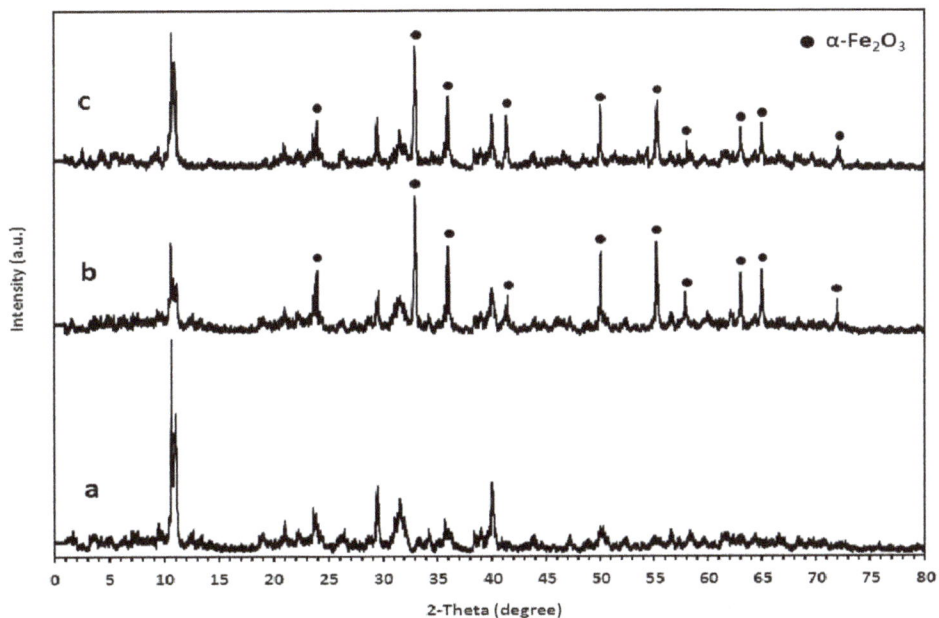

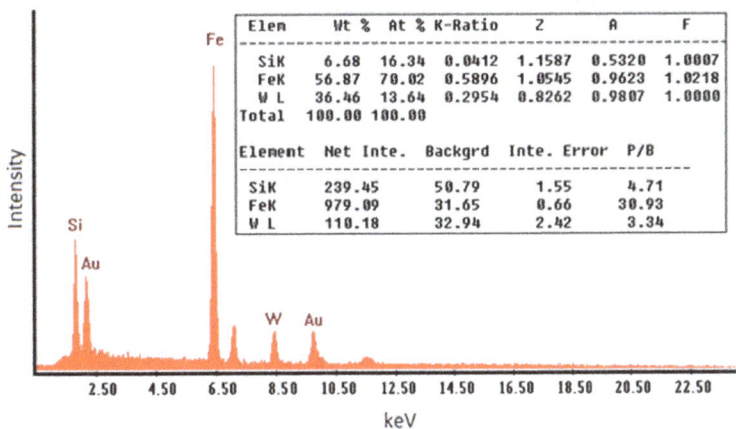

Fig. 7 SEM image and EDX results of α-Fe$_2$O$_3$/12-TSA·7H$_2$O prepared by SSD method

particles supported on the surface of 12-TSA·7H$_2$O were calculated using XRD and Warren–Averbach method (taking account of device errors) whose averages for SSD and FHRC methods were 50.5 and 70.82 nm, respectively. The BET surface area of catalyst prepared by SSD and FHRC methods were determined 57.53 and 39.84 (m^2/g), respectively. It seems that the high amount of iron oxide formed on the base has been increase the BET surface area of catalyst prepared with SSD method.

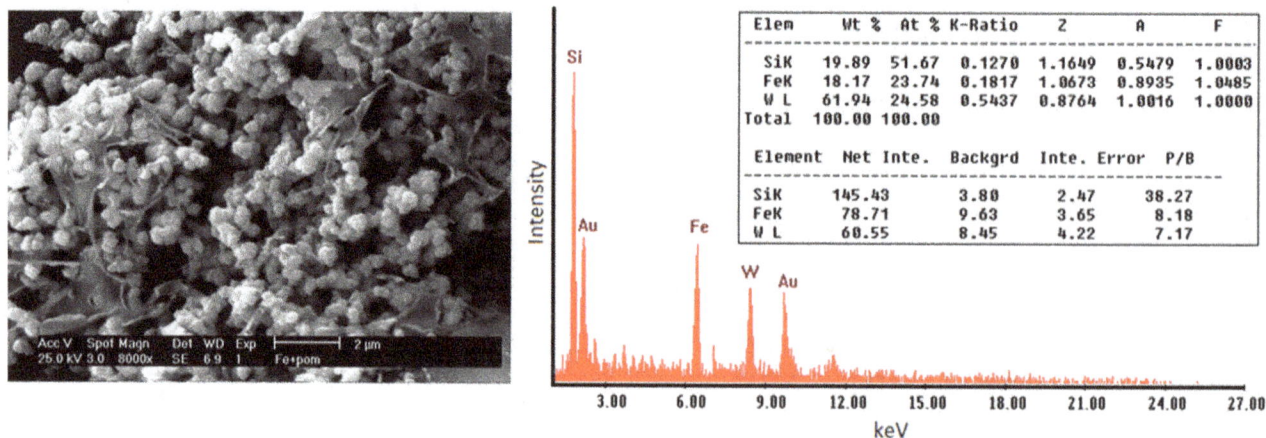

Elem	Wt %	At %	K-Ratio	Z	A	F
SiK	19.89	51.67	0.1270	1.1649	0.5479	1.0003
FeK	18.17	23.74	0.1817	1.0673	0.8935	1.0485
W L	61.94	24.58	0.5437	0.8764	1.0016	1.0000
Total	100.00	100.00				

Element	Net Inte.	Backgrd	Inte. Error	P/B
SiK	145.43	3.80	2.47	38.27
FeK	78.71	9.63	3.65	8.18
W L	60.55	8.45	4.22	7.17

Fig. 8 SEM image and EDX results of α-Fe$_2$O$_3$/12-TSA·7H$_2$O prepared by FHRC method

UV/Vis spectra

The absorbance of TC solutions during photocatalytic process (using α-Fe$_2$O$_3$/12-TSA·7H$_2$O prepared by SSD method and according to exp. no. 8) at initial and after 10, 20, 30, 40 and 50 min irradiation time verses wavelength are depicted in Fig. 9. In all experiments, $x\%$ was calculated at $\lambda_{max} = 357$ nm. The wavelength of maximum absorbance (in 357 nm) did not change with time, then this wavelength for measuring the concentration of pollutants was chosen. Furthermore, absorbance changes in 357 nm were completely regular and measurable.

Performance of photocatalysts

Having carried out all experiments based on Table 2, $x\%$ values were calculated at $\lambda_{max} = 357$ nm following 50 min after reaction which have been reported in Table 4. In general, comparing $x\%$ values reveals that the degree of TC photocatalytic degradation by pure α-Fe$_2$O$_3$ is lower than that of α-Fe$_2$O$_3$/12-TSA·7H$_2$O prepared through SSD and

FHRC methods. This means that supporting α-Fe$_2$O$_3$ NPs leads to increase their photocatalytic activity. Also, comparing the results of SSD and FHRC methods indicates that α-Fe$_2$O$_3$/12-TSA·7H$_2$O prepared through SSD method was effective from the aspect of TC photocatalytic degradation and has yielded more $x\%$ values. Comparing $x\%$ values in one series of experiments (1 through 19) shows that the highest degradation percentage has been obtained in exp. no. 7. To better compare the results, $x\%$ histogram versus experiment number for pure α-Fe$_2$O$_3$ and α-Fe$_2$O$_3$/12-TSA·7H$_2$O prepared through SSD and FHRC methods has been shown in Fig. 10. The histogram clearly indicates that in all experiments α-Fe$_2$O$_3$ NPs supported on the surface of 12-TSA·7H$_2$O (particularly through SSD method) had more photocatalytic efficacy and has degraded more TC.

Photocatalytic mechanism

According to exp. no. 7, the effects of UV irradiation, pure α-Fe$_2$O$_3$ NPs and α-Fe$_2$O$_3$/12-TSA·7H$_2$O prepared by two different SSD and FHRC methods on the photodegradation

Fig. 9 UV/Vis spectral absorption changes of TC solution photodegraded by α-Fe$_2$O$_3$/12-TSA·7H$_2$O prepared through SSD method (pH 4, Initial concentration of TC = 70 ppm, catalyst concentration = 50 ppm, H$_2$O$_2$ concentration = 0.5 ppm)

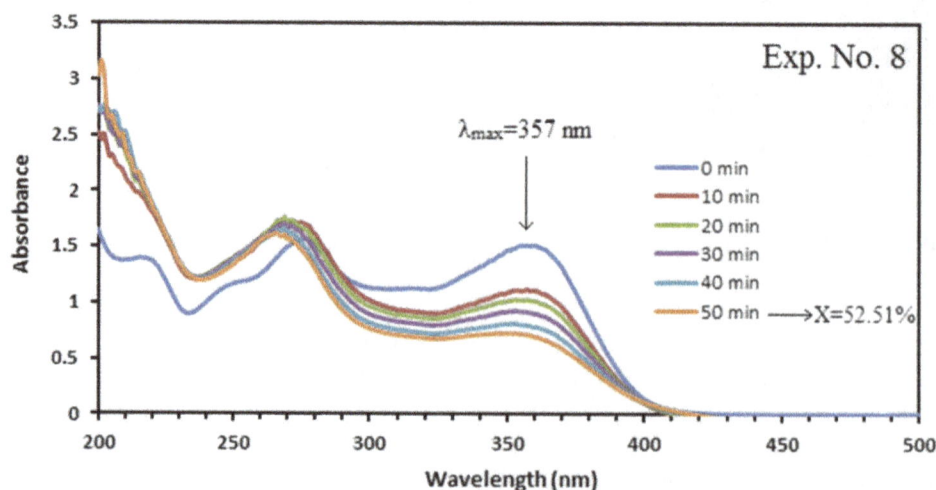

Table 4 $x\%$ values after 50 min photodegradation process at $\lambda_{max} = 357$ nm

Exp. no.	$x\%$		
	Pure α-Fe$_2$O$_3$ NPs	α-Fe$_2$O$_3$/12-TSA· 7H$_2$O prepared by SSD method	α-Fe$_2$O$_3$/12-TSA· 7H$_2$O prepared by FHRC method
1	64.11	78.59	66.32
2	75.39	93.61	85.95
3	67.56	85.35	73.29
4	48.83	62.74	53.22
5	37.14	60.51	45.07
6	36.97	69.61	60.38
7	82.17	97.39^a	88.44
8	37.95	52.51	42.66
9	65.46	84.34	74.61
10	44.37	58.94	47.92
11	32.84	47.91	38.91
12	62.00	91.48	74.28
13	29.84	47.06	37.71
14	40.69	62.45	48.72
15	39.31	55.34	45.62
16	37.02	69.75	59.79
17	66.83	82.21	77.13
18	36.83	69.44	60.13
19	65.83	87.17	81.84

^{a} Maximum value of $x\%$

of TC are presented in Fig. 11. This Figure designates that in the presence of α-Fe$_2$O$_3$/12-TSA·7H$_2$O prepared by SSD method and UV irradiation 97.39% of TC was degraded at the reaction time of 50 min while it was 88.44, 82.17 and 10.2% for α-Fe$_2$O$_3$/12-TSA·7H$_2$O prepared by FHRC method, pure α-Fe$_2$O$_3$ NPs and only UV, respectively. When α-Fe$_2$O$_3$ is illuminated by the light, electrons are promoted from the valence band (VB) to the conduction band (CB) of the semi conducting oxide to give electron–hole pairs. The VB potential (h_{VB}) is positive enough to generate hydroxyl radicals at the surface, and the CB potential (e_{CB}) is negative enough to reduce molecular oxygen. The hydroxyl radical is a powerful oxidizing agent and attacks TC molecules present at or near the surface of α-Fe$_2$O$_3$. It causes the photo-oxidation of TC according to the following reactions [45–50]:

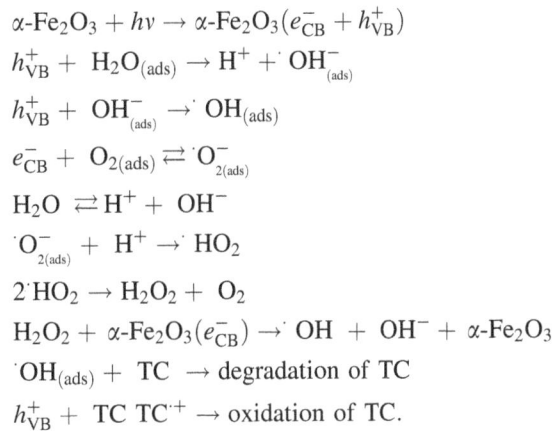

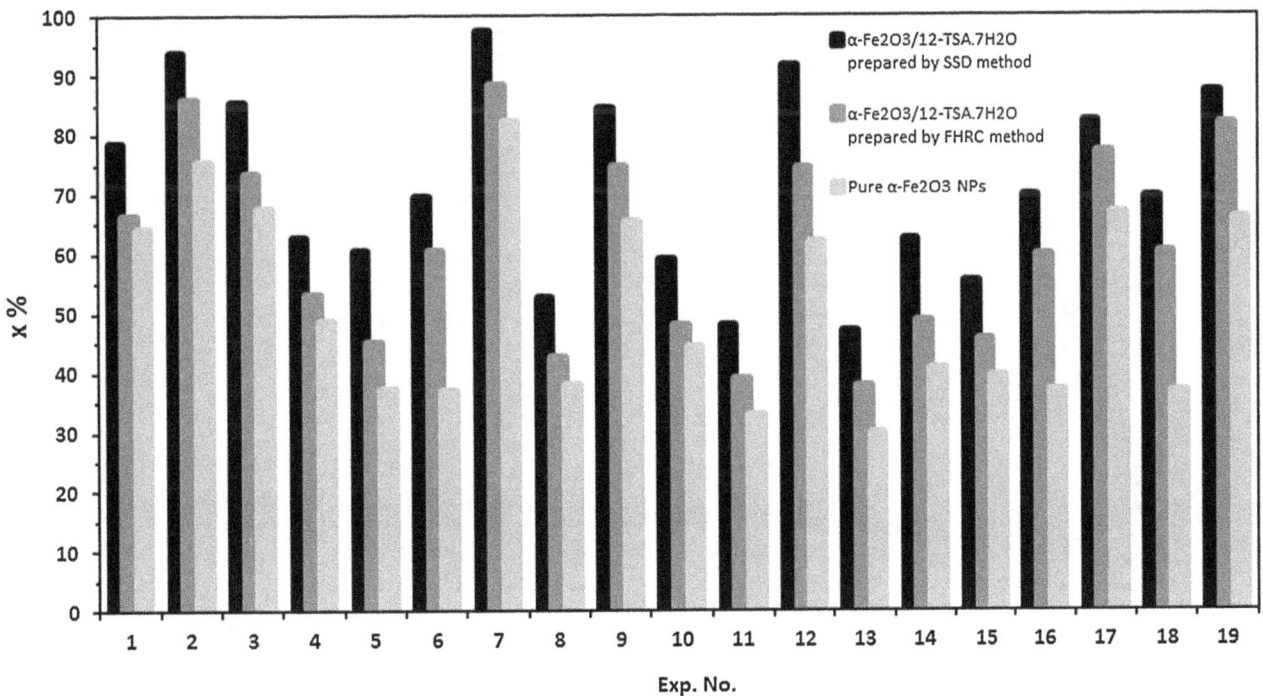

$$\alpha\text{-Fe}_2\text{O}_3 + h\nu \rightarrow \alpha\text{-Fe}_2\text{O}_3(e_{CB}^- + h_{VB}^+)$$

$$h_{VB}^+ + H_2O_{(ads)} \rightarrow H^+ + \cdot OH_{(ads)}^-$$

$$h_{VB}^+ + OH_{(ads)}^- \rightarrow \cdot OH_{(ads)}$$

$$e_{CB}^- + O_{2(ads)} \rightleftarrows \cdot O_{2(ads)}^-$$

$$H_2O \rightleftarrows H^+ + OH^-$$

$$\cdot O_{2(ads)}^- + H^+ \rightarrow \cdot HO_2$$

$$2 \cdot HO_2 \rightarrow H_2O_2 + O_2$$

$$H_2O_2 + \alpha\text{-Fe}_2\text{O}_3(e_{CB}^-) \rightarrow \cdot OH + OH^- + \alpha\text{-Fe}_2\text{O}_3$$

$$\cdot OH_{(ads)} + TC \rightarrow \text{degradation of TC}$$

$$h_{VB}^+ + TC \; TC^{\cdot +} \rightarrow \text{oxidation of TC.}$$

Fig. 10 $x\%$ values versus experiment number

The mechanism is summarized in Fig. 12. The main role of the foundation is creating the perfect conditions for putting the TC and hydroxyl radical beside each other. Photocatalytic activity increased after stabilizing iron oxide on 12-TSA·7H$_2$O. To comment on this result, we propose that the hydroxyl radicals, on the surface of iron oxide, are easily transferred onto the surface of 12-TSA·7H$_2$O. That means the organic pollutants such as TC, which have already been adsorbed on the nonphotoactive 12-TSA·7H$_2$O, have chances to be degraded due to the appearance of hydroxyl radicals, resulting in the enhancement of the photodegradation performance of α-Fe$_2$O$_3$/12-TSA·7H$_2$O (as shown in Fig. 12b).

Kinetics of photocatalytic degradation of TC

Figure 13 displays the plot of ln(C_0/C) versus reaction time for TC. The linearity of the plot suggests that the photodegradation reaction approximately follows the pseudo-first order kinetics with a rate coefficient $k = 0.0098$ min^{-1}.

The statistical analysis (optimum conditions)

Since α-Fe$_2$O$_3$ NPs supported through SSD method have shown more effective than other photocatalysts from the

view of the TC photocatalytic degradation, then in this section we carry out the statistical results analysis of the photocatalytic process in which α-Fe$_2$O$_3$/12-TSA·7H$_2$O prepared by SSD method has been utilized. Analysis of variance (ANOVA) is a set consists of a number of statistical methods used to analyze the differences among group means and their associated procedures. ANOVAs are useful for testing three or more means variables for statistical significance. ANOVA was used for graphical analyses of the data to obtain the interaction between the process variables and the responses. The quality of the fit polynomial model was expressed by the coefficient of determination R^2, and its statistical significance was checked by the Fisher's F test in the same program. Model terms were evaluated by the P value. In Table 5, the estimated effects and coefficients for $x\%$ have been listed. In this table, standard deviation (S), correlation coefficient, pried R^2 and adjusted R^2 values were also reported. The square of the correlation coefficient for each response was computed as the coefficient of determination (R^2). The accuracy and variability of the model can be evaluated by R^2. The R^2 value is always between 0 and 1. The closer the R^2 value to 1, the stronger the model is and the better the model predicts the response ($x\%$). R^2 value was reported to be 0.9915 in this paper. The "pried R^2" of 0.9662 is in reasonable agreement with the "adj R^2" of 0.9848,

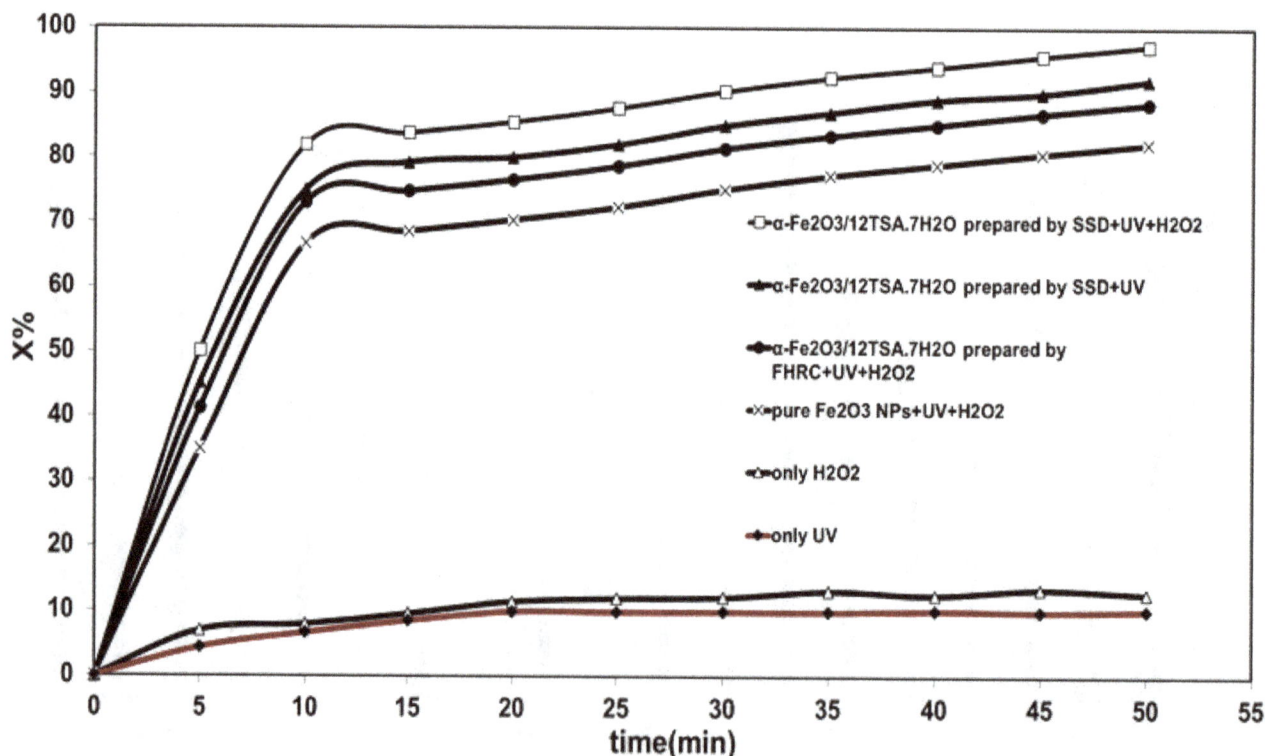

Fig. 11 Effect of UV light, pure α-Fe$_2$O$_3$ NPs and α-Fe$_2$O$_3$/12-TSA·7H$_2$O prepared by FHRC and SSD methods on TC degradation (pH 8, Initial concentration of TC = 30 ppm, catalyst concentration = 150 ppm, H$_2$O$_2$ concentration = 0.1 ppm)

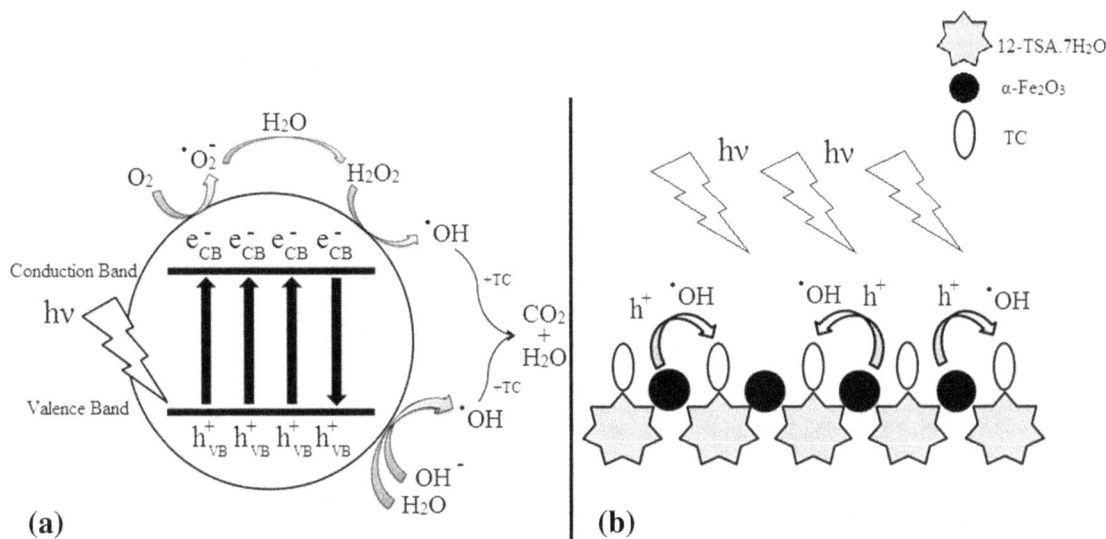

Fig. 12 General mechanism of the photocatalysis (**a**) and photocatalytic activity of α-Fe$_2$O$_3$/12-TSA·7H$_2$O (**b**)

confirming good predictability of the model. Due to Table 5 and the significant variables effects on the response, affect magnitudes of the initial concentration of TC, pH, H$_2$O$_2$ concentration and catalyst concentration equal to 31.59, 3.72, 2.48 and 7.35, respectively. Thus, the significant reaction parameters were (the most to the least significant): initial concentration of TC > catalyst concentration > pH > and H$_2$O$_2$ concentration. Of course, it is necessary to note that despite other three variables, the variable of the initial concentration of TC has a negative effect on the response (−31.59). This means that increasing the initial concentration of TC leads to decrease $x\%$ and conversely. In this way, the effects about the variables interaction were reported in Table 5. As can be seen from these results, it is the only interaction of variables, namely the initial concentration of TC and the catalyst concentration which have positive effects (3.10). The interaction of the initial concentration of TC with pH and the interaction of H$_2$O$_2$ concentration with catalyst concentration have

both negative and roughly the same effects on the $x\%$ value (−4.67 and −4.66, respectively). In Table 5, the coefficients of each term have been reported which are the same term coefficients in response function which they will be given in the following. It is vital to note that P values have been assessed considering Alpha $(\alpha) = 0.05$. Table 6 depicts the results of ANOVA. The effect on the response was increased by increasing the value of F parameter and decreasing P parameter. For main effects (with 4 degrees of freedom) including the initial concentration of TC, pH, H$_2$O$_2$ concentration and catalyst concentration, F and P values have obtained as 278.34 and <0.0001, respectively. Besides, these values were 18.41 and <0.0001 for 2-way interactions (with 3 freedom degree), respectively. In Table 7, complementary results have been listed which have been used for drawing residual plots. Residual values were calculated from subtracting experimental $x\%$ values and fitted values.

In order to compare the variables effect (from the viewpoint of magnitude) on the response, the Fig. 14a could be investigated which is one Pareto chart of the standardized effects. In this Figure, those variables whose effects on response is negative (−) or positive (+) have been marked. The results revealed that the effect of the initial concentration of TC on the $x\%$ is greater than other variables effect (at least three times) but the effect of this variable is negative i.e. increasing or decreasing the initial concentration of TC leads to decrease and increase $x\%$, respectively. In order to better investigate the residual values, residual plot versus exp. no. has been illustrated in Fig. 14b. As it is seen, eight points (residuals) are located under zero line (negative), nine points above zero line (positive) and two points roughly on the zero line. Due to this and comparing distance of points from zero line, it

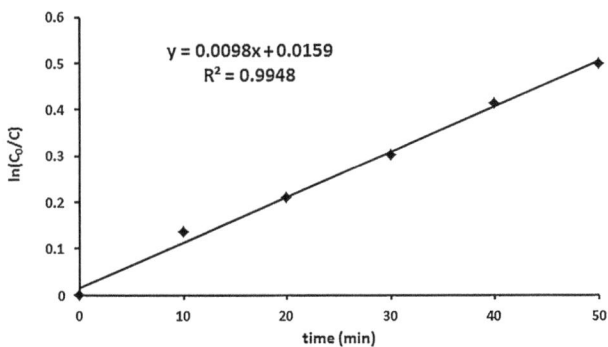

Fig. 13 Plot of reciprocal of pseudo-first order rate constant against initial concentration of TC = 70 ppm, catalyst concentration = 50 ppm, H$_2$O$_2$ concentration = 0.5 ppm and pH 4

Table 5 Estimated effects and coefficients for $x\%$

Term	Effect	Coef	SE Coef	T (Coef/SE Coef)	P value	Result
Constant	–	71.73	0.4905	146.22	<0.0001	Significant
Initial con. of TC	–31.59	–15.79	0.4905	–32.19	<0.0001	Significant
pH	3.72	1.86	0.4905	3.79	0.004	Significant
H_2O_2 con.	2.48	1.24	0.4905	2.52	0.03	
Catalyst con.	7.35	3.68	0.4905	7.49	<0.0001	Significant
Initial con. of TC × pH	–4.66	–2.33	0.4905	–4.75	0.001	Significant
Initial con. of TC × catalyst con.	3.10	1.55	0.4905	3.16	0.010	
H_2O_2 con. × catalyst con.	–4.67	–2.34	0.4905	–4.76	0.001	Significant
Center point	–	–2.13	1.2345	–1.72	0.116	

$S = 1.96219$, $R^2 = 99.15\%$, Pred $R^2 = 96.62\%$, Adj $R^2 = 98.48\%$

Table 6 ANOVA results

Source	Degree of freedom	Seq SS	Adj SS	Adj MS	F value
Initial con. of TC	1	3990.68	3990.68	3990.68	1036.49
pH	1	55.25	55.25	55.25	14.35
H_2O_2 con.	1	24.54	24.54	24.54	6.37
Catalyst con.	1	216.12	216.12	216.12	56.13
Initial con. of TC × pH	1	86.73	86.73	86.73	22.53
Initial con. of TC × catalyst con.	1	38.55	38.55	38.55	10.01
H_2O_2 con. × catalyst con.	1	87.36	87.36	87.36	22.69

could be said that residual distribution is normal. An extremely useful procedure is to construct a normal probability plot of the residuals. If the underlying error distribution is normal, this plot will resemble a straight line. Figure 14c shows normal probability plot. In this plot, it is fully clear that residuals distribution is normal because points (especially central points) are close to straight line. If the model is correct and if the assumptions are satisfied, the residuals should be structureless; in particular, they should be unrelated to any other variable including the predicted response. A simple check is to plot the residuals versus the fitted values. Figure 14d displays plot of residuals versus fitted values. Mathematical model representing TC photocatalytic degradation in the range studied can be expressed by the following equation:

$$\text{Response} = x\%$$
$$= 71.73 - 15.79A + 1.86B + 1.24C$$
$$+ 3.68D - 2.33AB + 1.55AD - 2.34CD$$

where A, B, C and D are the initial concentration of TC, pH, H_2O_2 concentration and catalyst concentration, respectively.

In Fig. 15, the plots of main effects have been shown. These plots indicate that of four main effects, only the

Table 7 Residual values

Exp. no.	$x\%$	Fit	SE fit	Residual ($x\%$–fit)	St resid
1	78.59	77.6338	1.3875	0.9562	0.69
2	93.61	92.7296	1.3875	0.8804	0.63
3	85.35	86.5531	1.3875	–1.1977	–0.86
4	62.74	62.7282	1.3875	0.0118	0.01
5	60.51	61.7882	1.3875	–1.2782	–0.92
6	69.61	69.6000	1.1329	0.0100	0.01
7	97.39	94.9260	1.3875	2.4640	1.78
8	52.51	54.7502	1.3875	–2.2402	–1.61
9	84.34	86.0067	1.3875	–1.6667	–1.20
10	58.94	59.5918	1.3875	–0.6518	–0.47
11	47.91	47.5998	1.3875	0.3102	0.22
12	91.48	93.1572	1.3875	–1.6778	–1.21
13	47.06	46.6598	1.3875	0.4002	0.29
14	62.45	60.5318	1.3875	1.9182	1.38
15	55.34	53.8102	1.3875	1.5298	1.10
16	69.75	69.6000	1.1329	0.1500	0.09
17	82.21	84.3567	1.3875	–2.1467	–1.55
18	69.44	69.6000	1.1329	–0.1600	–0.10
19	87.17	84.7843	1.3875	2.3882	1.72

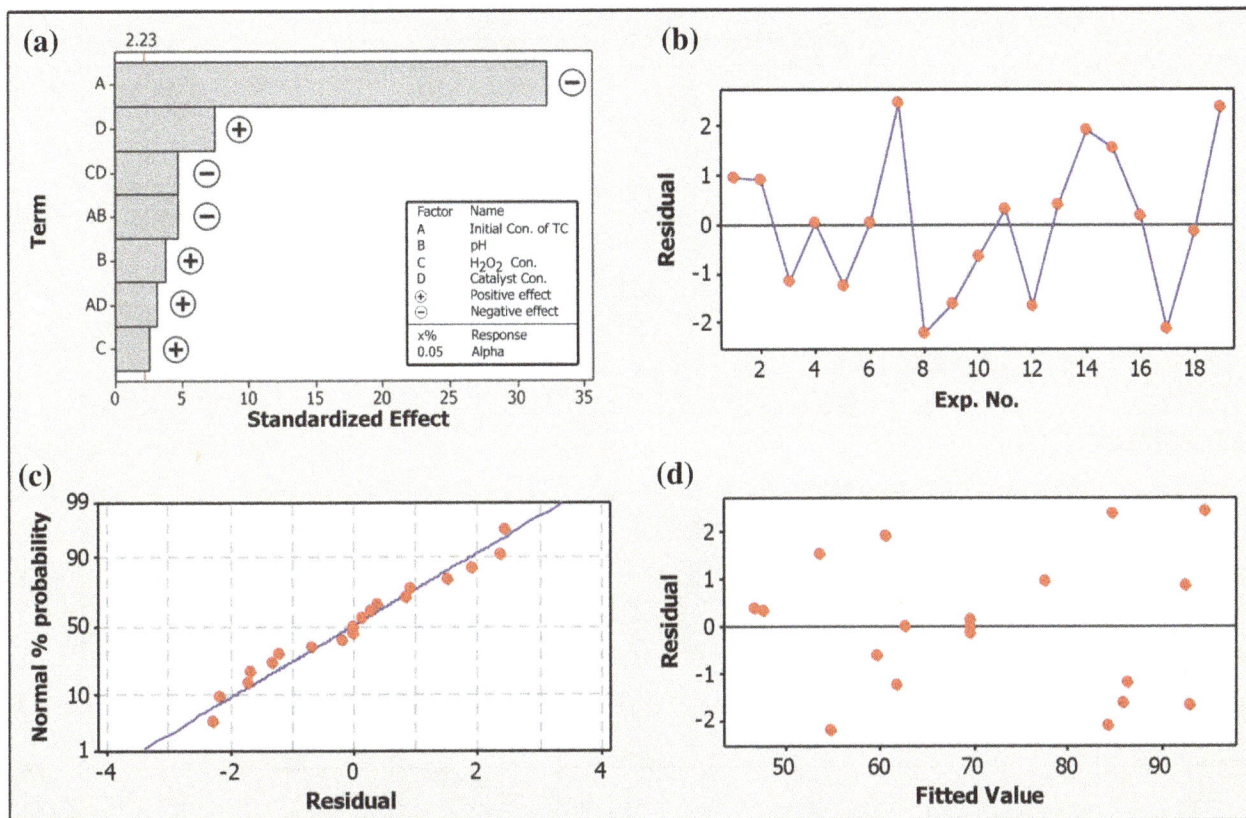

Fig. 14 a Pareto chart of the standardized effects, **b** plot of residuals versus exp. no., **c** Normal probability plot and **d** plot of residuals versus fitted values

variable of the initial concentration of TC has a negative effect on response ($x\%$); effects of other variables on response were positive. In effect, increasing the initial concentration of TC and decreasing pH, H_2O_2 concentration and catalyst concentration will be caused to decrease and increase $x\%$, respectively (if the interaction effect of variable is ignored). The slope of line in main effect plots is one indicator of magnitude related to the variable effect on the response. Therefore, the order of affecting variables from magnitude viewpoint is as initial concentration of TC > catalyst concentration > pH > H_2O_2 concentration which confirm the results of Fig. 14a.

In Fig. 16, interaction plots for $x\%$ have been presented. Generally, in such plots the more parallel the lines, the lower the interaction effect would be and the more intersecting the lines, the higher the interaction effect would be. As it is observed, there is a significant interaction effect among catalyst concentration and H_2O_2 concentration variables. This effect is slightly found at interaction among pH and catalyst concentration variables. Figure 17 shows a cube plot for $x\%$. Using this plot, one could easily identify the conditions for reaching the desirable $x\%$. For example,

in order to reach maximum degradation ($x\% = 97.39$) the variables of pH, the initial concentration of TC, catalyst concentration and H_2O_2 concentration should be at levels of +1(8), −1(30 ppm), +1(150 ppm) and −1(0.1 ppm), respectively. Generally, considering the interaction effects is very important because it may place the unpredictable effects on the response. For example, based on the results of Fig. 15 even though H_2O_2 concentration had simply a positive effect on $x\%$, the maximum $x\%$ was achieved in those conditions where H_2O_2 concentration was at its minimum level (see exp. no 7 in Table 4). For the same reason, the interaction effect of variables should not be ignored in studying variables for reaching optimal conditions.

Finally, to determine the stability of the catalyst after 5 steps photocatalytic decomposition process, catalyst separation and then drying it, the FTIR spectrum of the sample showed that the catalyst structure have not changed. To determine the reusability of catalyst, 5 times was repeated experiment in the optimal conditions. Results, respectively, are as follows: X1 = 97.39, X2 = 97.32, X3 = 97.24, X4 = 97.20, X5 = 97.21. These results show that reusability of catalyst is appropriate.

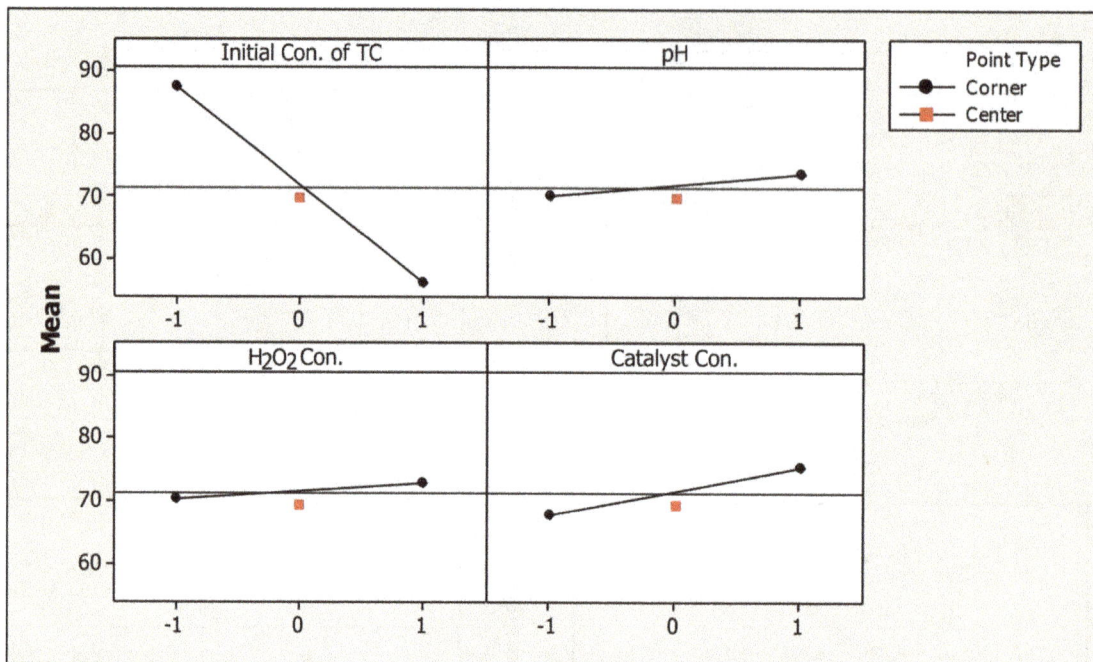

Fig. 15 Main effects plot for *x*%

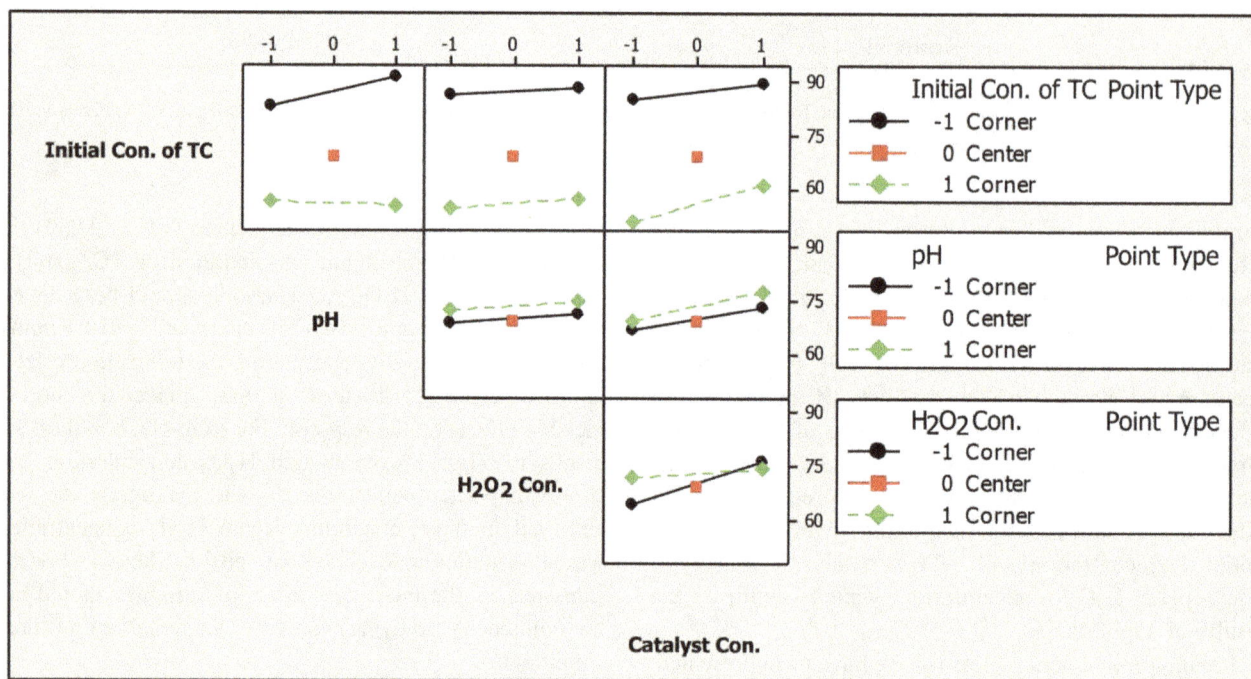

Fig. 16 Interaction plot for *x*%

Conclusions

The results revealed that:

1. Spherical α-Fe$_2$O$_3$ NPs had been successfully synthesized and supported on the surface of 12-TSA·7H$_2$O through SSD and FHRC methods with no decrease of NPs photocatalytic efficiency and chemical change of 12-TSA·7H$_2$O which are indicative of being effective these supporting methods.

2. While supporting α-Fe$_2$O$_3$ NPs on the surface of 12-TSA·7H$_2$O help to recover them from the medium and reusing them, it causes to enhance their photocatalytic activities.

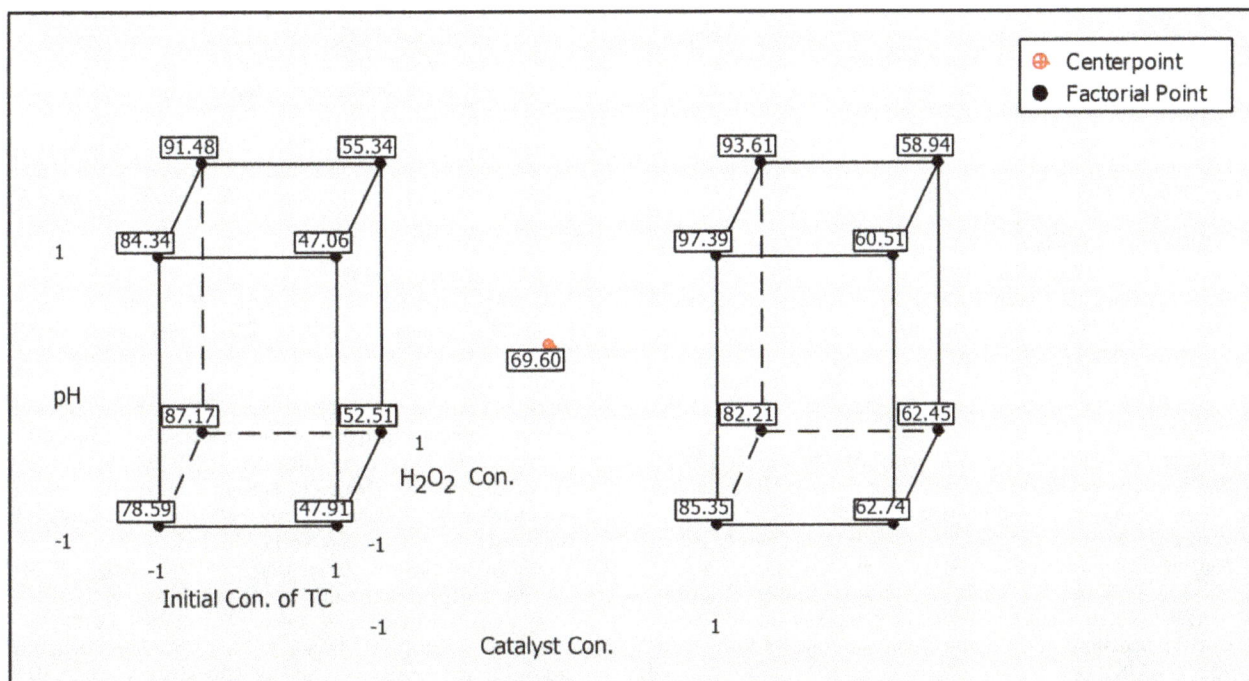

Fig. 17 Cube plot (data means) for $x\%$

3. Nanophotocatalytic effect of α-Fe$_2$O$_3$/12-TSA·7H$_2$O prepared through SSD and FHRC methods on the TC degradation is greater than pure α-Fe$_2$O$_3$ NPs.

4. As shown in the analysis of EDX, amount of iron oxide supported on the 12-TSA·7H$_2$O using SSD method is greater than the FHRC method. Then photo-activity of catalyst that prepared with SSD method is higher than FHRC method. Therefore, SSD method is more suitable.

5. The statistical analysis results indicated that the model used in this paper is significantly reliable and valid.

6. In the process of the TC photocatalytic degradation using α-Fe$_2$O$_3$/12-TSA·7H$_2$O prepared though SSD method, four parameters of pH, the initial concentration of TC, catalyst concentration and H$_2$O$_2$ concentration are effective on $x\%$. If interaction effects of variables are ignored, only the initial concentration of TC has a negative effect on the $x\%$.

7. The interaction effects of variables are very important and should be considered for optimizing the conditions because it significantly affects the $x\%$. For example, even though H$_2$O$_2$ concentration has simply a positive effect on $x\%$, interaction effects cause to yield maximum $x\%$ at conditions where H$_2$O$_2$ concentration is at minimum level.

8. The optimum conditions for the TC degradation process by α-Fe$_2$O$_3$/12-TSA·7H$_2$O prepared through SSD is as pH 8, initial concentration of TC = 30 ppm, catalyst concentration = 150 ppm and H$_2$O$_2$

concentration = 0.1 ppm so that they cause to reach maximum degradation (97.39%).

9. The kinetics of photocatalytic degradation of TC is of the pseudo-first order with $k = 0.0098$ min^{-1}.

Acknowledgements The authors wish to thank the Islamic Azad University of Arak, Iran, for financial support.

References

1. Balcioglu IA, Otker M (2003) Treatment of pharmaceutical wastewater containing antibiotics by O$_3$ and O$_3$/H$_2$O$_2$ processes. Chemosphere 50:85–95
2. Novo A, Andre S, Viana P, Nunes OC, Manaia CM (2013) Antibiotic resistance, antimicrobial residues and bacterial community composition in urban wastewater. Water Res 47:1875–1887
3. Hou J, Wang C, Mao D, Luo Y (2016) The occurrence and fate of tetracyclines in two pharmaceutical wastewater treatment plants of Northern China. Environ Sci Pollut Res Int 23:1722–1731
4. Kümmerer K (2009) Antibiotics in the aquatic environment—a review-part I. Chemosphere 75:417–434
5. Brown KD, Kulis J, Thomson B, Chapman TH, Mawhinney DB (2006) Occurrence of antibiotics in hospital, residential, and dairy effluent, municipal wastewater, and the Rio Grande in New Mexico. Sci Total Environ 366:772–783

6. Pailler JY, Krein A, Pfister L, Hoffmann L, Guignard C (2009) Solid phase extraction coupled to liquid chromatography-tandem mass spectrometry analysis of sulfonamides, tetracyclines, analgesics and hormones in surface water and wastewater in Luxembourg. Sci Total Environ 407:4736–4743

7. Homem V, Santos L (2011) Degradation and removal methods of antibiotics from aqueous matrices—a review. J Environ Manag 92:2304–2347

8. Xia Y, Xiong Y, Lim B, Skrabalak SE (2009) Shape-controlled synthesis of metal nanocrystals: simple chemistry meets complex physics? Angew Chem Int Ed 48:60–103

9. Jun YW, Choi JS, Cheon J (2007) Heterostructured magnetic nanoparticles: their versatility and high performance capabilities. Chem Commun 12:1203–1214

10. Chen J, Xu L, Li W, Gou X (2005) α-Fe$_2$O$_3$ nanotubes in gas sensor and lithium-ion battery applications. Adv Mater 17:582–586

11. Raming TP, Winnubst AJA, Van Kats CM, Philipse AP (2002) The synthesis and magnetic properties of nanosized hematite (α-Fe$_2$O$_3$) particles. J Colloid Interface Sci 249:346–350

12. Walter D (2006) Characterization of synthetic hydrous hematite pigments. Thermochim Acta 445:195–199

13. Shekhah O, Ranke W, Schüle A, Kolios G, Schlögl R (2003) Styrene synthesis: high conversion over unpromoted iron oxide catalysts under practical working conditions. Angew Chem Int Ed 42:5760–5763

14. Mishra M, Chun DM (2015) α-Fe$_2$O$_3$ as a photocatalytic material: a review. Appl Catal A 498:126–141

15. Farahmandjou M, Soflaee F (2015) Synthesis and characterization of α-Fe$_2$O$_3$ nanoparticles by simple co-precipitation method. Phys Chem Res 3:191–196

16. Liang H, Liu K, Ni Y (2015) Synthesis of mesoporous α-Fe$_2$O$_3$ via sol–gel methods using cellulose nano-crystals (CNC) as template and its photo-catalytic properties. Mater Lett 159:218–220

17. Diab M, Mokari T (2014) Thermal decomposition approach for the formation of α-Fe$_2$O$_3$ mesoporous photoanodes and an α-Fe$_2$O$_3$/CoO hybrid structure for enhanced water oxidation. Inorg Chem 53:2304–2309

18. Jiang T, Poyraz AS, Iyer A, Zhang Y, Luo Z, Zhong W, Miao R, El-Sawy AM, Guild CJ, Sun Y, Kriz DA, Suib SL (2015) Synthesis of mesoporous iron oxides by an inverse micelle method and their application in the degradation of orange II under visible light at neutral pH. J Phys Chem C 119:10454–10468

19. Askarinejad A, Bagherzadeh M, Morsali A (2011) Sonochemical fabrication and catalytic properties of α-Fe$_2$O$_3$ nanoparticles. J Exp Nanosci 6:217–225

20. Tadic M, Panjan M, Damnjanovic V, Milosevic I (2014) Magnetic properties of hematite (α-Fe$_2$O$_3$) nanoparticles prepared by hydrothermal synthesis method. Appl Surf Sci 320:183–187

21. Bharathi S, Nataraj D, Mangalaraj D, Masuda Y, Senthil K, Yong K (2010) Highly mesoporous α-Fe$_2$O$_3$ nanostructures: preparation, characterization and improved photocatalytic performance towards Rhodamine B (RhB). J Phys D Appl Phys 43:1–9

22. Chen M, Liu J, Chao D, Wang J, Yin J, Lin J, Fan HJ, Shen ZX (2014) Porous α-Fe$_2$O$_3$ nanorods supported on carbon nanotubes-graphene foam as superior anode for lithium ion batteries. Nano Energy 9:364–372

23. Rancourt DG, Julian SR, Daniels JM (1985) Mössbauer characterization of very small superparamagnetic particles; application to intra-zeolitic α-Fe$_2$O$_3$ particles. J Magn Magn Mater 49:305–316

24. Nikazar M, Gholivand K, Mahanpoor K (2008) Photocatalytic degradation of azo dye acid red 114 in water with TiO$_2$ supported on clinoptililite as a catalyst. Desalination 219:293–300

25. Hill CL (1998) Polyoxometalates. Chem Rev 98:1–387

26. Pope MT (1983) Heteropoly and Isopoly Oxometalates. Springer-Verlag, Berlin, Heidelberg

27. Kozhevnikov IV (2012) Catalysis by heteropoly acids and multicomponent polyoxometalates in liquid-phase reactions. Chem Rev 98:171–198

28. Es'haghi Z, Hooshmand S (2015) Dispersive solid-liquid phase microextraction based on nanomagnetic preyssler heteropolyacid: a novel method for the preconcentration of nortriptyline. J Sep Sci 38:1610–1617

29. Wang L, Zhou B, Liu J (2013) Anticancer Polyoxometalates. Prog Chem 25:1131–1141

30. Judd DA, Netlles HJ, Nevis N, Snyder JP, Liotta DC, Tang J, Ermolieff JJ, Schinazi FR, Hill CL (2001) Polyoxometalate HIV-1 protease inhibitors. A new mode of proteas inhibition. J Am Chem Soc 123:886–897

31. Wang X, Liu J, Li J, Liu J (2001) Synthesis, characterization and in vitro antitumor activity of diorganometallo complexes γ-Keggin anions. Inorg Chem Commun 4:372–374

32. Lin Z, Zhongqun L, Wenjun C, Shaojin C (1996) Removing Cs from nuclear waste liquid by crown ether and heteropoly acid: simulated tests. J Radioanal Nucl Chem 205:49–56

33. Heylen S, Smeekens S, Kirschhock CEA, Parac-Vogt TN, Martens JA (2010) Temperature swing adsorption of NOx over Keggin type heteropolyacids. Energy Environ Sci 3:910–916

34. Lihua B, Qizhuang H, Qiong J, Enbo W (2001) Synthesis, properties and crystal structure of (Gly)$_2$H$_4$SiW$_{12}$O$_{40}$·5·5H$_2$O. J Mol Struct 597:83–91

35. Soled S, Miseo S, McVicker G, Gates WE, Gutierrez A, Paes J (1996) Preparation and catalytic properties of supported heteropolyacid salts. Chem Eng J Biochem Eng J 64:247–254

36. Abolghasemi MM, Hassani S, Rafiee E, Yousefi V (2015) Nanoscale-supported heteropoly acid as a new fiber coating for solid-phase microextraction coupled with gas chromatography-mass spectrometry. J Chromatogr A 1381:48–53

37. Chen F, Ma J, Dong Z, Liu R (2014) Characterization and catalytic performance of heteropoly acid H$_4$SiW$_{12}$O$_{40}$ supported on nanoporous materials. J Nanosci Nanotechnol 14:7293–7299

38. Badday AS, Abdullah AZ, Lee KT (2014) Transesterification of crude Jatropha oil by activated carbon-supported heteropolyacid catalyst in an ultrasound-assisted reactor system. Renewable Energy 62:10–17

39. Taylor DB, McMonagle JB, Moffat JB (1985) Cation effects on the surface and bulk structure of the salts of 12-tungstosilicic acid. J Colloid Interface Sci 108:278–284

40. Nabizadeh R, Jahangiri Rad R (2016) Nitrate adsorption by pan-oxime-nano Fe$_2$O$_3$ using a two-level full factorial design. Res J Nanosci Nanotechnol 6:1–7

41. Jiao H, Wang J (2013) Single crystal ellipsoidal and spherical particles of a-Fe$_2$O$_3$: hydrothermal synthesis, formation mechanism, and magnetic properties. J Alloys Compd 577:402–408

42. North EO, Bailar JC, Jonelis FG (2007) Silicotungstic Acid. Inorg Synth 1:129–132

43. Bamoharram FF (2009) Vibrational spectra study of the interactions between Keggin heteropolyanions and amino acids. Molecules 14:3214–3221

44. Kozhevnikov IV, Sinnema A, Jansen RJJ, Bekkum HV (1994) ^{17}O NMR determination of proton sites in solid heteropoly acid H$_3$PW$_{12}$O$_{40}$. ^{31}P, ^{29}Si and ^{17}O NMR, FT-IR and XRD study of H$_3$PW$_{12}$O$_{40}$ and H$_4$SiW$_{12}$O$_{40}$ supported on carbon. Catal Lett 27:187–197

45. Zhao D, Sheng G, Chen C, Wang X (2012) Enhanced photocatalytic degradation of methylene blue under visible irradiation on graphene@TiO$_2$ dyade structure. Appl Catal B 111:303–308

46. Zhang S, Li J, Zeng M, Zhao G, Xu J, Hu W, Wang X (2013) In situ synthesis of water-soluble magnetic graphitic carbon

Photocatalytic degradation of tetracycline aqueous solutions by nanospherical α-Fe2O3 supported...

199

nitride photocatalyst and its synergistic catalytic performance. Appl Mater Interfaces 5:12735–12743

47. Yao K, Basnet P, Sessions H, Larsen GK, Hunyadi Murph SE, Zhao Y (2016) Fe_2O_3–TiO_2 core–shell nanorod arrays for visible light photocatalytic applications. Catal Today 270:51–58

48. Zhang S, Fan Q, Gao H, Huang Y, Liu X, Li J, Xu X, Wang X (2016) Formation of Fe_3O_4@MnO_2 ball-in-ball hollow spheres as a high performance catalyst with enhanced catalytic performances. J Mater Chem A 4:1414–1422

49. Guo S, Zhang G (2016) Green synthesis of a bifunctional Fe–montmorillonite composite during the Fenton degradation process and its enhanced adsorption and heterogeneous photo-Fenton catalytic properties. RSC Adv 6:2537–2545

50. Mehraj O, Pirzada BM, Mir NA, Khan MZ, Sabir S (2016) A highly efficient visible-light-driven novel p-n junction Fe_2O_3/BiOI photocatalyst: surface decoration of BiOI nanosheets with Fe_2O_3 nanoparticles. Appl Surf Sci 387:642–651

16

Investigation of corrosion resistance of steel used in beet sugar processing juices

H. D. Ada[1] · S. Altanlar[2] · F. Erdem[2] · G. Bereket[3]

Abstract In this study, corrosion behaviors of materials used in diffusion units and equipment used in juice clarification steps and tubes in evaporators at Ankara Sugar Factory were investigated in terms of juice production and juice clarification processes as well as juice medium at evaporation stages. The measurements have been performed by comparing steel types used in these units and alternative types of steels that can also be used during the study. For this purpose, pH and Brix (Bx, refractometric dry matter) values of raw juice, thin juice and juice taken from evaporator have been measured during 2009–2010, 2010–2011, and 2011–2012 campaign periods of Ankara Sugar Factory. In addition to these measurements, traditional weight loss and electrochemical tests such as Linear Polarization Resistance (LP), Tafel Extrapolation (TP), Electrochemical Impedance Spectroscopy (EIS) were performed to measure and compare the corrosion rate of the metals used in different juice mediums. The metals included in the study were AISI 316L, AISI 304L grade stainless steel, St 37.2 grade carbon steel and nickel-coated St 37.2. The highest and the lowest corrosion rates were recorded for raw juice and thin juice, respectively. St 37.2 steel had the fastest corrosion rate, whereas the stainless steel AISI 316L has the slowest corrosion rate. However, AISI 316L shows only slightly higher corrosion resistance compared to the corrosion resistance of AISI 304L in different juices. Therefore, AISI 304L steel, which is cheaper than AISI 316L, can be selected as a substation of St 37.2 steel.

Keywords Beet sugar juice · Alloys · Corrosion · Electrochemical Methods

Introduction

Along with the development of sugar industry, corrosion has drawn attention in sugar industry and in terms of corrosion, various stages of corrosion are analyzed with regards to material and environment by many researchers. Common types of corrosion seen in processing of sugar are bacterial corrosion, erosion corrosion, pitting corrosion, stress corrosion and high temperature corrosion [1–3]. In Ankara Sugar Factory, most of the equipment parts are made of St 37.2 type steel because of its low cost, good mechanical properties and ease of fabrication. However, corrosion is a major problem in Ankara Sugar Factory due to the relatively low corrosion resistance of St 37.2 type steel. Maintenance, replacement and repair of equipment due to corrosion and abrasion have diverse effect on the operating cost. Therefore, the use of stainless steel such as AISI 304L and AISI 316L that have better corrosion resistance for various processing industries as a construction material in sugar factory can minimize these operating costs. For this reason, most sugar mills around the world have switched over the components made of various grades of stainless steel [4]. Corrosion resistance of steels in juice medium should be evaluated to choose the proper materials that can be used in a sugar factory. Corrosion in sugar factories can arise mainly due to the presence of some acids, impurities in juice and high temperature [5]. Some

✉ G. Bereket
gbereket@ogu.edu.tr

1 Dumlupınar University, 43100 Kütahya, Turkey

2 Technological Research Division, Turkish Sugar Factories Co, 06930 Ankara, Turkey

3 Deparment of Chemistry, Faculty of Science and Letters, Eskişehir Osmangazi University, 26480 Eskişehir, Turkey

process parameters such as pH, temperature, dissolved oxygen content, brix and conductivity values of juices also affect the corrosion rate of steel [6–9]. The aim of performing corrosion tests in equipment is determining the best specifications of the materials, probable service life of the equipment or products, most economical means for reducing corrosion and studying the corrosion mechanism [10]. Corrosion rate measurement by weight loss of coupon specimens especially for the uniform corrosion is the simplest and the best service condition technique. For this reason, weight loss corrosion tests are performed regularly in different types of juices produced in Ankara Sugar Factory and in boiler water mediums. However, electrochemical corrosion rate measurement techniques such as Tafel Analysis, Linear Polarization and Electrochemical Impedance Spectroscopic measurements are not conducted regularly. Though numerous studies are made on the corrosion resistance of stainless steel in different corrosive mediums including sugarcane juice, investigation on the comparative corrosion resistance study of AISI 304L, AISI 316L types stainless steel, nickel-coated St 37.2 steel and St 37.2 steel in different types of juices taken from Ankara Sugar Factory does not exist. In the present study, it was aimed to measure corrosion resistance of St 37.2, nickel-coated St 37.2, AISI 304L, AISI 316L stainless steel in different types of juices produced in Ankara Sugar Factory using weight loss immersion test, Tafel Analysis, Linear Polarization Resistance and Electrochemical Impedance Spectroscopy.

Materials and methods

Materials

The studied steels were stainless steel AISI 316L, AISI 304L and carbon steel St 37.2. They were received from Ankara Sugar Factory. Table 1 provides the details of their chemical composition.

Weight loss corrosion test

ASTMD 3263 was followed as a guide for weight loss immersion tests [11]. Two specimens of each materials were prepared in a rectangular size of $10 \times 3.5 \times 0.15$ cm. Corrosion rate of the steel samples was calculated by:

Corrosion rate (mm/year): $(W_i - W_f)$ 24.365/A.T.7830
W_i: initial weight (g) A: surface area (m^2)
W_f: final weight (g) T: time (hour)
(mpy = 0.0254 mm/year)

Electrochemical measurements

Electrochemical tests were performed to determine the corrosion rate of AISI 304L, AISI 316L stainless steels, St 37.2 low carbon steel and nickel-coated St37.2 in different types of juices produced during three campaign periods in Ankara Sugar Factory. St 37.2 electrode was electroless nickel-coated according to the procedure given in the literature Surface and Coating Technology 201 (2006)90–101 [12]. For this purpose, electrochemical experiments were conducted using Gamry Reference 600 potentiostat galvanostat/ZRA system with Gamry Framework/Echem Analyst (Version 5.50) software. A platinum electrode that has a surface area of 1.5 cm^2 was used as the counter electrode and a saturated calomel electrode (SCE) was used as the reference electrode. Working electrodes were AISI 316L, AISI 304L, St 37.2 and nickel-coated St 37.2 steels with an area of 4.9 cm^2. The surfaces of the working electrodes were polished to mirror brightness with a polishing cloth using 3, 1, 0.05 microns of aluminum polishing solutions, respectively. The electrolytic cell was made of a 750 ml Pyrex glass flask with four entrances for reference electrode, working electrode, counter electrode and aeration/deaeration. Measurements were conducted in aerated and unstirred solutions. The mounted samples were immersed in juice medium to obtain a constant potential, which is referred as the Open Circuit Potential. About 30 min was sufficient to attain constant potential condition. In all cases, Open Circuit Potential was established first, and then the experiments were carried out. Electrochemical Impedance Spectroscopy was performed using a potential amplitude of 10 mV and the frequency ranging from 100 kHz to 5 mHz. EIS data were analyzed using the ZSimpwin 3.10 program which provided accurate information about the circuit. The impedance data were analyzed using the electrical equivalent circuit R(QR) presented in Fig. 2. The measure of goodness of fit of the model was the χ^2 parameter; during the analysis, χ^2 did not exceed 1×10^{-4}, attesting to a very high fit of received impedance spectra to the proposed electrical equivalent circuit. The same circuit has also been used in similar aggressive environments [13].

Table 1 Composition of steels (wt %)	Grade of steel	C	Cr	Ni	Mo	Mn	P	Si	S	Cu
	St 37.2	0.063	0.016	0.018	0.014	0.528	–	0.181	0.012	0.047
	AISI 304L	0.03	18–20	8–12	–	0.045	1	2	0.03	–
	AISI 316L	0.03	16–18	10–14	2–3	2	0.045	0.75	0.03	–

Table 2 The results of weight loss corrosion test obtained in juice samples taken from various units during 2009–2010 campaign period

Juice	Metals	Briks	pH	Corrosion rate determined by weight loss test (mpy)
Raw juice	AISI 304L	16.52	6.75	0.0042
Raw juice	AISI 316L	16.52	6.75	0.0037
Thin juice	AISI 304L	15.82	8.75	0.0025
Thin juice	AISI 316L	15.82	8.75	0.0025
2A Evap.	AISI 304L	35.50	8.79	0.0021
2A Evap.	AISI 316L	35.50	8.79	0.0020

Fig. 1 The *Bode curves* of AISI 304L and AISI 316L electrodes in raw juice

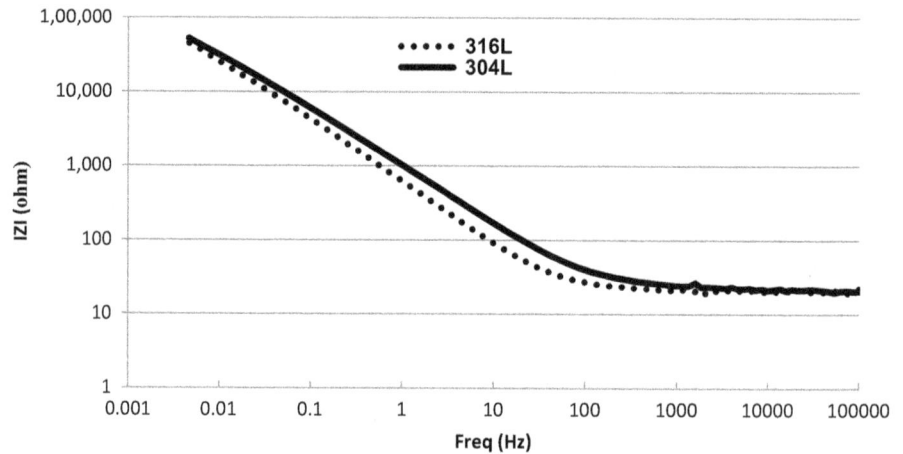

The Linear Polarization Technique was applied using a scan rate of 0.125–1 mV/s and a polarization of ±10 mV in relation to the Open Circuit Potential. The parameters of Tafel analysis were a scan rate of 1 mVs and a polarization of ±600 mV (Open Circuit Potential).

Results and discussion

Corrosion behavior of St 37.2, nickel-coated St 37.2, AISI 304L and AISI 316L stainless steel in different types of juices produced in Ankara Sugar Factory was studied during 2009–2010, 2010–2011, and 2011–2012 campaign periods. For this purpose, pH, Brix (Bx, refractometric dry mater) values of thin juice, juice from diffusion unit, juice from clarification stages have been measured. In addition, weight loss, EIS, LP and Tafel analysis were performed to determine the corrosion rates of the studied materials in juice medium.

Results of measurements conducted during 2009–2010 campaign period

In this campaign period, corrosion resistance of AISI 304L and AISI 316L steels for substitution of St 37.2 using weight loss, EIS, LP, and Tafel analysis was evaluated.

In Table 2, corrosion rates of steels determined by weight loss measurement technique are presented. It can be observed that highest corrosion rates were recorded in raw juice followed by thin juice and the lowest corrosion rates were recorded in juice taken from 2A evaporator, respectively. In the factory, juice temperature in 2A evaporators reaches up to a degree ranged between 130 and 140 °C. In fact, corrosion rates in such high temperature levels should be much higher than the corrosion rates measured in the laboratory conditions (e.g., 25 °C). However, it was not possible to generate such a high temperature in the laboratory conditions. As it can be seen from Table 2, the highest brix content is in juice taken from 2A evaporators and this causes measured corrosion rates in this medium at room temperature to be the lowest.

Figure 1 shows the Bode plots of AISI 304L and AISI 316L steels obtained in raw juice. Slightly higher impedance modulus was recorded for AISI 316L which implies that AISI 316L has slightly higher corrosion resistance compared to AISI 304L in raw juice. Similar Bode plots were also obtained for EIS measurements conducted in thin juice and juice taken from 2A evaporators (measurements were taken at room temperature). In these mediums, AISI 316L shows slightly higher corrosion resistance compare to the corrosion resistance of AISI 304L. Impedance data were analyzed using an electrical equivalent circuit R(QR) illustrated in Fig. 2.

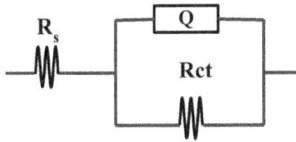

Fig. 2 Equivalent circuit R(QR) used to simulate EIS data

In the equivalent circuit used for analyzing impedance plot, R_s denotes solution resistance, R_{ct} denotes charge transfer resistance of corroding metal and Q reflects constant phase element of corroding metal–solution interface. The values of impedance parameters derived from the fitting of equivalent circuits to experimental data are given in Table 3.

Polarization resistance measurement was also conducted for AISI 304L and AISI 316L type steels for the corrosion measurements in various juices. However, nonlinear behavior of E–I curves in the potential region ±20 mV with respect to corrosion potentials were recorded. Relatively, high corrosion rate or active dissolution condition is needed to observe linear behavior in E–I plots during polarization resistance measurements. Departure from linearity appears when absolute values of the Tafel constants are low as $\beta_a = \beta_c = 30$ mV. Unequal Tafel constants further compress the range of apparent linearity and also result in asymmetry about origin when $\beta_a = 30$ mV and $\beta_c = 118$ mV [11].

Potentiodynamic polarization measurements were conducted for AISI 316L and AISI 304L steels in different juice mediums and Fig. 3 shows the polarization curves recorded in raw juice. However, Tafel analysis was not conducted for determination of the corrosion rates since at least one decade of linearity on E–log I curves is desirable for maximum accuracy to determine i_{corr} by Tafel extrapolation [14]. Lower current densities and more anodic E_{ocp} values were recorded for AISI 316L. Corrosion resistance of AISI 316L and AISI 304L in different types of juices obtained from weight loss, EIS and potentiodynamic measurements shows that AISI 316L has only slightly higher resistance. Thus, the use of AISI 304L which is cheaper than AISI 316L can be envisaged as substation of St 37.2 steel.

Table 3 The EIS circuit analysis results obtained in juice samples taken from various units during 2009–2010 campaign period by fitting the impedance spectra to the R(QR) circuit

	Metals	Briks	pH	R_s (kΩ cm^2)	Q (Ss^{-n}/cm^2)	n	R_{ct} (kΩ cm^2)	χ^2
Raw juice	AISI 304L	16.5	6.8	0.0207	0.00034	0.85	259.5	1.67×10^{-4}
Raw juice	AISI 316L	16.5	6.8	0.0022	0.00023	0.79	261.5	2.58×10^{-4}
Thin juice	AISI 304L	15.8	8.8	0.0168	0.00017	0.71	294.8	3.55×10^{-4}
Thin juice	AISI 316L	15.8	8.8	0.0172	0.00025	0.80	295.2	1.11×10^{-4}
2A Evap.	AISI 304L	35.5	8.8	0.0149	0.00025	0.79	365.0	9.05×10^{-4}
2A Evap.	AISI 316L	35.5	8.8	0.0149	0.00036	0.82	376.5	5.34×10^{-4}

Fig. 3 The current–potential *curves* of AISI 304L and AISI 316L steels in raw juice

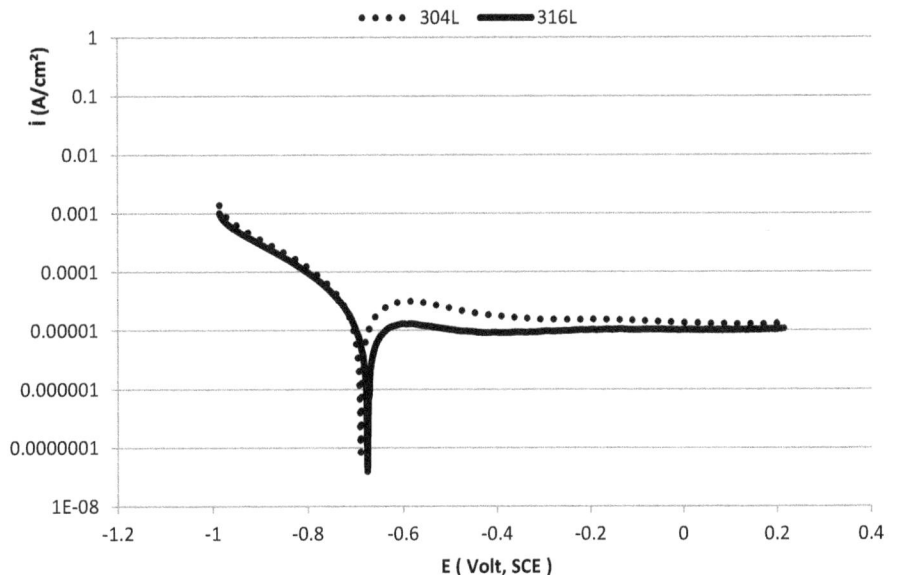

Results of measurements conducted during 2010–2011 campaign period

In this campaign period, corrosion resistance of AISI 304L, electroless Ni-coated St 37.2 and St 37.2 samples in different types of juices were determined by weight loss, EIS, LP and potentiodynamic measurement methods. Electroless Ni deposits are widely used in different industries for their unique combination of properties such as wear resistance, corrosion resistance and higher hardness [15]. The anticorrosive behavior of these deposits gained greater importance and applicability [16]. Therefore, in this campaign period, corrosion resistance of nickel-coated St 37.2 sample was also evaluated for considering replacement of the materials used in Ankara Sugar Factory.

Table 4 shows corrosion rates of St 37.2 and AISI 304L steels determined by weight loss measurement technique. Highest corrosion rates were recorded in raw juice and the lowest corrosion rates were recorded in juices obtained from 2A evaporators in 2009–2010 campaign periods.

Figure 4 shows the Bode plots of AISI 304L, nickel-coated St 37.2 and St 37.2 samples obtained from raw juice. Similar Bode plots were also obtained for EIS measurements conducted in clarified juice (juice obtained after liming), in thin juice and in juice taken from 2A evaporator. Impedance data were analyzed using R(QR) equivalent circuit.

The values of impedance parameters derived from the fitting of equivalent circuit to experimental data are given in Table 5 which shows that corrosion resistance of electroless nickel-coated St 37.2 steel was increased compared to the uncoated St 37.2 steel. However, corrosion resistance of electroless nickel-coated St 37.2 steel is much lower than corrosion resistance of AISI 304L steel. This implies that electroless nickel coating is not feasible in juice medium as substation of St 37.2 steel. According to the results given in Table 5, corrosion rates of the studied materials depend on pH and brix (solid concentration in the juice). To evaluate the effect of pH on the corrosion, measured corrosion rates in raw juice, clarified juice and

Table 4 The results of weight loss corrosion test obtained in juice samples taken from various units during 2010–2011 campaign period

Juice	Metals	Briks	pH	Corrosion rate determined by weight loss test (mpy)
Raw juice	AISI 304L	14.7	6.8	0.0056
Raw juice	St 37.2	14.7	6.8	0.2100
Raw juice after liming	AISI 304L	14.9	9.0	0.0040
Raw juice after liming	St 37.2	14.9	9.0	0.0900
Thin juice	AISI 304L	14.1	8.8	0.0042
Thin juice	St 37.2	14.1	8.8	0.0081
2A Evap.	AISI 304L	30.2	8.9	0.0029
2A Evap.	St 37.2	30.2	8.9	0.0370

Fig. 4 The *Bode curves* of AISI 304L, St 37.2 and Ni-coated St 37.2 electrodes in raw juice

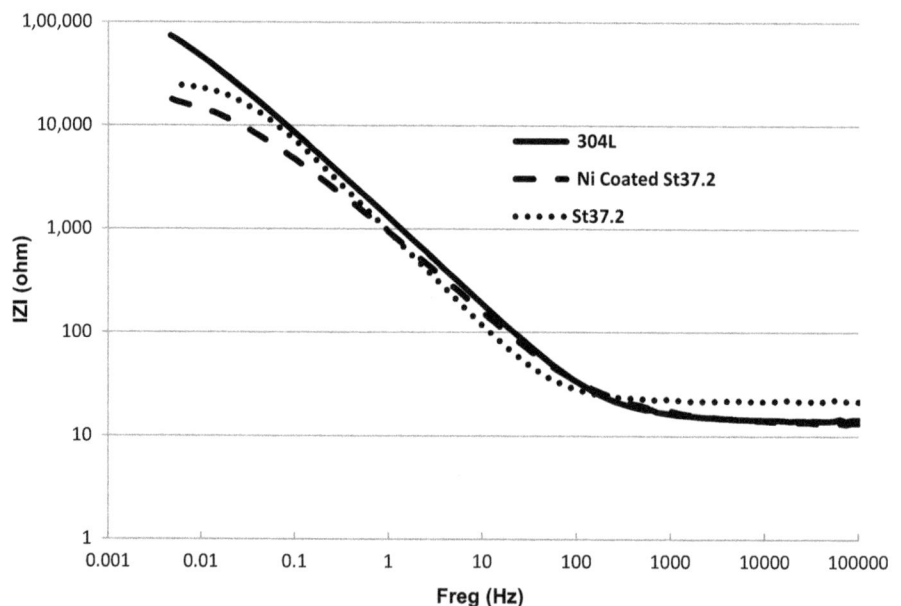

Table 5 The EIS circuit analysis results obtained in juice samples taken from various units during 2010–2011 campaign period by fitting the impedance spectra to the R(QR) circuit

	Metals	Briks	pH	R_s (kΩ cm^2)	Q_1 (Ss^{-n}/cm^2)	n	R_{ct} (kΩ cm^2)	χ^2
Raw juice	AISI 304L	14.7	6.8	0.0326	0.00004	0.72	218	6.21×10^{-4}
Raw juice	Ni coat.	14.7	6.8	0.0280	0.00046	0.87	8.1	5.96×10^{-4}
Raw juice	St 37.2	14.7	6.8	0.0340	0.00074	0.73	4.9	6.13×10^{-4}
Raw juice after liming	AISI 304L	14.9	9.0	0.0240	0.00013	0.82	254	2.71×10^{-4}
Raw juice after liming	Ni coat.	14.9	9.0	0.0213	0.00003	0.88	13.4	3.12×10^{-4}
Raw juice after liming	St 37.2	14.9	9.0	0.0189	0.00034	0.78	11.7	4.52×10^{-4}
Thin juice	AISI 304L	14.1	8.8	0.0019	0.00008	0.81	256	2.67×10^{-4}
Thin juice	Ni coat.	14.1	8.8	0.0171	0.00012	0.89	25	3.12×10^{-4}
Thin juice	St 37.2	14.1	8.8	0.0867	0.00003	0.66	125	4.52×10^{-4}
2A Evap.	AISI 304L	30.2	8.9	0.0136	0.00023	0.83	273	6.13×10^{-4}
2A Evap.	Ni coat.	30.2	8.9	0.0220	0.00020	0.91	26.6	5.21×10^{-4}
2A Evap.	St 37.2	30.2	8.9	0.0139	0.00027	0.76	20.5	4.67×10^{-4}

Table 6 The result of linear polarization measurements of juice samples taken from various units obtained during 2010–2011 campaign period

Juice	Metals	Briks	pH	E_{corr} (mV)	i_{corr} (μA)	R_p (kΩ cm^2)
Raw juice	Ni coat.	14.7	6.8	−332.6	394.0	6.4
Raw juice	St 37.2	14.7	6.8	−705.3	770.0	2.2
Raw Juice after liming	Ni coat.	14.9	9.0	−280.6	189.4	13.8
Raw juice after liming	St 37.2	14.9	9.0	−313.5	195.0	11.3
Thin juice	Ni coat.	14.0	8.8	−335.9	124.1	21.0
Thin juice	St 37.2	14.0	8.8	−257.8	19.8	136.0
2A Evap.	Ni coat.	30.2	8.9	−337.6	98.1	26.6
2A Evap.	St 37.2	30.2	8.9	−401.3	120.8	21.6

thin juice should be compared with each other. In these juice mediums, brix content can be considered to be almost equal. Highest corrosion rates were recorded in raw juice, which has the lowest pH value. Our results are consistent with those obtained by Gupta et al. [17]. They performed a study to evaluate the effect of pH on corrosion rate of mild steel in raw, draft and clarified juice. They have found that the corrosion rate was highest for raw cane juice followed by clarified juice and draft juice [18], respectively. When the measured corrosion rates of AISI 304L and nickel-coated steels in juice taken from 2A evaporator (measured at room temperature), and in clarified juice are compared with each other, it can be seen that the lowest corrosion rates are recorded in juice taken 2A evaporator. pH values of these three juice mediums are almost same. However, the juice taken from 2A evaporator has the highest brix and highest corrosion resistance. This result is consistent with the result obtained by Bajpai et al. [19]. They reported that corrosion rate linearly increases with decrease in brix. Brix contains organic substances such as sucrose, amino acids, betaine, glucose, fructose that acts as in inhibitors for the corrosion of steel in corrosive medium. Similarly Date

Palm Fruit Juice due to presence of mentioned organic substances acts as inhibitor for the corrosion Aluminum Alloy in 3.5 % NaCl [20]. However, corrosion resistance of St 37.2 in juice taken from 2A evaporator was found to be lower than the corrosion rate in thin juice. Evaporators are working with high vacuum which is an oxygen-free environment. Therefore, electrochemical experiments conducted in juice taken from 2A evaporator have a medium that has low oxygen content. Formation of porous nature of oxide film on St 37.2, which has a protective character, is not much feasible. So, lower corrosion rate was recorded in juice taken from 2A evaporator although it has the highest brix content.

Polarization resistance measurements were also conducted for St 37.2 and nickel-coated St 37.2 for the corrosion measurements in various juices. Table 6 shows the result of Linear Polarization measurements. As it can be seen from the table, the highest corrosion rates were recorded for samples in raw juice. On the other hand, St 37.2 steel has the least corrosion rate. Table 6 also shows that measured E_{corr} values of St 37.2 in juice mediums (except in thin juice) are more negative than that of

Fig. 5 The current–potential *curves* of AISI 304L, St 37.2 and Ni-coated St 37.2 steels in 2A evaporator

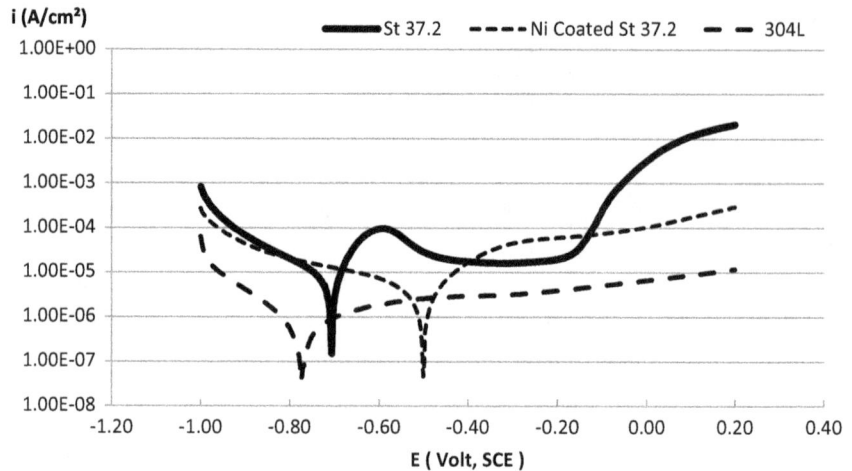

Fig. 6 The current–potential *curves* of AISI 304L, St 37.2 and Ni-coated St 37.2 steels in clarified juice

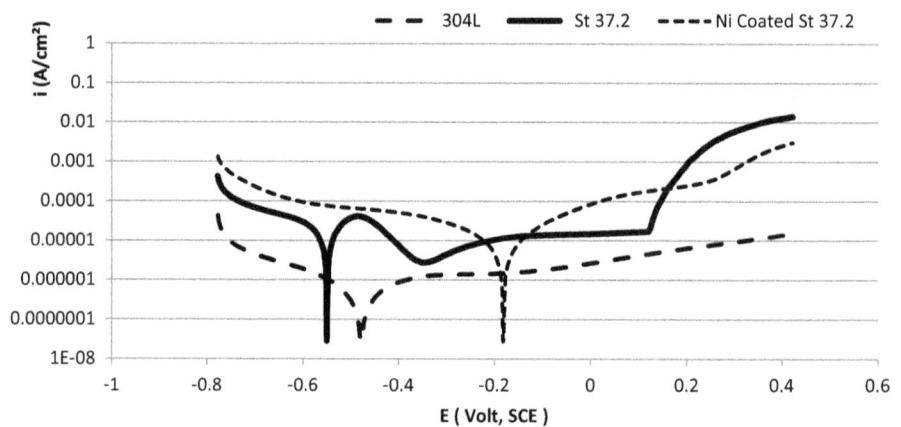

measured E_{corr} values of nickel-coated St 37.2 indicating lower corrosion resistance of St 37.2. Similarly, for carbon steel negative E_{corr} value was more thus higher corrosion rate than the stainless steel alloys was recorded in raw sugar juice environment [21]. Potentiodynamic polarization measurements were also conducted for St 37.2 and nickel-coated St 37.2 steels. Lowest current densities and most noble Open Circuit Potential were recorded for AISI 304L steel indicating highest corrosion resistance in the studied juice mediums. However, for St 37.2 drastic current increase in the polarization curves shown in Figs. 5 and 6 was observed at more anodic potentials from Open Circuit Potential in juice taken from 2A evaporator and in clarified juice. pH values of these mediums are about 9 and obtained Open Circuit Potential values are close to −300 mVversus SHE. Thus, this region in Fe–H_2O system in Pourbaix Diagram corresponds to $Fe_2O_3 \cdot H_2O$ phase [22]. Due to non-adherent porous nature of this formed oxide film on St 37.2, destruction of this film at more noble potentials was observed as current increase [23].

Results of measurements conducted during 2011–2012 campaign period

In this campaign period, corrosion resistance of AISI 304L and St 37.2 steels in various juice mediums was evaluated using weight loss EIS, LP, and potentiodynamic measurements. This comparative study of corrosion of AISI 304L steel and St 37.2 steel in sugar juices is to probe the feasibility of replacing St 37.2 with AISI 304L in different parts of Ankara Sugar Factory.

Table 7 shows corrosion rates of St 37.2 and AISI 304L steels determined by weight loss measurement technique. Highest corrosion rates for AISI 304L were recorded in raw juice and the lowest corrosion were recorded in juice taken from 2A evaporators as in 2009–2010 and 2010–2011 campaign periods.

Figure 7 shows the Bode plots of AISI 304L, St 37.2 samples obtained from raw juice. Similar Bode plots were also obtained for EIS measurements conducted in clarified juice, thin juice and the juice taken from 2A evaporator.

Table 7 The results of weight loss corrosion test obtained in juice samples taken from various units during 2011–2012 campaign period

Juice	Metals	Briks	pH	Corrosion rate with coupon test (mpy)
Raw juice	AISI 304L	15.6	5.9	0.0061
Raw juice	St 37.2	15.6	5.9	0.3700
Raw juice after liming	AISI 304L	14.9	9.0	0.0038
Raw juice after liming	St37.2	14.9	9.0	0.0840
Thin juice	AISI 304L	14.8	8.9	0.0038
Thin juice	St 37.2	14.8	8.9	0.0075
2A Evap.	AISI 304L	30.3	9.1	0.0037
2A Evap.	St 37.2	30.3	9.1	0.0410

Fig. 7 The *Bode curves* of AISI 304L and St 37.2 electrodes in raw juice

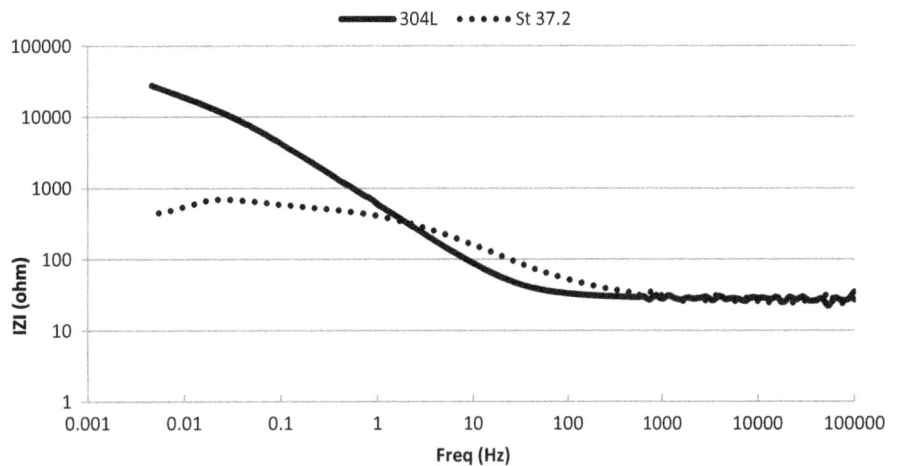

Table 8 The EIS circuit analysis results obtained in juice samples taken from various units during 2011–2012 campaign period (stabilization made with R(QR) circuit

	Metals	Briks	pH	R_s (kΩ cm^2)	Q_1 (Ss^{-n}/cm^2)	n	R_{ct} (kΩ cm^2)	χ^2
Raw juice	AISI 304L	15.6	5.9	0.1370	0.00007	0.87	178.0	4.88×10^{-4}
Raw juice	St37.2	15.6	5.9	0.1240	0.00010	0.76	3.1	5.08×10^{-4}
Raw juice after liming	AISI 304L	14.9	9.0	0.0860	0.00006	0.76	267.0	1.67×10^{-4}
Raw juice after liming	St 37.2	14.9	9.0	0.1040	0.00006	0.81	11.5	3.55×10^{-4}
Thin juice	AISI 304L	14.8	8.9	0.0910	0.00003	0.75	283.0	1.19×10^{-4}
Thin juice	St 37.2	14.8	8.9	0.0824	0.00009	0.66	145.0	9.10×10^{-4}
2A Evap.	AISI 304L	30.3	9.1	0.0590	0.00002	0.73	288.0	3.84×10^{-4}
2A Evap.	St 37.2	30.3	9.1	0.0700	0.00031	0.65	26.1	3.90×10^{-4}

The values of impedance parameters derived from fitting of R(QR) equivalent circuits are given in Table 7. Potentiodynamic polarization measurements were also conducted for AISI 304L and St 37.2 in clarified juice (juice obtained after liming), in thin juice and in juice taken from 2A evaporator. The polarization curves obtained show similar behaviors to those obtained in 2010–2011 campaign period.

Polarization resistance measurements were also conducted for St 37.2 for the corrosion measurements in various juice mediums. Examination of polarization resistance

values given in Table 8 shows that the highest corrosion resistance was recorded in thin juice and the lowest corrosion resistance values are recorded in raw juice medium.

Conclusion

In the present study, corrosion resistance of AISI 304L, AISI 316L, St37.2, and nickel-coated St 37.2 steel in different juice mediums of Ankara Sugar Factory was evaluated using weight loss and electrochemical techniques such as Electrochemical Impedance Spectroscopy, Linear Polarization and Tafel Extrapolation methods. TP method is not found to be applicable for obtaining quantitative corrosion rates; whereas LP method was applicable for obtaining qualitative corrosion rates of St 37.2 and nickel-coated St 37.2 steel. However, EIS method was found to be suitable for determining quantitative corrosion rates for all steel types included in the study. Results obtained from EIS measurements and weight loss coupon test had the same trends. According to these rates, it can be concluded that corrosion rates were dependent on pH and brix content of the sugar juice. Thus, the highest and the lowest corrosion rates were recorded for raw juice and thin juice, respectively. St 37.2 steel had the fastest corrosion rate while stainless steel AISI 316L corroded has the least juices extracted from beet sugar as evaluated by weight loss and electrochemical techniques. For economic reasons, the low-cost AISI 304L steel can be used in Ankara Sugar Factory since it has slightly lower corrosion resistance compared to AISI 316L steel.

Acknowledgments The authors are thankful to Technological Research Division of Sugar Institute belonging to Turkish Sugar Factories Co. for providing laboratory facilities, scientific and technical cooperation as well as for financial support.

References

1. Zumelzu E, Goyos I, Cabezas C, Opitz O, Parada A (2002) Wear and corrosion behaviour of high-chromium (14–30% Cr) cast iron alloys. J Mater Prosess Technol 128:250–255
2. Zumelzu E, Cabezas C, Opitz O, Quiroz E, Goyos L, Parada A (2003) Microstructural characteristics and corrosion behaviour of high-chromium cast iron alloys in sugar media. Prot Met 39:183–188
3. Panigrahi BK, Srikanth S, Singh J (2007) Corrosion failure in the sugar industry: a case study. J Fail Anal Preven 7:187–191
4. R.K. Goyal, Rajesh Khosla and Pravin Goel (2006). In: Proceeding of the 67th Annual Convention of STAI., Ahmadabad
5. P.Goel, R.Khosla, L.K. Singhal and R.K.Goyal (2007) Role of stainless steel to combat corrosion in the Indian sugar industry, National federation of Cooperative sugar factories ltd. Vol 38 No:12 pp. 39–46
6. Bajpai A, Shukla NP (1992) Corrosion and its prevention in cane sugar industry. Int Sugar J 94:76–80. doi:10.1007/s40090-016-0077-9
7. Farias CA, Lins VFC (2011) Corrosion Resistance of Steels Used in Alcohol and Sugar Industry. Chem Eng Technol 34:1393–1401
8. Durmoo S, Richard C, Beranger G, Moita Y (2008) Biocorrosion of stainless steel grade 304L (SS304L) in sugar cane juice. Electrochim Acta 54:74–79
9. Wesley SB, Goyal HS, Mishra SC (2012) Corrosion behavior of ferritic steel, austenitic steel and low carbon steel grades in sugarcane juice. J Mater Metall Eng 2:9–22
10. D.O. Sprowls (1986) Metal Handbook, Corrosion, 9th edn.,vol 13. ASM International, Metals Park Ohio, p. 624
11. John Denny H (1996) Princ Prev Corros 2:148
12. Singh DDN, Gosh R (2006) Surf Coat Technol 201:90–101
13. Gerengi H, Bereket G, Kurtay M (2016) A morphological and electrochemical comparison of the corrosion process of aluminum alloys under simulated acid rain conditions. J Taiwan Inst Chem Eng 58:509–516
14. Palaniappa SK Seshadri (2007) Structural and phase transformation behaviour of electroless Ni-P, Ni-W-P deposits. Mater Sci Eng 460–461:638–644
15. Bai A, Chuang PY, Hu CC (2003) The corrosion behaviour of Ni–P deposits with high Phosphorous content in brine media. Mater Chem Phys 82:93
16. S.K. Gupta, A. Bajpai and V. Sharma (2002). In: Proceedings of the 64th Annual Convention of STAI. Cochin pp. 23–34
17. S.K. Gupta, A. Bajpal, and V. Sharma (2002) Pipeline corrosion in cane sugar industry indian sugar vol. 52. No. 9, pp. 681–687
18. Ashutosh Bajpai and N.P. Shukla (1992) Int Sugar J 94 pp. 76–80
19. Pierre, R.R. (2006) Handbook of Corrosion Eng.,McGraw-Hill Company p. 117
20. Gerengi H (2012) Anticorrosive properties of date palm (*Phoenix dactylifera* L.) fruit juice on 7075 type aluminum alloy in 3.5 % NaCl solution. Ind Eng Chem Res 51(39):12835–12843
21. Wesley SB, Goyal HS, Mishra SC (2012) J Mater Metall Eng 2(1):9–22
22. Pourbaix M (1974) Atlas of electrochemical equilibria in aqueous solutions. NNACE, Houston
23. Pohlman SL (1987) Corrosion and erosion based materials selection for the M242 autocannon barrel in a marine operating environment general corrosion.ASM International, Metal Park

Permissions

All chapters in this book were first published in IJIC, by Springer International Publishing AG.; hereby published with permission under the Creative Commons Attribution License or equivalent. Every chapter published in this book has been scrutinized by our experts. Their significance has been extensively debated. The topics covered herein carry significant findings which will fuel the growth of the discipline. They may even be implemented as practical applications or may be referred to as a beginning point for another development.

The contributors of this book come from diverse backgrounds, making this book a truly international effort. This book will bring forth new frontiers with its revolutionizing research information and detailed analysis of the nascent developments around the world.

We would like to thank all the contributing authors for lending their expertise to make the book truly unique. They have played a crucial role in the development of this book. Without their invaluable contributions this book wouldn't have been possible. They have made vital efforts to compile up to date information on the varied aspects of this subject to make this book a valuable addition to the collection of many professionals and students.

This book was conceptualized with the vision of imparting up-to-date information and advanced data in this field. To ensure the same, a matchless editorial board was set up. Every individual on the board went through rigorous rounds of assessment to prove their worth. After which they invested a large part of their time researching and compiling the most relevant data for our readers.

The editorial board has been involved in producing this book since its inception. They have spent rigorous hours researching and exploring the diverse topics which have resulted in the successful publishing of this book. They have passed on their knowledge of decades through this book. To expedite this challenging task, the publisher supported the team at every step. A small team of assistant editors was also appointed to further simplify the editing procedure and attain best results for the readers.

Apart from the editorial board, the designing team has also invested a significant amount of their time in understanding the subject and creating the most relevant covers. They scrutinized every image to scout for the most suitable representation of the subject and create an appropriate cover for the book.

The publishing team has been an ardent support to the editorial, designing and production team. Their endless efforts to recruit the best for this project, has resulted in the accomplishment of this book. They are a veteran in the field of academics and their pool of knowledge is as vast as their experience in printing. Their expertise and guidance has proved useful at every step. Their uncompromising quality standards have made this book an exceptional effort. Their encouragement from time to time has been an inspiration for everyone.

The publisher and the editorial board hope that this book will prove to be a valuable piece of knowledge for researchers, students, practitioners and scholars across the globe.

List of Contributors

Roland Tolulope Loto
Department of Mechanical Engineering, Covenant University, Ota, Ogun, Nigeria

Abimbola Patricia Popoola and Akanji Lukman Olaitan
Department of Chemical, Metallurgical and Materials Engineering, Tshwane University of Technology, Pretoria, South Africa

Somaia G. Mohammad
Central Agricultural Pesticides Laboratory, Agriculture Research Center (ARC), Dokki, Giza, Egypt

Sahar M. Ahmed
Egyptian Petroleum Research Institute, Ahmed El-Zomor St., Nasr City, Cairo, Egypt

M. Benahmed and N. Djeddi
Laboratoire des Molécules Bioactives et Applications, Université Larbi Tébessi, Route de Constantine, 12000 Tébessa, Algeria

S. Akkal
Laboratoire de Phytochimie et Analyses physicochimiques et Biologiques, Département de Chimie, Faculté de Sciences exactes, Université Mentouri Constantine, Route d'Ain el Bey, 25000 Constantine, Algeria

H. Laouar
Laboratoire de Valorisation des Ressources Naturelles Biologiques, Département de Biologie, Universite´ Ferhat Abbas de Sétif, Sétif, Algeria

Chynthia Devi Hartono, Kevin Jonathan Marlie, Felycia Edi Soetardjo and Suryadi Ismadji
Department of Chemical Engineering, Widya Mandala Surabaya Catholic University, Kalijudan 37, Surabaya 60114, Indonesia

Jindrayani Nyoo Putro and Yi Hsu Ju
Department of Chemical Engineering, National Taiwan University of Science and Technology, No. 43, Sec. 4, Keelung Rd, Taipei 106, Taiwan, People's Republic of China

Dwi Agustin Nuryani Sirodj
Department of Industrial Engineering, Widya Mandala Surabaya Catholic University, Kalijudan 37, Surabaya 60114, Indonesia

A. S. Fouda, A. S. Abousalem and G. Y. EL-Ewady
Chemistry Department, Faculty of Science, Mansoura University, Mansoura 35516, Egypt

Loutfy H. Madkour
Chemistry Department, Faculty of Science and Arts, Baljarashi, Al-Baha University, P.O. Box 1988, Al-Baha, Saudi Arabia

I. H. Elshamy
Chemistry Department, Faculty of Science, Tanta University, Tanta 31527, Egypt

Aref Shokri
Young Researchers and Elite Club, Arak Branch, Islamic Azad University, Arak, Iran

Kazem Mahanpoor
Department of Chemistry, Faculty of Science, Arak Branch, Islamic Azad University, Arak, Iran

Farid Safari1 and Abtin Ataei
Department of Energy Engineering, Science and Research Branch, Islamic Azad University, Tehran, Iran

Ahmad Tavasoli
Department of Energy Engineering, Science and Research Branch, Islamic Azad University, Tehran, Iran
School of Chemistry, College of Science, University of Tehran, Tehran, Iran

Vajjiravel Murugesan and Elumalai Marimuthu
Department of Chemistry, B S Abdur Rahman Crescent University, Vandalur, Chennai 600 048, India

K. S. Yoganand and M. J. Umapathy
Department of Chemistry, College of Engineering, Anna University, Chennai 600 025, India

Turuvekere K. Chaitra and Kikkeri N. Mohana
Department of Studies in Chemistry, Manasagangotri, University of Mysore, Mysuru 570006, Karnataka, India

Harmesh C. Tandon
Department of Chemistry, Sri Venkateswara College, Dhaula kuan, New Delhi 110021, India

Noureddine Boudechiche, Hassiba Mokaddem and Zahra Sadaoui
Laboratory of Engineering Reaction, Faculty of Engineering Mechanic and Engineering Processes, USTHB, BP 32, Algiers, Algeria

Mohamed Trari
Laboratory of Storage and Valorization of Renewable Energies, Faculty of Chemistry, USTHB, BP 32, Algiers, Algeria

Sirajo Abubakar Zauro and B. Vishalakshi
Department of Post-Graduate Studies and Research in Chemistry, Mangalore University, Mangalagangothri, Dakshina Kannada, Mangalore, Karnataka 574199, India

Liping Zhang and Shaohai Fu
Key Laboratory of Eco-Textile, Jiangnan University, Ministry of Education, Wuxi 214122, Jiangsu, China

Benjamin Tawiah
Key Laboratory of Eco-Textile, Jiangnan University, Ministry of Education, Wuxi 214122, Jiangsu, China

Department of Industrial Art (Textiles), Kwame Nkrumah University of Science and Technology, Private Mail Bag, Kumasi, Ghana

Benjamin K. Asinyo and William Badoe
Department of Industrial Art (Textiles), Kwame Nkrumah University of Science and Technology, Private Mail Bag, Kumasi, Ghana

Shi Yunhai, Liang Shan, Li Wei, Gatabazi Remy and Du JianPeng
Research Centre of Chemical Engineering, East China University of Science and Technology, 130 Meilong Road, Box 368, Shanghai 200237, China

Luo Dong
Chongqing General Gas Purification Plant, Southwest Oil and Gas Field Branch Company, 44 Taohua Road, Chongqing 401220, China

Majid Saghi and Kazem Mahanpoor
Department of Chemistry, Islamic Azad University, Arak Branch, Arāk, Iran

H. D. Ada
Dumlupınar University, 43100 Kütahya, Turkey

S. Altanlar and F. Erdem
Technological Research Division, Turkish Sugar Factories Co, 06930 Ankara, Turkey

G. Bereket
Deparment of Chemistry, Faculty of Science and Letters, Eskişehir Osmangazi University, 26480 Eskişehir, Turkey

Index

www.ingramcontent.com/pod-product-compliance
Lightning Source LLC
Chambersburg PA
CBHW082033190326
41458CB00010B/3356